Springer Series on
Wave Phenomena
5

Edited by L.B. Felsen

Springer
Berlin
Heidelberg
New York
Barcelona
Budapest
Hong Kong
London
Milan
Paris
Singapore
Tokyo

Springer Series on
Wave Phenomena
Editors: L. M. Brekhovskikh   L. B. Felsen   H. A. Haus
Managing Editor: H. K.V. Lotsch

L.M. Brekhovskikh   O.A. Godin

Acoustics of Layered Media I

Plane and Quasi-Plane Waves

With 44 Figures

 Springer

Professor Leonid M. Brekhovskikh, Academician
Russian Academy of Sciences, P.P. Shirshov Institute of Oceanology,
36 Nakhimovsky Av., Moscow 117851, Russia

Dr. Oleg A. Godin
School of Earth and Ocean Sciences, University of Victoria, P.O. Box 1700,
Victoria, B.C. V8W 2Y2, Canada and
Russian Academy of Sciences, P.P. Shirshov Institute of Oceanology,
36 Nakhimovsky Av., Moscow 117851, Russia

Series Editors:

Professor Leonid M. Brekhovskikh, Academician
Russian Academy of Sciences, P.P. Shirshov Institute of Oceanology,
36 Nakhimovsky Av., Moscow 117851, Russia

Professor Leopold B. Felsen, Ph.D.
Boston University, Boston, MA 02215, USA

Professor Hermann A. Haus
Department of Electrical Engineering & Computer Sciences, MIT, Cambridge, MA 02139, USA

Managing Editor: Helmut K.V. Lotsch
Springer-Verlag, Tiergartenstrasse 17,
D-69121 Heidelberg, Fed. Rep. of Germany

Second, Updated Printing 1998

ISSN 0931-7252
ISBN 3-540-64724-4 Springer-Verlag Berlin Heidelberg New York (softcover)

ISBN 3-540-51038-9 Springer-Verlag Berlin Heidelberg New York (1st Edition; hardcover)

Library of Congress Cataloging-in-Publication Data applied for.

Die Deutsche Bibliothek-CIP-Einheitsaufnahme
Brechovskich, Leonid M.: Acoustics of layered media / L. M. Brekhovskikh; O. A. Godin.
Berlin ; Heidelberg ; New York ; London ; Paris ; Tokyo ; Hong Kong ; Barcelona ; Budapest : Springer
1. Plane and quasi-plane waves. - 1998
(Springer series on wave phenomena ; Vol. 5)
ISBN 3-540-64724-4

© Springer-Verlag Berlin Heidelberg 1990
Printed in Germany

SPIN 10687498   54/3144 – 5 4 3 2 1 0 – Printed on acid-free paper

Preface

This second printing of *Acoustics of Layered Media. I: Plane and Quasi-Plane Waves* incorporates a number of improvements and additions. The modifications have been made to reflect recent progress in research on the subject as well as experience gathered in teaching from the first edition. However, the structure of the monograph remains unchanged.

Based on *Acoustics of Layered Media. I: Plane and Quasi-Plane Waves* and its sequel *Acoustics of Layered Media. II. Point Sources and Bounded Beams* senior undergraduate courses on wave propagation in inhomogeneous media and on ocean acoustics have been taught at Moscow Institute of Physics and Technology for a number of years. The authors are indebted to their students and many colleagues for numerous comments which helped to correct typographical errors and to improve the text of the first edition.

Most of the preparation of the second printing has been done by one of the authors (OG); the other (LB) is grateful to him for that.

Moscow, Russia   *L.M. Brekhovskikh*
Victoria, B.C., Canada   *O.A. Godin*
December 1997

Preface to the First Edition

This monograph is devoted to the systematic presentation of the theory of sound-wave propagation in layered structures. These structures can be man-made, such as ultrasonic filters, lenses, surface-wave delay lines, or natural media, such as the ocean and the atmosphere, with their marked horizontal stratification. A related problem is the propagation of elastic (seismic) waves in the earth's crust.

These topics have been treated rather completely in the book by L.M. Brekhovskikh, *Waves in Layered Media,* the English version of the second edition of which was published by Academic Press in 1980. Due to progress in experimental and computer technology it has become possible to analyze the influence of factors such as medium motion and density stratification upon the propagation of sound waves. Much attention has been paid to propagation theory in near-stratified media, i.e., media with small deviations from strict stratification. Interesting results have also been obtained in the fields of acoustics which had been previously considered to be "completely" developed. For these reasons, and also because of the inflow of researchers from the related fields of physics and mathematics, the circle of persons and research groups engaged in the study of sound propagation has rather expanded. Therefore, the appearance of a new summary review of the field of acoustics of layered media has become highly desirable.

Since *Waves in Layered Media* became quite popular, we have tried to retain its positive features and general structure. However, a large part of the present book is devoted to new material which has previously not been considered in the mono-graphical literature. The major part reviews old topics treated from the new points of view afforded by recent theoretical methods. Most of the material presented stems from our own work and has been previously published in various journals. Results obtained by other researchers are also presented, but with our own interpretations. A bibliography is included to enable the reader to find additional material.

Although the book is devoted to acoustical waves, most of the developed theoretical approaches and many results can be easily transferred to the domain of electromagnetic waves. It should be mentioned that, unlike *Waves in Layered Media*, only the theory of plane and quasi-plane waves is treated here. The problem of point sources in layered media is the topic of our forthcoming volume. Although the results of the present work will be exploited there widely, the publisher and authors have agreed that cramping all the material into one book would be unwise.

The authors are grateful to V.V. Vavilova and T.I. Tzyplakova for their great help in preparing the manuscript.

Moscow   *L.M. Brekhovskikh*
September, 1989   *O.A. Godin*

Contents

1. Basic Equations for Wave Processes in Fluids and Solids

In this chapter the equations and boundary conditions for elastic waves in liquids and solids, with particular attention to layered media, are obtained. Emphasis is placed on *quasi-plane waves* which are defined to be waves with a harmonic dependence on time and two spatial Cartesian coordinates. This type of wave motion is particularly important because studying quasi-plane waves provides a deep physical insight into sound propagation in layered media, and quite general acoustic fields can be represented as a superposition of these waves.

1.1 Sound in Layered Fluids

1.1.1 Derivation of Wave Equations

Let v_0, p_0, ϱ be the particle velocity, pressure, and the medium density in the absence of the wave and v, p, ϱ' disturbances of these properties caused by the sound wave. The sums of these quantities $\tilde{v} = v_0 + v$, $\tilde{p} = p_0 + p$, and $\tilde{\varrho} = \varrho + \varrho'$ obey hydrodynamic equations, the Euler and continuity [1.1]:

$$\frac{\partial \tilde{v}}{\partial t} + (\tilde{v} \cdot \nabla)\tilde{v} = -\frac{1}{\tilde{\varrho}}\nabla\tilde{p} \quad , \tag{1.1.1}$$

$$\frac{\partial \tilde{\varrho}}{\partial t} + \nabla \cdot (\tilde{\varrho}\tilde{v}) = 0 \quad , \tag{1.1.2}$$

as well as the state equation

$$\tilde{p} = \tilde{p}(\tilde{\varrho}, \tilde{S}) \quad , \tag{1.1.3}$$

where density and entropy $\tilde{S}$ are chosen as independent variables. In (1.1.1 and 2) t is time and $\partial/\partial t + \tilde{v} \cdot \nabla$ is the convective or material time derivative, i.e., the rate of change of the physical quantity under consideration at a chosen fluid particle. In multicomponent fluids, $\tilde{p}$ in (1.1.3) also depends on the concentrations of the various components (humidity of the atmosphere, salinity of sea water, etc.). By neglecting the diffusion of components in such fluids and the thermal conductivity, we may treat the propagation of sound as an adiabatic process. The entropy density of chosen particle is then constant and we have

$$\left(\frac{\partial}{\partial t} + \tilde{v} \cdot \nabla\right)\tilde{p} = \tilde{c}^2 \left(\frac{\partial}{\partial t} + \tilde{v} \cdot \nabla\right)\tilde{\varrho} \quad , \tag{1.1.4}$$

where

$$\tilde{c} = \tilde{c}(\tilde{\varrho}, \tilde{S}) = \sqrt{(\partial\tilde{p}/\partial\tilde{\varrho})_{\tilde{S}}} \tag{1.1.5}$$

is the sound velocity, as we shall see below.

The linear set of acoustic equations can be obtained if we substitute $\tilde{v} = v_0 + v$, $\tilde{p} = p_0 + p$, $\tilde{\varrho} = \varrho + \varrho'$ into (1.1.1), (1.1.2), and (1.1.4) and separate the first-order terms in v, p, ϱ' (zero-order terms satisfy the hydrodynamic equations separately):

$$\frac{d}{dt}v + (v \cdot \nabla)v_0 = -\frac{1}{\varrho}\nabla p + \frac{\varrho'}{\varrho^2}\nabla p_0 \quad , \tag{1.1.6}$$

$$\frac{d}{dt}\varrho' + \varrho' \operatorname{div} v_0 + \operatorname{div}(\varrho v) = 0 \quad , \tag{1.1.7}$$

$$(v \cdot \nabla)p_0 + \frac{d}{dt}p = c^2 \frac{d}{dt}\varrho' + (c^2)' \frac{d}{dt}\varrho + c^2(v \cdot \nabla)\varrho \quad . \tag{1.1.8}$$

Here $d/dt \equiv \partial/\partial t + v_0 \cdot \nabla$, c is the sound velocity not influenced by the wave, $(c^2)' \equiv \tilde{c}^2 - c^2$. Conditions under which this linearization is possible and the assumption of adiabaticity is justified were thoroughly discussed in [1.2], see also [Ref. 1.18, Chap. 1]. These conditions are satisfied in rather broad ranges of wave amplitudes and frequencies.

Using (1.1.6–8), we consider two most important cases when the set of linear acoustic equations reduces to a single expression:

a) Let us suppose that in the absence of the wave our medium is at rest: $v_0 = 0$. It follows from (1.1.1) that for this case $\nabla p_0 = 0$ and (1.1.6–8) become much simpler. Eliminating $d\varrho'/dt$ from (1.1.7) and (1.1.8) we obtain

$$\frac{\partial v}{\partial t} = -\frac{1}{\varrho}\nabla p \quad , \tag{1.1.9}$$

$$\operatorname{div} v + \frac{1}{\varrho c^2} \cdot \frac{\partial p}{\partial t} = 0 \quad . \tag{1.1.10}$$

After applying the divergence operator to (1.1.9) and $\partial/\partial t$ to (1.1.10) and subtracting one from the other we get

$$\frac{\partial}{\partial t}\left(\frac{1}{\varrho c^2}\frac{\partial p}{\partial t}\right) - \operatorname{div}\left(\frac{\nabla p}{\varrho}\right) = 0 \quad , \tag{1.1.11}$$

i.e., the closed-form equation describing sound-wave propagation in inhomogeneous, possibly nonsteady-state medium at rest. Other characteristics of the sound field can be found if the sound pressure $p(r, t)$ is known. For example, the particle velocity v, can be found by the use of (1.1.9).

b) In many instances real media (the ocean, the earth crust and atmosphere, some media in technical applications, etc.) can be approximated by layered ones. The theory of sound wave propagation in such media is rather thoroughly developed. In these cases the properties of the medium depend on only one space coordinate and, possibly, on time.

In the following discussion, we assume that the medium's parameters in the unperturbed state do not depend on "horizontal" Cartesian coordinates x, y but only on the "vertical" coordinate z. Such is the case in horizontally stratified media which are the primary subject of the treatment below. We further suppose that the velocity of unperturbed flow v_0 is horizontal and does not depend on time. It follows from (1.1.1) that under these conditions $\nabla p_0 = 0$. Taking into account the obvious relations div $v_0 = 0$, and $v_0 \cdot \nabla \varrho = 0$ and eliminating the unknown ϱ' from (1.1.6–8) yields [compare with (1.1.9, 10)]:

$$\frac{dv}{dt} + w\frac{dv_0}{dz} = -\frac{\nabla p}{\varrho} \quad , \tag{1.1.12}$$

$$\frac{1}{\varrho c^2}\frac{dp}{dt} + \text{div } v = 0 \quad , \tag{1.1.13}$$

where w is vertical component of v. To eliminate the horizontal components of v from these equations we apply the div operator to (1.1.12) and d/dt to (1.1.13). By substracting one from the other and noting the relation

$$\text{div}\frac{dv}{dt} - \frac{d}{dt}\text{div } v = \left(\frac{dv_0}{dz} \cdot \nabla\right)w$$

(which can be obtained by direct differentiation), we get

$$\frac{d}{dt}\left(\frac{1}{\varrho c^2}\frac{dp}{dt}\right) - \text{div}\left(\frac{\nabla p}{\varrho}\right) - 2\left(\frac{dv_0}{dz} \cdot \nabla\right)w = 0 \quad . \tag{1.1.14}$$

To eliminate w and obtain a closed equation for p we differentiate (1.1.14) by t and substitute dw/dt from (1.1.12) into the result. Thus we obtain the acoustic wave equation for moving layered media

$$\frac{d}{dt}\left[\frac{d}{dt}\left(\frac{1}{\varrho c^2}\frac{dp}{dt}\right) - \text{div}\left(\frac{\nabla p}{\varrho}\right)\right] + 2\left(\frac{dv_0}{dz} \cdot \nabla\right)\left(\frac{1}{\varrho}\frac{\partial p}{\partial z}\right) = 0 \quad . \tag{1.1.15}$$

In layered media at rest (1.1.15) reduces to (1.1.11).

For the general case of a moving, inhomogeneous in three dimensions medium, the wave equation for p can be obtained in the important case of slow flows ($v_0 \ll c$) [1.3]. By assuming that the motion of the particles in the sound wave is a potential one, that is curl $v_0 \equiv 0$, curl $v \equiv 0$, it is possible to reduce (1.1.1–3) to a single wave equation without assumptions about the magnitude of the ratio v_0/c [1.4]. The reader will find a more detailed discussion of acoustic wave equations for moving fluid in [Ref. 1.19, Sect. 4.1].

We have not yet taken into account possible external forces acting upon the medium. Presence of such forces gives rise to additional terms in the right-hand sides of (1.1.1, 6) and the wave equations do not reduce to (1.1.11, 15) if $\nabla p_0 \neq 0$. One external force which always acts upon a fluid is gravity. It induces the fluid's stratification $c(z)$, $\varrho(z)$ in the unperturbed state and also influences the sound propagation. However, this effect is important only at frequencies $f \lesssim 10\,\text{Hz}$ in the atmosphere and at even lower frequencies in the ocean. The simultaneous presence of elastic and gravitational forces in the atmosphere gives rise to special acoustic-gravity waves with typical periods of 5 to 10 minutes, which are rather important in the dynamics of the atmosphere [1.5–7]. These problems are beyond the scope of this book, however.

1.1.2 Plane Waves and Spherical Waves

Equations (1.1.11 and 15) describe sound propagation in the general case of inhomogeneous and nonsteady-state liquids. We begin the analysis of these equations, however, by considering several simple but important limiting cases. The simplest is the homogeneous, steady-state medium at rest (ϱ = const, c = const, $v_0 \equiv 0$). In this case (1.1.11) reduces to

$$\frac{\partial^2 p}{\partial t^2} - c^2 \Delta p = 0 \quad , \tag{1.1.16}$$

which is the wave equation in a narrow sense of this term usually used in mathematics [1.8]. Two important solutions to (1.1.16) are worth mentioning:

$$p_1 = f(\boldsymbol{n} \cdot \boldsymbol{r}/c - t) \quad , \quad n^2 = 1 \quad , \quad \boldsymbol{n} = \text{const} \quad , \tag{1.1.17}$$

$$p_2 = r^{-1} F(r/c \pm t) \quad , \quad r \equiv |\boldsymbol{r}| \quad , \tag{1.1.18}$$

where f and F are arbitrary smooth functions and the unit vector $\boldsymbol{n}$ is normal to those planes where the argument of the function f, and hence p_1, are constants. Waves which are described by (1.1.17) are called *plane waves*. They propagate along $\boldsymbol{n}$ at the speed c without changing their shape or amplitude. The wave p_2 is called *spherical wave*. It is spherically symmetric, that is, the sound pressure and other characteristics of the sound field are constants at spheres r = const at fixed moments of time.

1.1.3 Boundary Conditions

Equations (1.1.6–8) describe acoustic fields in unbounded, inhomogeneous liquids if their parameters are smooth functions of time and positional coordinates, so that all the derivatives in these equations have finite values. If the liquids are bounded or their parameters are discontinuous at some surfaces, *boundary conditions* must be specified for these equations. The simplest conditions arise in two cases: that of absolutely rigid and absolutely soft (pressure-release or free) boundaries. In the first case a given boundary S is not deformed by the wave, hence

$$v_n(\boldsymbol{r}) = 0 \quad , \quad \boldsymbol{r} \in S \quad , \tag{1.1.19}$$

where v_n is the component of the particle velocity vector in the wave which is normal to S [of course, $(v_0)_n = 0$, too]. For an absolutely soft surface the total pressure is zero, i.e.,

$$p + p_0 = 0 \quad , \quad \boldsymbol{r} \in \tilde{S} \quad , \tag{1.1.20}$$

where $\tilde{S}$ is the position of the surface under the action of the sound wave. This boundary condition is nonlinear with respect to the amplitude of the wave. Its linearization in the case of small amplitudes is the simplest one when the pressure in an undisturbed liquid is constant ($\nabla p_0 = 0$). Since p is of the first order of the magnitude and p_0 satisfies the boundary condition in the absence of the wave, we have for the linear approximation

$$p = 0 \quad , \quad \boldsymbol{r} \in S \quad . \tag{1.1.21}$$

In the general case of an interface between two fluids at rest two boundary conditions have to be satisfied:

4

a) Equality of normal components of particle velocities at both sides of the interface (kinematic condition), and

b) equality of the forces acting upon each part of the interface from both sides; otherwise a part of the surface would move with infinitely great acceleration (dynamic condition).

Denoting the difference of function f values at both sides of S by $[f]_S$, these two conditions may be written as

$$[v_n]_S = 0 \quad \text{and} \quad [p]_S = 0 \ . \tag{1.1.22}$$

This first condition may also be written in terms of p by using (1.1.9). Then we will have, instead of (1.1.22):

$$\left[\frac{1}{\varrho} \frac{\partial p}{\partial n} \right]_S = 0 \quad , \quad [p]_S = 0 \quad , \tag{1.1.23}$$

where $\partial/\partial n$ is the derivative of p along the normal to S.

The boundary condition (1.1.19) at an absolutely rigid boundary can also be written in terms of a pressure field:

$$\frac{\partial p}{\partial n} = 0 \quad , \quad r \in S \ . \tag{1.1.24}$$

Conditions (1.1.21, 24) valid at ideal boundaries are referred to as boundary conditions of the first and the second kind.

In the case of a nonsteady-state acoustic field initial conditions must also be taken into account. If a sound wave is generated by a source initiated at time $t = t_0$, these conditions are $p(r, t_0) = 0$, $(\partial p/\partial t)_{t=t_0} = 0$ at any r.

1.2 Harmonic Waves

1.2.1 Conditions at Infinity

By supposing that the medium parameters are independent of time, we can reduce the number of independent variables in the wave equations by using the spectral representation

$$p(r, t) = \int_{-\infty}^{+\infty} p(r, \omega) \exp(-\mathrm{i}\omega t) d\omega \quad , \quad \mathrm{i} = \sqrt{-1} \quad , \tag{1.2.1}$$

where ω is the frequency.

To simplify the calculations we use the complex form for the wave-field description, keeping in mind that only their real part has physical meaning. The time derivative of an elementary harmonic wave $p(r, \omega) \exp(-\mathrm{i}\omega t)$ corresponds to just multiplication by $-\mathrm{i}\omega$. Then (1.1.11 and 15) are considerably simplified. Equation (1.1.11) now becomes

$$\Delta p(r, \omega) - \nabla \ln \varrho(r) \cdot \nabla p(r, \omega) + k^2 p(r, \omega) = 0 \quad , \tag{1.2.2}$$

where $k = \omega/c(r)$ is the wave number. Waves of fixed frequency ω are called *monochromatic*, or *continuous*. It is worth noting that (1.2.2) can also be used for describing sound fields in dispersive media, in which some parameters (for example the sound velocity) depend on frequency and when (1.1.11) has no sense.

Monochromatic waves are unlimited in time. Therefore, one has no such initial conditions as mentioned at the end of previous section. Instead, usually conditions at $r \to \infty$ are specified. These conditions serve to distinguish fields generated by sources located in the finite region from waves propagating from infinity.

One of the ways to attain a "physical" solution to our equations is to assume that a small absorption of waves occurs in the medium. The solutions of interest are those that vanish at infinity ($r \to \infty$). When these equations are solved, one can obtain the true solution for a nondissipative medium by letting the absorption coefficient tend to zero. This method of addressing real physical problems is termed the *principle of limiting absorption*.

Another way to obtain such a solution in a nondissipative medium is to use the so-called *radiation condition*. This uses the fact that at $r \to \infty$, the sound field under consideration consists only of the waves propagating to infinity. Concrete forms of this condition can be different in different cases. For the case of an infinite, homogeneous medium at rest it is given by [1.8]:

$$\lim_{r \to \infty} r \left[\frac{\partial}{\partial r} p(r, \omega) - \mathrm{i}kp(r, \omega) \right] = 0 \quad . \tag{1.2.3}$$

This form suggests that the wave's energy flux has the same direction as its phase gradient. In dispersive media, when $c = c(\omega)$, another situation is possible when the group and phase velocities have opposite directions. In this case a wave carrying energy *from the source* will propagate *to the source*. In this case the minus sign between the brackets in (1.2.3) must be changed to a plus sign. In considering projections of phase and group velocities on some direction, in the following discussion we will suppose (unless otherwise noted) that these projections have the same sign.

1.2.2 Waves with Harmonical Dependence on Horizontal Coordinates and Time

In a layered, steady-state medium, the sound field equations can be further simplified and reduced to an ordinary differential equation. The spectral representation with respect to horizontal coordinates x, y can be used for this purpose:

$$p(r, \omega) = \int\!\!\!\int_{-\infty}^{+\infty} p(z, \xi, \omega) \exp{(\mathrm{i}\xi \cdot r)} d^2\xi \quad , \quad \xi = (\xi_1, \xi_2, 0) \quad . \tag{1.2.4}$$

In this case, taking the partial derivatives (except for $\partial/\partial z$) of the *quasi-plane* wave $p(r, t) = p(z, \xi, \omega) \exp{(\mathrm{i}\xi \cdot r - \mathrm{i}\omega t)}$ is simply equivalent to multiplication by some factors and (1.1.15) becomes

$$\frac{\partial^2}{\partial z^2} p(z, \xi, \omega) - \left(\frac{\partial}{\partial z} \ln \varrho \beta^2 \right) \left[\frac{\partial}{\partial z} p(z, \xi, \omega) \right]$$
$$+ (k^2 \beta^2 - \xi^2) p(z, \xi, \omega) = 0 \quad , \tag{1.2.5}$$

where

$$\beta = 1 - \boldsymbol{\xi} \cdot \boldsymbol{v_0}/\omega \quad . \tag{1.2.6}$$

The factor before $\partial p/\partial z$ in (1.2.5) has a singularity at $\beta = 0$. In the neighborhood of $\beta = 0$ a strong interaction of sound and flow take place which is sometimes called the "resonant interaction" [1.9, 10]. In the case of slow flows ($v_0 \ll c$), β is almost unity.

When the fluid is at rest we obtain from (1.2.2, 5)

$$\frac{\partial^2}{\partial z^2} p(z, \boldsymbol{\xi}, \omega) - \left(\frac{\partial}{\partial z} \ln \varrho \right) \left[\frac{\partial}{\partial z} p(z, \boldsymbol{\xi}, \omega) \right]$$

$$+ (k^2 - \xi^2) p(z, \boldsymbol{\xi}, \omega) = 0 \quad . \tag{1.2.7}$$

Solutions to this equation must obey the boundary condition (1.1.23).

Let us now turn to the boundary conditions of solutions of (1.2.5). Obviously, continuity of normal to boundary particle displacements serves as the kinematic boundary condition. To write it in terms of sound pressure we note that:

1.   the normal velocity is the material time derivative of the normal displacement,
2.   taking the material time derivative

$$\frac{d}{dt} \equiv \frac{\partial}{\partial t} + \boldsymbol{v_0} \cdot \nabla = -i\omega\beta$$

is equivalent to multiplying by $-i\omega\beta$,
3.   in layered media unperturbed by sound, the boundaries are horizontal, hence the normals are parallel to z-axis,
4.   the z-component of the particle velocity is related to $\partial p/\partial z$ by the formula

$$w = \frac{1}{i\omega\beta\varrho} \frac{\partial p}{\partial z} \quad ,$$

which follows easily from (1.1.12).

After putting this chain of arguments together, we obtain for the kinematic boundary condition

$$\left[\frac{1}{\varrho\beta^2} \cdot \frac{\partial p}{\partial z} \right]_S = 0 \quad . \tag{1.2.8}$$

It is important to note that the normal to the boundary displacement, but not the vertical component of the particle velocity must be continuous across the boundary. Disregard for this fact was the cause of several mistakes [1.11, 12]. It can be shown that condition (1.2.8) is simply the continuity of $v_n + v_{0n}$, i.e., the normal to the perturbed boundary component of the total velocity. The difference between $\tilde{v}_n$ and w is linear with respect to the amplitude of the sound wave if $v_0 \neq 0$ and can not be neglected.

The dynamic boundary condition is again the continuity of the sound pressure

$$[p]_S = 0 \quad . \tag{1.2.9}$$

Equations (1.2.8, 9) become equivalent to (1.1.23) if $v_0 = 0$.

Note that one can obtain the boundary conditions (1.2.8,9) from (1.2.5) by regarding a given boundary, say $z = z_1$, as an extreme case of media with smooth variation of parameters in the region $(z_1 - \varepsilon, z_1 + \varepsilon)$. Indeed, (1.2.5) can be written in integral form:

$$\frac{1}{\varrho\beta^2}\cdot\frac{\partial p}{\partial z}\Big|_{z=z_1-\varepsilon}^{z=z_1+\varepsilon} = -\int_{z_1-\varepsilon}^{z_1+\varepsilon}\frac{k^2\beta^2 - \xi^2}{\varrho\beta^2}p(z',\xi,\omega)dz' \quad.$$

Here, the right-hand side tends to zero when $\varepsilon \to 0$, since the integrand is a finite function. Hence, when $\varepsilon \to 0$,

$$\frac{1}{\varrho\beta^2}\cdot\frac{\partial p}{\partial z}\Big|_{z=z_1+\varepsilon} - \frac{1}{\varrho\beta^2}\cdot\frac{\partial p}{\partial z}\Big|_{z=z_1-\varepsilon} \to 0 \quad,$$

and one obtains (1.2.8). Condition (1.2.9) can be obtained by an analogous method.

In a homogeneous medium there exists a general solution to (1.2.7)

$$p(z,\xi,\omega) = A(\xi,\omega)\exp(\mathrm{i}\mu z) + B(\xi,\omega)\exp(-\mathrm{i}\mu z) \quad, \tag{1.2.10}$$

where

$$\mu = \sqrt{k^2 - \xi^2} \quad. \tag{1.2.11}$$

By restoring the factor $\exp(\mathrm{i}\xi\cdot r - \mathrm{i}\omega t)$ in (1.2.10) which was omitted in the equations subsequent to (1.2.4), we shall see that this solution describes two plane waves propagating in directions symmetric with respect to the horizontal plane. Wave vectors of these waves are $kn = (\xi_1, \xi_2, \pm\mu)$, where n is normal to the wave front, as in (1.1.17).

In a homogeneous medium moving at constant speed ($v_0 =$ const), (1.2.5) also has a general solution of the type (1.2.10), but the *dispersion relation*, i.e., the relation between wave vector and frequency, appears to be more complicated than in (1.2.11):

$$\mu = \sqrt{k^2\beta^2 - \xi^2} = \sqrt{\omega^2 c^{-2}(1 - \xi\cdot v_0/\omega)^2 - \xi^2} \quad. \tag{1.2.12}$$

Consider solution (1.2.10) when $B = 0$. Let θ be the angle between the wave vector and the z-axis. We choose the direction of the x-axis to be along v_0. Then

$$\mu = q\cos\theta \quad, \quad \xi = (q\sin\theta\cos\varphi, q\sin\theta\sin\varphi, 0) \quad, \tag{1.2.13}$$

where $q = qn$ is the wave vector, φ is the angle between the x-axis and projection of n upon the horizontal plane xy. Solution (1.2.10) can now be written as

$$p(r,t) = A\exp(\mathrm{i}q\cdot r - \mathrm{i}\omega t) \quad, \quad q = k\frac{1}{1 + (v_0\sin\theta\cos\varphi)/c} \quad. \tag{1.2.14}$$

Here, the dispersion relation (1.2.12) was also taken into account.

The phase velocity of the wave

$$c_{\mathrm{ph}} = |\omega/q| = |c + v_0\sin\theta\cos\varphi| = |c + v_0\cdot q/|q|| \tag{1.2.15}$$

is equal in magnitude to the sum of the sound velocity in a medium at rest and the projection of the flow's velocity v_0 on the wave's direction of propagation.

The phase velocity as well as the wavelength are at a maximum when the wave propagates along the flow and at a minimum in the opposite case.

The dispersion relation (1.2.12) can be written in vector form

$$\omega = qc + \boldsymbol{q} \cdot \boldsymbol{v}_0 \quad , \quad q = \pm |\boldsymbol{q}| \quad . \tag{1.2.16}$$

Here we have taken into account that

$$\boldsymbol{\xi} \cdot \boldsymbol{v}_0 = \boldsymbol{q} \cdot \boldsymbol{v}_0 \quad , \quad \mu^2 + \xi^2 = q^2 \quad .$$

The equation for the group velocity follows from (1.2.16):

$$c_g \equiv \frac{\partial \omega}{\partial \boldsymbol{q}} = c \frac{\boldsymbol{q}}{q} + \boldsymbol{v}_0 \quad . \tag{1.2.17}$$

We see that the direction of the group velocity or that of the energy transportation is different from the phase velocity direction $c_{ph} = c_{ph} \boldsymbol{q} / |\boldsymbol{q}|$. The magnitudes of the vectors c_{ph} and c_g are also different. These vectors are equal only in cases when the wave propagates either parallel or antiparallel to the flow, that is when $\pm \boldsymbol{v}_0 \parallel \boldsymbol{q}$.

1.2.3 Modified Wave Equations

In the medium at rest with $c = c(z)$, $\varrho = \text{const}$, (1.2.7) reduces to the one-dimensional Helmholtz equation

$$\frac{\partial^2}{\partial z^2} p(z, \xi, \omega) + (k^2 - \xi^2) p(z, \xi, \omega) = 0 \quad . \tag{1.2.18}$$

The monochromatic sound field $p(\boldsymbol{r}, \omega)$ in three-dimensional representation, under the same conditions, obeys the three-dimensional Helmholtz equation

$$\Delta p(\boldsymbol{r}, \omega) + k^2 p(\boldsymbol{r}, \omega) = 0 \quad , \tag{1.2.19}$$

which follows from (1.2.2). Exact or approximate solutions of (1.2.18 or 19) are the topics of a great number of papers in acoustics and in electro-magnetic theory. These solutions are rather well studied indeed. For this reason we would also like to transform (1.2.5), which describes the sound field in general types of layered media, to the Helmholtz equation.

The simplest way to achieve this goal is to introduce a new dependent variable

$$\Psi(z, \xi, \omega) = \frac{p(z, \xi, \omega)}{\beta(z, \xi, \omega) \sqrt{\varrho(z)}} \quad . \tag{1.2.20}$$

Equation (1.2.5) then becomes

$$\frac{\partial^2}{\partial z^2} \Psi + \left\{ k^2 \beta^2 - \xi^2 + \frac{1}{2 \varrho \beta^2} \frac{\partial^2}{\partial z^2} (\varrho \beta^2) \right.$$

$$\left. - \frac{3}{4} \left[\frac{1}{\varrho \beta^2} \frac{\partial}{\partial z} (\varrho \beta^2) \right]^2 \right\} \Psi = 0 \quad . \tag{1.2.21}$$

The last equation is of the same type as (1.2.18), but with a new, "effective" wave number. The use of this equation appears to be rather effective for studying sound-wave propagation in layered media when the functions $\varrho(z)$ and $\boldsymbol{v}_0(z)$ are sufficiently smooth. Difficulties arise when strong or abrupt changes of one or both of these functions take place at some z. Large and strongly varying terms occur in the coefficients in (1.2.21) and (1.2.5) and the use of approximate and numerical methods become difficult.

To eliminate this difficulty and to obtain a general equation that can be applied to smooth as well as abrupt changes in the parameters of the medium, we must find an equation where the derivatives of $\varrho(z)$ and $v_0(z)$ do not appear in the coefficients. This can be achieved by transforming the independent variable z [1.13, 20].

First we note that (1.2.5) has the desired form of (1.2.18) if $\varrho = $ const and $v_0 = 0$. At $\omega = 0$, $\xi = 0$ its solution is obviously

$$p = Az + B \quad , \tag{1.2.22}$$

where A and B are constants. If $\omega = 0$, $\xi = 0$ but $\varrho \neq$ const, $v_0 \not\equiv 0$ the general solution to (1.2.5) is

$$p = A\zeta(z) + B \quad , \tag{1.2.23}$$

where

$$\zeta(z) = \varrho_0^{-1} \int\limits_{z_0}^{z} \varrho(z')\beta^2(z')dz' \quad , \quad z_0 = \text{const} \quad , \tag{1.2.24}$$

and $\varrho_0 > 0$ is an arbitrary factor with the same dimension as density. The function $\zeta(z)$ monotonically increases with z. Comparing (1.2.22 and 23) suggests a method for obtaining the new independent vertical variable $\zeta(z)$ in media with density and flow stratification. Indeed, by changing the independent variable in (1.2.5) according to relation (1.2.24) we obtain the Helmholtz equation

$$\frac{\partial^2}{\partial \zeta^2}p + (k^2\beta^2 - \xi^2)\left(\frac{\varrho_0}{\varrho\beta^2}\right)^2 p = 0 \quad , \tag{1.2.25}$$

where the effective wave number is the function of sound velocity, density, and flow velocity but not of their derivatives. Therefore, the coefficients in (1.2.25) are finite unless $\beta \neq 0$. The use of (1.2.25) is also convenient when the medium's parameters are specified by tables (being obtained in experiment for example) since the derivatives of ϱ and v_0 need not be estimated.

In the new coordinate system (x, y, ζ) the boundary conditions appear to be the same as those in the original coordinates when there is no density and flow stratification:

$$[p]_S = 0 \quad , \quad \left[\frac{\partial p}{\partial \zeta}\right]_S = 0 \quad . \tag{1.2.26}$$

This form of the boundary conditions permits us to describe the sound field by (1.2.25) in the entire medium including the boundaries, where derivatives of $\varrho(z)$ and $v_0(z)$ do not exist. We do not need to consider solutions to the equation on different sides of a boundary and account for their mutual adjustment as satisfying the conditions of (1.2.26) is automatically established by (1.2.25).

The transformations of the dependent variable in (1.2.20), to deduce (1.2.21), and the independent variable used for obtaining (1.2.25) both contain the horizontal wave vector ξ. Hence, it appears impossible to obtain differential equations which are analogous to (1.2.21 and 25) for $p(r, \omega)$ in the general three-dimensional case. Only when a fluid is at rest ($\beta \equiv 1$) do transformations (1.2.20, 24) not include the frequency or sound wave vector. We can then perform the transformation of the wave equation without assuming that the dependence of sound pressure on the horizontal coordintes is harmonic. Namely, by using the transformation

$$\Psi = \frac{p(r, \omega)}{\sqrt{\varrho(r)}} \quad , \tag{1.2.27}$$

we obtain from (1.2.2)

$$\Delta\Psi + \left[k^2 + \frac{1}{2\varrho}\Delta\varrho - \frac{3}{4}\left(\frac{1}{\varrho}\nabla\varrho\right)^2 \right]\Psi = 0 \quad . \tag{1.2.28}$$

Note that (1.2.28) describes the sound field in media with any given three-dimensional inhomogeneities (i.e., without any assumption about stratification).

Using the transformation of the vertical coordinate

$$\zeta(z) = \varrho_0^{-1}\int\limits_{z_0}^{z}\varrho(z')dz' \quad , \tag{1.2.29}$$

(1.1.11) can be rewritten as

$$\frac{\partial^2}{\partial x^2}p + \frac{\partial^2}{\partial y^2}p + \frac{\varrho^2}{\varrho_0^2}\frac{\partial^2}{\partial\zeta^2}p - \frac{\partial}{\partial t}\left(\frac{1}{c^2}\frac{\partial p}{\partial t}\right) = 0 \quad . \tag{1.2.30}$$

Here the density must be independent of the horizontal coordinates as well as of time. In contrast, the sound velocity can depend on t as well as on r. Equation (1.2.30), as (1.2.25), does not contain space derivatives of the medium parameters and describes the sound field in media where discontinuous changes of these parameters can occur (that is, where c and ϱ are piecewise-continuous functions).

1.3 Elastic Waves in Isotropic Solids

1.3.1 General Relations

The main features of the elasticity theory can be found in [1.5, 14]. Solids, in contrast to liquids, resist shear. Hence, in addition to longitudinal waves, shear waves can also propagate in solids. A deformed state of a solid can be completely specified by the *displacement vector* $u(r, t)$, that is, the displacement of the particle at a time t from its initial position r. For convenience, we label the vector components by numerical indices, e.g., $r \equiv (x, y, z) \equiv (x_1, x_2, x_3)$, $u \equiv (u_1, u_2, u_3)$. Forces generated in the process of deformation are specified by the *stress tensor* $\sigma_{ij}(r, t)$, $i, j = 1, 2, 3$. The component σ_{ij} is the projection of the force acting on the unit area perpendicular to the axis i upon the axis j. The relation between stresses and deformations in the linear theory of elasticity and in the most simple case of locally isotropic solids, is given by *Hooke's law*

$$\sigma_{ij} = \lambda \frac{\partial u_k}{\partial x_k} \delta_{ij} + \mu \left(\frac{\partial u_i}{\partial x_j} + \frac{\partial u_j}{\partial x_i} \right) \quad . \tag{1.3.1}$$

Here and below we use the convention that whenever the same index appears in the same term twice, the term must be automatically summed over this (so-called dummy) index. The Kronecker symbol δ_{ij} is 1 for $i = j$ and 0 for $i \neq j$. Note that the stress tensor is symmetric: $\sigma_{ij} = \sigma_{ji}$. The elastic constants λ and μ are the so-called Lamé constants, where μ is the shear modulus. When $\mu = 0$ we have a liquid with no resistance to shear. In this case, the stress tensor is related to the pressure by the formula $\sigma_{ij} = -p\delta_{ij}$. Newton's second law for a particle in a solid in terms of the stress tensor is:

$$\varrho \frac{\partial^2}{\partial t^2} u_j = \frac{\partial}{\partial x_i} \sigma_{ij} \quad . \tag{1.3.2}$$

The parameters of the medium are assumed to be independent of time. The physical meaning of (1.3.2) is that a particle in the medium is accelerated by the sum of the forces applied to its boundaries. By substituting (1.3.1) into (1.3.2), we obtain the equation describinig propagation of elastic waves in locally isotropic solid:

$$\varrho \frac{\partial^2}{\partial t^2} u_j = \frac{\partial}{\partial x_j} \left(\lambda \frac{\partial u_k}{\partial x_k} \right) + \frac{\partial}{\partial x_i} \left[\mu \left(\frac{\partial u_i}{\partial x_j} + \frac{\partial u_j}{\partial x_i} \right) \right] \quad . \tag{1.3.3}$$

The vector form of this equation is

$$\varrho \frac{\partial^2 u}{\partial t^2} = (\lambda + \mu) \operatorname{grad} (\operatorname{div} u) + \mu \Delta u + \operatorname{grad} \lambda \operatorname{div} u$$
$$+ \operatorname{grad} \mu \times \operatorname{curl} u + 2(\operatorname{grad} \mu \cdot \nabla)u \quad , \tag{1.3.4}$$

where $\times$ indicates a vector product. In (1.3.3,4) the Lamé constants must be continuous in space and have finite derivatives.

At interfaces between solids solutions of these equations must obey some boundary conditions. These can be rather diverse and depend on the kind of contact between the solids. The most important case is that of contact of solids without slip (welded contact). In this problems, the kinematic boundary condition is simply that continuity of the displacement u at the boundary must be maintained. The dynamic boundary condition is that we must have continuity in the components of the stress tensor: σ_{nj}, $j = 1, 2, 3$, where n denotes the axis normal to the boundary. Note that now we have altogether six boundary conditions instead of two as in case of contact between two liquids. This is due to the fact that in solids there are more types of waves which can interact at the boundaries.

At a welded contact with an infinitely rigid wall, the condition $u = 0$ must be fulfilled; no conditions are imposed upon the stress tensor. If the solids can slip along the boundary (contact with slip, also called perfectly lubricated or unbonded contact) then four conditions take place: $[u_n]_S = 0$, $[\sigma_{nn}]_S = 0$, and $\sigma_{nj} = 0$, $j \neq n$. These conditions are also true at the boundary between a solid and a nonviscous fluid. At the boundary of a solid and vacuum (free, or traction-free, boundary) three conditions exist: $\sigma_{nj} = 0$, $j = 1, 2, 3$.

The elastic wave equation in inhomogeneous solids is much more complicated than the sound wave equation (1.1.11) in liquids. In fact, (1.3.4) is a set of three coupled scalar equations; each of these equations is not less complicated than (1.1.11).

The coupling of these scalar equations relates to the transformation of one type of wave into another at each point in an inhomogeneous medium.

Since (1.3.4) is rather complicated, analysis of simple limiting cases is important to gain a physical understanding of the problem. Some of these cases, where (1.3.4) reduces to a set of independent scalar wave equations, were considered in [1.15]. The special cases of layered media where longitudinal and shear waves are generally interrelated, but one can propagate without generating the other, were considered in [1.16].

1.3.2 Elastic Waves in Homogeneous Solids

In a homogeneous solid, (1.3.4) can be transformed into

$$\frac{\partial^2 u}{\partial t^2} = \frac{\lambda + 2\mu}{\varrho} \operatorname{grad}(\operatorname{div} u) - \frac{\mu}{\varrho} \operatorname{curl}(\operatorname{curl} u) \quad . \tag{1.3.5}$$

In this transformation, the identity

$$\Delta a = \operatorname{grad}(\operatorname{div} a) - \operatorname{curl}(\operatorname{curl} a)$$

must be taken into account.

In general, the displacement vector u can be related to the scalar φ and vector ψ potentials by

$$u = u_l + u_t \quad , \quad u_l = \operatorname{grad} \varphi \quad , \quad u_t = \operatorname{curl} \psi \quad . \tag{1.3.6}$$

Note that the relative change in the volume of the particle during deformation is div u. Hence in (1.3.6) the deformation is divided into two parts: one connected with change in the particle volume (u_l) and another without change in volume (the pure shear component u_t).

By introducing (1.3.6) into (1.3.5) we obtain

$$\frac{\partial^2 u_l}{\partial t^2} - \frac{\lambda + 2\mu}{\varrho} \Delta u_l + \frac{\partial^2 u_t}{\partial t^2} - \frac{\mu}{\varrho} \Delta u_t = 0 \quad . \tag{1.3.7}$$

After successively applying the operators div and curl to the last equation we find

$$\operatorname{div}\left(\frac{\partial^2 u_l}{\partial t^2} - \frac{\lambda + 2\mu}{\varrho} \Delta u_l\right) = 0 \quad , \quad \operatorname{curl}\left(\frac{\partial^2 u_l}{\partial t^2} - \frac{\lambda + 2\mu}{\varrho} \Delta u_l\right) = 0 \quad ,$$

$$\operatorname{div}\left(\frac{\partial^2 u_t}{\partial t^2} - \frac{\mu}{\varrho} \Delta u_t\right) = 0 \quad , \quad \operatorname{curl}\left(\frac{\partial^2 u_t}{\partial t^2} - \frac{\mu}{\varrho} \Delta u_t\right) = 0 \quad .$$

It follows from these relations that the quantities in brackets are vectors which depend only on t. Supposing that u_l and u_t are defined to within an accuracy of additive vectors depending only on t, we obtain the equations

$$\frac{\partial^2 u_l}{\partial t^2} - c_l^2 \Delta u_l = 0 \quad , \quad c_l = \sqrt{(\lambda + 2\mu)/\varrho} \quad , \tag{1.3.8}$$

$$\frac{\partial^2 u_t}{\partial t^2} - c_t^2 \Delta u_t = 0 \quad , \quad c_t = \sqrt{\mu/\varrho} \quad . \tag{1.3.9}$$

Hence, longitudinal and shear waves propagate independently in a homogeneous solid.

According to (1.3.6), the potentials φ and ψ are defined, respectively, to within an accuracy of an arbitrary function of t and the gradient of an arbitrary scalar $\psi_1 = \mathrm{grad}\, A(\mathbf{r}, t)$. Taking this into account one can easily obtain from (1.3.8–11) the equations for the potentials

$$\frac{\partial^2 \varphi}{\partial t^2} - c_l^2 \Delta \varphi = 0 \quad , \tag{1.3.10}$$

$$\frac{\partial^2 \psi}{\partial t^2} - c_t^2 \Delta \psi = 0 \quad . \tag{1.3.11}$$

Equations (1.3.8, 9) are analogous to (1.1.16). The simplest solutions of the latter we considered in Sect. 1.1. Plane wave solutions of (1.3.10, 11) are, compare with (1.1.17),

$$\varphi = f(\mathbf{n} \cdot \mathbf{r}/c_l - t) \quad , \quad n^2 = 1 \quad , \quad \mathbf{n} = \mathrm{const} \quad , \tag{1.3.12}$$

$$\psi = f(\mathbf{n} \cdot \mathbf{r}/c_t - t) \quad , \quad n^2 = 1 \quad , \quad \mathbf{n} = \mathrm{const} \quad . \tag{1.3.13}$$

where c_l and c_t are the corresponding wave velocities. Since $\mu \geq 0$ always and $\lambda > 0$ for real solids [1.5, 14] it follows from (1.3.8, 9) that

$$c_l > \sqrt{2} c_t \quad . \tag{1.3.14}$$

The plane wave in (1.3.12) is longitudinal, as is the wave in (1.1.17). Indeed, we have for the displacement

$$\mathbf{u}_l = \mathrm{grad}\, \varphi = \mathbf{n} \frac{\dot{f}(\mathbf{n} \cdot \mathbf{r}/c_l - t)}{c_l} \quad , \quad \dot{f}(\xi) \equiv \frac{df(\xi)}{d\xi} \quad ,$$

that is, displacement occurs along the direction of propagation. The plane wave in (1.3.13) is transverse. The displacement is normal to the wave propagation direction as one can see from the relations

$$\mathbf{n} \cdot \mathbf{u}_t = \mathbf{n} \cdot \mathrm{curl}\, \psi = c_t^{-1} \mathbf{n} \cdot [\mathbf{n} \times \dot{f}(\mathbf{n} \cdot \mathbf{r}/c_t - t)] = 0 \quad .$$

1.3.3 Elastic Wave Equations in Layered Solids

Waves in layered media with spherical symmetry were treated in [Ref. 1.17; Chap. 9]. We now consider horizontally stratified solids. We assume harmonical dependence of the waves on the horizontal coordinates and on time

$$\mathbf{u}(\mathbf{r}, \omega) = \mathbf{u}(z, \xi, \omega) \exp(i\xi \cdot \mathbf{r} - i\omega t) \tag{1.3.15}$$

and choose that the x-axis is directed along the vector ξ. Then the field of the wave will not depend on the coordinate y. Under these conditions (1.3.4) can be written as a set of three ordinary scalar differential equations:

$$-\omega^2 \varrho u_1 = i\xi \left[(\lambda + \mu) \frac{\partial u_3}{\partial z} + \frac{\partial \mu}{\partial z} u_3 \right] + \frac{\partial}{\partial z} \left(\mu \frac{\partial u_1}{\partial z} \right)$$
$$- \xi^2 (\lambda + 2\mu) u_1 \quad , \tag{1.3.16}$$

$$-\omega^2 \varrho u_2 = \frac{\partial}{\partial z}\left(\mu \frac{\partial u_2}{\partial z}\right) - \xi^2 \mu u_2 \quad , \tag{1.3.17}$$

$$-\omega^2 \varrho u_3 = \mathrm{i}\xi\left[\frac{\partial}{\partial z}(\lambda u_1) + \mu\frac{\partial u_1}{\partial z}\right] + \frac{\partial}{\partial z}\left[(\lambda + 2\mu)\frac{\partial u_3}{\partial z}\right] - \xi^2 \mu u_3 \quad . \tag{1.3.18}$$

Equations (1.3.16–18) cease to be coupled if $\xi = 0$, that is, when the propagation of the wave is normal to the layers. We also note that in the case of arbitrary incidence, (1.3.17) is not coupled with the two others. Hence, the waves with displacement along the y-axis propagate independently of the waves polarized in the xz-plane. In seismology, the case when $u_1 = u_3 = 0$ is usually referred to as an SH wave (Shear or Secondary wave, Horizontal polarization). In the second case (when displacement lies in the vertical plane), the wave may be shear and is referred to as an SV wave (Shear or Secondary wave, Vertical polarization) or longitudinal which is then called a P wave (Primary wave, due to its greater speed).

We shall first consider the SH wave omitting, for simplicity, the index "2" in the displacement $\boldsymbol{u} = (0, u_2, 0)$. According to Hooke's law (1.3.1), only four components of the stress tensor are not zero in this case:

$$\sigma_{12} = \sigma_{21} = \mathrm{i}\xi\mu u \quad , \qquad \sigma_{23} = \sigma_{32} = \mu\frac{\partial u}{\partial z} \quad . \tag{1.3.19}$$

Equation (1.3.17) for an SH wave can be written in the form

$$\frac{\partial^2}{\partial z^2}u - \frac{\partial}{\partial z}\left(\ln\frac{1}{\mu}\right)\frac{\partial}{\partial z}u + (k_t^2 - \xi^2)u = 0 \quad , \tag{1.3.20}$$

where

$$k_t = \omega/c_t = \omega\sqrt{\varrho/\mu} \tag{1.3.21}$$

is the wave number for a shear wave. Note that (1.3.20) is analogous to (1.2.7) for acoustic waves; here we have the displacement u in place of the sound pressure p, and the inverse shear modulus $1/\mu$ in place of the density ϱ. Moreover, the boundary conditions for an SH wave at a free boundary, an absolutely rigid wall, and at a boundary of solids with welded contact appear to be the same as the conditions in (1.1.24, 21, 23) for a sound wave at an absolutely rigid wall, a free boundary, and at a boundary of two liquids, respectively, if the substitution $u \to p$, $1/\mu \to \varrho$ is used.

Equation (1.3.20), just as (1.2.7) can be reduced to the Helmholtz equation by the appropriate transformation of dependent or independent variables. Thus, the substitution

$$\Psi = \sqrt{\mu}u \tag{1.3.22}$$

yields from (1.3.20) the equation

$$\frac{\partial^2}{\partial z^2}\Psi + \left[k_t^2 - \xi^2 + \frac{1}{4}\left(\frac{1}{\mu}\frac{\partial \mu}{\partial z}\right)^2 - \frac{1}{2\mu}\frac{\partial^2 \mu}{\partial z^2}\right]\Psi = 0 \quad , \tag{1.3.23}$$

which is analogous to (1.2.21).

Introducing the new vertical coordinate

$$\zeta(z) = \mu_0\int_{z_0}^{z}\mu^{-1}(z')dz' \quad , \qquad \mu_0 = \text{const} \quad , \qquad z_0 = \text{const} \quad , \tag{1.3.24}$$

we transform (1.3.20) to, compare with (1.2.25),

$$\partial^2 u / \partial \zeta^2 + (k_t^2 - \xi^2)(\mu/\mu_0)^2 u = 0 \quad , \tag{1.3.25}$$

which can be applied to media with piecewise-continuous parameters.

Thus, we see that the theory of shear horizontally polarized wave propagation in layered solids is quite similar to the theory of sound-wave propagation in layered liquids.

The propagation theory of vertically polarized elastic waves, which obey (1.3.16 and 18), appears to be much more complicated. First, we note that in the case of welded contact between two solids we have four boundary conditions

$$[\sigma_{13}]_S = 0 \quad , \quad [\sigma_{33}]_S = 0 \quad , \quad [u_1]_S = 0 \quad , \quad [u_3]_S = 0 \quad . \tag{1.3.26}$$

The u_2 component of the displacement vector and components σ_{2j}, $j = 1, 2, 3$ of the stress tensor are zero in the case of vertical polarization. It is also important to note that the potential ψ can be chosen in such a way that only component ψ_2 is not zero. Indeed, the displacement u_t lies in the vertical plane and does not depend on the x and z components of ψ, according to (1.3.6). Therefore, we can assume

$$\psi = (0, \psi_2, 0) \tag{1.3.27}$$

without loss of generality.

Discussion of P and SV waves propagation in layered solids is continued first assuming media consisting of homogeneous layers in Chap. 4 and then more generally in Chaps. 6 and 7.

2. Plane Waves in Discretely Layered Fluids

A discretely layered medium is a set of homogeneous layers in contact. Such a model is widely used because it is a good approximation of many real geophysical and technical systems. The theory of sound-wave propagation in discretely layered media has been elaborated rather thoroughly. Media with continuously varying parameters can be approximated by discrete media by assuming an increasing number of layers with decreasing thickness.

In this chapter, waves of the form $p(z, \xi, \omega) \exp(i\xi \cdot r - i\omega t)$ with harmonic dependence on time and horizontal coordinates will be discussed. For brevity, the arguments in $p(z, \xi, \omega)$ and often the exponential will be omitted. In addition, the dissipation of the wave will be neglected. This aspect will be discussed in Chap. 7. We will begin this chapter with a very important generalization in the concept of a plane wave.

2.1 Inhomogeneous Plane Waves. Energy of Sound Waves

In the definition of a plane wave (1.1.17), the vector n was assumed to be real. In the case of *monochromatic* plane waves, however, the assumption about the realness of the wave vector kn is not necessary. Let us look for the solution to the wave equation (1.1.16) in homogeneous media at rest in the form

$$p = A \exp(iq \cdot r - i\omega t) \quad , \quad A = \text{const} \quad . \tag{2.1.1}$$

Substitution of this expression into (1.1.16) yields

$$q \cdot q = k^2 \quad . \tag{2.1.2}$$

This condition can be satisfied by the complex vector $q = q_1 + iq_2$, where q_1 and q_2 are real vectors, if

$$q_1 \cdot q_2 = 0 \quad , \quad q_1^2 - q_2^2 = k^2 \quad . \tag{2.1.3}$$

The expression

$$p = A \exp[-q_2 \cdot r + i(q_1 \cdot r - \omega t)] \quad , \tag{2.1.4}$$

with q_1 and q_2 satisfying these conditions, describes an *inhomogeneous plane wave*. Its fronts (planes of constant phase) are orthogonal to the vector q_1. The amplitude of the wave varies exponentially along its fronts (in contrast to the case of a homogeneous plane wave). The amplitude is constant at a plane orthogonal to the vector q_2. According to (2.1.3), constant phase planes and constant amplitude planes are orthogonal to each other. The phase velocity of an inhomogeneous wave is

$$c_{ph} = \omega/q_1 = \omega(k^2 + q_2^2)^{-1/2} < c \quad ,$$

and is less than the velocity of homogeneous plane sound waves.

An inhomogeneous plane wave can not exist in infinite homogeneous space since, at infinity, the sound pressure in such a wave would be infinite. In layered media, however, inhomogeneous waves occur rather often.

Suppose that the wave's direction of propagation lies in the xz-plane. If θ is the angle between this direction and the z-axis, the homogeneous plane wave (2.1.1) may be written as

$$p = A \exp [\mathrm{i}(kz \cos \theta + kx \sin \theta - \omega t)] \quad . \tag{2.1.5}$$

An inhomogeneous plane wave may be written in the same form but θ is then complex. By assuming in (2.1.5), for example, $\theta = \pi/2 - \mathrm{i}\alpha$ we obtain

$$p = A \exp (\mathrm{i}kx \cdot \cosh \alpha - kz \sinh \alpha - \mathrm{i}\omega t) \quad . \tag{2.1.6}$$

Such a wave propagates in the x- and decays in the z-direction. The phase velocity $c_{ph} = c/\cosh \alpha$ decreases and the attenuation coefficient in the z-direction increases when α increases.

The energy density E and specific energy flux I in homogeneous and inhomogeneous plane sound waves are, according to [2.1, 2],

$$E = E_K + E_I \quad , \quad E_K = \frac{\varrho v^2}{2} \quad , \quad E_I = \frac{p^2}{2\varrho c^2} \quad , \tag{2.1.7}$$

$$I = pv \quad . \tag{2.1.8}$$

Acoustic energy consists of two parts: kinetic E_K that is the energy of a particle's motion, and internal E_I, representing the work done against the sound pressure. The quantities E and I, when averaged over the half period $T/2 = \pi/\omega$, we will denote by E_T and I_T. To calculate these values by the use of (2.1.7, 8) we must take into account that only the real parts of the complex quantities p and v must be substituted into these relations. Note that if $a = |a| \exp (\mathrm{i}\alpha)$ and $b = |b| \exp (\mathrm{i}\beta)$ are two complex quantities then the identity

$$[\mathrm{Re} \{a \exp (-\mathrm{i}\omega t)\} \mathrm{Re} \{b \exp (-\mathrm{i}\omega t)\}]_T$$
$$= \tfrac{1}{2}|ab| \cos (\alpha - \beta) = \tfrac{1}{2}\mathrm{Re} \{ab^*\} \tag{2.1.9}$$

holds, where the asterisk denotes a complex conjugate. Now we obtain from (2.1.7–9)

$$E_T = \frac{\varrho|v|^2}{4} + \frac{|p|^2}{4\varrho c^2} = \frac{|\nabla p|^2 + k^2|p|^2}{4\varrho\omega^2} \quad , \tag{2.1.10}$$

$$I_T = 0.5 \, \mathrm{Re} \{p^* v\} = (2\omega\varrho)^{-1} \mathrm{Im} \{p^* \nabla p\} \quad . \tag{2.1.11}$$

The average sound energy density in an inhomogeneous wave (2.1.4) is

$$E_T = (2\varrho\omega^2)^{-1} q_1^2 |A|^2 \exp (-2q_2 \cdot r) \quad . \tag{2.1.12}$$

It decays exponentially along the direction q_2. This density and that in a homogeneous wave

$$E_T = (2\varrho c^2)^{-1}|A|^2 \tag{2.1.13}$$

do not vary in the direction of the wave propagation.

It is also worth noting that the energy density in an inhomogeneous wave is greater than in an ordinary plane wave if the sound pressure is the same in both waves. This is because under this condition the particle velocity is greater in the inhomogeneous wave. The average specific energy flux in inhomogeneous and in ordinary plane waves are, correspondingly,

$$I_T = (2\omega\varrho)^{-1}q_1|A|^2 \exp(-2q_2 \cdot r) \quad , \tag{2.1.14a}$$

$$I_T = (2\varrho c)^{-1})n|A|^2 \quad , \quad n = q_1/k \quad , \quad q_2 = 0 \quad . \tag{2.1.14b}$$

Although the inhomogeneous wave is slow, the energy flux is greater than in the ordinary wave, however, if the sound pressure $|p|$ is the same since E_T is greater. Note also that the instantaneous energy flux is not zero along the q_2-direction whereas the average flux I_T occurs only in the q_1-direction which is orthogonal to q_2.

The wave equation for sound pressure in homogeneous moving media ($v_0 =$ const) follows from (1.1.15):

$$c^{-2}\frac{d^2p}{dt^2} - \Delta p = 0 \quad . \tag{2.1.15}$$

Solutions to this equation also include homogeneous (1.2.14) and inhomogeneous plane waves. The real and imaginary parts of the wave vector in an inhomogeneous wave are related to each other by

$$k^2\left[\left(1 - \frac{q_1 \cdot v_0}{\omega}\right)^2 - \left(\frac{q_2 \cdot v_0}{\omega}\right)^2\right] = q_1^2 - q_2^2 \quad ,$$

$$k^2 q_2 \cdot v_0 \frac{(q_1 \cdot v_0 - \omega)}{\omega^2} = q_1 \cdot q_2 \quad . \tag{2.1.16}$$

Presence of flow does not change the results obtained above if v_0 is orthogonal to q_1 as well as to q_2. The situation becomes more complicated in the case of arbitrary direction in the flow velocity. In particular the planes of constant phase and constant amplitude are not orthogonal if $q_2 \cdot v_0 \neq 0$ and $q_1 \cdot v \neq \omega$. Readers interested in the acoustic energy in homogeneous and inhomogeneous moving media are referred to [2.16].

2.2 Reflection at the Interface of Two Homogeneous Media

Let the plane $z = 0$ be the interface between the "upper" ($z > 0$, sound velocity c, density ϱ) and "lower" ($z < 0$, c_1, ϱ_1) homogeneous media at rest. The plane sound wave is assumed to be incident from the half-space $z > 0$ upon this interface (Fig. 2.1). Let the xz-plane coincide with the *incidence plane,* that is, with the plane including the normal to the interface as well as the wave vector of the incident wave. The amplitude of the incident wave is assumed to be unity. Then the incident and the reflected waves can be written correspondingly:

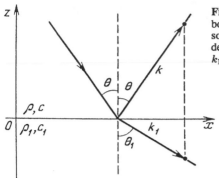

Fig. 2.1. Reflection and refraction of a plane wave at the boundary between two media. *Upper medium*: speed of sound c, density ϱ, *lower medium*: speed of sound c_1, density ϱ_1. Reflected wave vector k, refracted wave vector k_1

$$p_i = \exp\left[ik(x \sin\theta - z \cos\theta)\right] \quad ,$$
$$p_r = V \exp\left[ik(x \sin\theta + z \cos\theta)\right] \quad , \quad k = \omega/c \quad , \tag{2.2.1}$$

where θ is the *incidence angle,* that is, the angle between the wave vector and the z-axis, V is the *reflection coefficient,* the ratio between the complex amplitudes of the reflected and incident waves. The total sound pressure in the upper medium is

$$p = p_i + p_r = \left[\exp\left(-ikz \cos\theta\right) + V\right.$$
$$\left. \times \exp\left(ikz \cos\theta\right)\right] \exp\left(ikx \sin\theta\right) \quad . \tag{2.2.2}$$

The refracted wave in the lower medium can be written as

$$p_1 = W \exp\left[ik_1(x \sin\theta_1 - z \cos\theta_1)\right] \quad , \quad k_1 = \omega/c_1 \quad , \tag{2.2.3}$$

where θ_1 is *refraction angle* and W is *transmission (refraction) coefficient.*

The quantities V, W, and θ_1 must be determined by the use of the boundary conditions (1.1.23). Since these conditions must be fulfilled simultaneously at any point (x, y, z) at the boundary, waves with different horizontal wave vectors must satisfy the conditions independently. Hence, incident reflected and refracted waves must have the same horizontal component of the wave vector. This fact was already accounted for in (2.2.1). Obviously this is an immediate consequence of the fact that the medium parameters are invariant with respect to horizontal translations. In addition, all these waves have the same frequency if the media are steady state. The concept of invariability of the horizontal component of the wave vector when applied to the refracted wave immediately yields *Snell's refraction law*

$$k \sin\theta = k_1 \sin\theta_1 \quad , \tag{2.2.4}$$

which can also be written as

$$\frac{\sin\theta}{\sin\theta_1} = n \quad , \tag{2.2.5}$$

where $n \equiv k_1/k = c/c_1$ is a *refraction index.* The following form of Snell's law is mnemonically convenient

$$c^{-1}(z) \sin\theta(z) = \text{const} \quad . \tag{2.2.6}$$

We will frequently use the concept of the impedance[1] of the wave

$$Z = -\frac{p}{v_z} = -\frac{i\omega \varrho p}{\partial p/\partial z} \quad . \tag{2.2.7}$$

In the general case of layered media and waves with harmonic dependence on time and horizontal coordinates, Z is a function of z, ω and ξ. Boundary conditions (1.1.23) can be written in terms of the impedance as

$$[p]_{z=0} = 0 \quad , \quad [Z]_{z=0} = 0 \quad . \tag{2.2.8}$$

The first of these conditions yields the relation between the reflection and refraction coefficients:

$$1 + V = W \quad . \tag{2.2.9}$$

To use the second condition of (2.2.8) we first find the impedance of the wave in the lower medium with the help of (2.2.3,7)

$$Z_1 = \frac{\varrho_1 c_1}{\cos \theta_1} \quad , \tag{2.2.10}$$

i.e., the quantity independent of z. In the same way we find the impedance of the total (incident plus reflected) field in the upper medium:

$$Z = \frac{\varrho c}{\cos \theta} \frac{\exp(-2ikz\cos\theta) + V}{\exp(-2ikz\cos\theta) - V} \quad . \tag{2.2.11}$$

The condition of the equality of (2.2.10 and 11) at $z = 0$ yields for the reflection coefficient

$$V = \frac{Z_1 \cos \theta - \varrho c}{Z_1 \cos \theta + \varrho c} \quad . \tag{2.2.12}$$

By the same reasoning, it follows that for the reflection coefficient (2.2.12) holds at the boundary of the homogeneous half-space $z > 0$ and at the *arbitrary layered half-space* $z < 0$, if the "input" impedance Z_1 of the latter is known at $z = 0$.

In the simplest case (homogeneous lower half-space) we find, by using relation (2.2.10) for Z_1:

$$V = \frac{m \cos \theta - n \cos \theta_1}{m \cos \theta + n \cos \theta_1} \quad , \quad m \equiv \frac{\varrho_1}{\varrho} \quad , \tag{2.2.13}$$

or

$$V = \frac{m \cos \theta - \sqrt{n^2 - \sin^2 \theta}}{m \cos \theta + \sqrt{n^2 - \sin^2 \theta}} \quad , \tag{2.2.14}$$

if (2.2.4) is taken into account.

If $\sin \theta > n$, the sign of the root $\sqrt{n^2 - \sin^2 \theta}$ is determined from the conditions of finiteness of the refracted wave given by (2.2.3) at infinity $z = -\infty$, namely:

$$\mathrm{Im}\{\cos \theta_1\} = \mathrm{Im}\{\sqrt{n^2 - \sin^2 \theta}/n\} \geq 0 \quad . \tag{2.2.15}$$

---

[1] To be more exact, the *impedance* defined by (2.2.7) is a *normal* one. If one uses in this definition the entire value of the velocity v instead of only its normal component v_z one obtains the *characteristic* impedance. In our discussion, however, the latter is rarely used.

Relation (2.2.12) takes its simplest and the most symmetrical form by substituting the impedance of incident wave $Z = \varrho c / \cos \theta$:

$$V = \frac{Z_1 - Z}{Z_1 + Z} \quad . \tag{2.2.16}$$

The transmission coefficient can be found by the use of (2.2.9):

$$W = \frac{2Z_1 \cos \theta}{Z_1 \cos \theta + \varrho c} = \frac{2Z_1}{Z_1 + Z} \quad , \tag{2.2.17}$$

or

$$W = \frac{2m \cos \theta}{m \cos \theta + n \cos \theta_1} = \frac{2m \cos \theta}{m \cos \theta + \sqrt{n^2 - \sin^2 \theta}} \quad . \tag{2.2.18}$$

Expressions (2.2.13, 14, 18) are called the *Fresnel formulas*.

Let us consider some particular cases. In the case of normal incidence ($\theta = \theta_1 = 0$) we have

$$V = \frac{m - n}{m + n} = \frac{\varrho_1 c_1 - \varrho c}{\varrho_1 c_1 + \varrho c} \quad ; \quad W = \frac{2m}{m + n} = \frac{2\varrho_1 c_1}{\varrho_1 c_1 + \varrho c} \quad .$$

We see that c, c_1, ϱ, and ϱ_1 enter into these expressions only as products ϱc and $\varrho_1 c_1$, called the *wave* or *characteristic impedances*.

In another case, where the sound velocities in both media are equal ($c = c_1$, $n = 1$) the reflection and transmission coefficients do not depend on the incidence angle:

$$V = \frac{\varrho_1 - \varrho}{\varrho_1 + \varrho} = \frac{m - 1}{m + 1} \quad ; \quad W = \frac{2\varrho_1}{\varrho_1 + \varrho} = \frac{2m}{m + 1} \quad . \tag{2.2.19}$$

If $n \neq 1$, $\theta \to \pi/2$ (grazing incidence) Eqs. (2.2.16, 17), yield $V \to -1$, $W \to 0$.

If θ satisfies the equation $m \cos \theta = (n^2 - \sin^2 \theta)^{1/2}$, the reflection coefficient becomes zero. This is the case of a completely transparent interface. We then find

$$\tan^2 \theta = \frac{m^2 - n^2}{n^2 - 1} \quad . \tag{2.2.20}$$

The complete transparency angle is real if $(m^2 - n^2)/(n^2 - 1) > 0$. This occurs when $1 < n < m$ or when $1 > n > m$.

One can show by simple differentiation of (2.2.14) that the reflection and transmission coefficients are monotonic functions of θ if $n > 1$. If $n < 1$ the functions are monotonic at $0 \leq \theta \leq \delta \equiv \arcsin n$. The reflection coefficient is complex if $\delta < \theta < \pi/2$. In this case, we have total reflection and (2.2.14) may be written as

$$V = \frac{m \cos \theta - i(\sin^2 \theta - n^2)^{1/2}}{m \cos \theta + i(\sin^2 \theta - n^2)^{1/2}} \equiv \exp(i\varphi) \quad ,$$

$$\varphi = -2 \arctan \frac{(\sin^2 \theta - n^2)^{1/2}}{m \cos \theta} \quad . \tag{2.2.21}$$

Here, the condition of (2.2.15) was also taken into account. In this case, we see that $|V| = 1$ and that φ is a monotonic function of θ. The phase shift $\varphi(\theta)$ is the cause of rather interesting phenomena when reflection of bounded beams or sound pulses are considered (Chap. 5 and [Ref. 2.3; Sect. 14; Ref. 2.17, Chap. 2]).

As for the transmission coefficient, its modulus decreases from 2 down to zero when θ increases from the *critical angle of total reflection* δ up to $\pi/2$. The phase difference between the refracted and incident waves is half of that between the reflected and incident waves: $W = |W| \exp(i\varphi/2)$.

In the case of total reflection, the field in the lower medium is, according to (2.2.3, 5), an inhomogeneous plane wave

$$p_1 = W \exp(\mu z + ikx \sin\theta) \quad , \quad \mu = k\sqrt{\sin^2\theta - n^2} \quad , \tag{2.2.22}$$

where the amplitude decreases exponentially as the distance $|z|$ from the boundary increases. From (2.2.7) we see that impedance $Z_1 = -i\omega\varrho_1/\mu$ is purely imaginary in this case. The average energy flux in the x-direction in the refracted wave can be calculated by the use of (2.1.14a). Such a flux in the z-direction is zero, hence the energy carried by the incident wave returns to the upper medium completely by the reflected wave.

It is instructive to discuss the behavior of the reflection coefficient V on its complex plane. By letting the abscissa represent the real part and the ordinate represent the imaginary part of V, for various relationships between the parameters of the two media we obtain curves such as those shown in Fig. 2.2. In cases a and b ($n > 1$) the reflection coefficient is real. Therefore, for various values of θ, this coefficient is confined to a segment of a straight line lying on the real axis. In case a the reflection coefficient is zero for θ defined by (2.2.20). In case b it is negative for all values $0 \le \theta \le \pi/2$ and is never zero.

In cases c and d ($n < 1$) total reflection occurs. Here, when $\theta > \delta = \arcsin n$, the points corresponding to the complex values of the reflection coefficient lie on a semicircle of unit radius. This shows graphically that the modulus of the reflection coefficient is unity and that only its phase changes as θ varies. In case d the reflection coefficient becomes zero at the angle θ which satisfies (2.2.20).

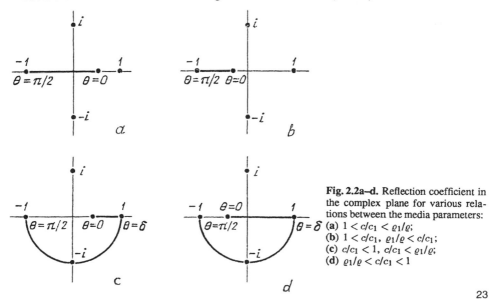

Fig. 2.2a–d. Reflection coefficient in the complex plane for various relations between the media parameters: (a) $1 < c/c_1 < \varrho_1/\varrho$; (b) $1 < c/c_1$, $\varrho_1/\varrho < c/c_1$; (c) $c/c_1 < 1$, $c/c_1 < \varrho_1/\varrho$; (d) $\varrho_1/\varrho < c/c_1 < 1$

23

We also discuss some symmetry properties of reflection and transmission processes. Let the wave be incident from the upper medium at an angle θ and its refraction angle in the lower one be θ_1. Now if we reverse the propagation direction and let the wave be incident from below at the angle θ_1, then according to (2.2.4) the refraction angle in the upper medium will be θ. Moreover, if the reflection coefficient for the wave incident from the upper medium at the angle θ is V, then according to (2.2.16) it will be $-V$ for the wave incident from below at the angle θ_1. In particular, if θ is the angle of complete transmission for the wave incident from above [$V(\theta) = 0$], then θ_1 is that angle for the wave incident from below. Such a simple rule does not hold for the transmission coefficient W, however. The quantity $W \cos \theta / \varrho c$, according to (2.2.18), is invariant when one reverses the propagation direction of the wave (θ, ϱ, and c are the refraction angle, density, and sound velocity of the medium where the refracted wave propagates).

It further follows from (2.2.3 and 9) that the pressure at the boundary in the refracted wave is $1+V$ times that in the incident one. Consider, for example, reflection at an air-water boundary assuming normal wave incidence from the air. In this case, one has $\varrho = 1.3 \times 10^{-3}$ g/cm^3, $\varrho_1 = 1$ g/cm^3, $c = 333$ m/s, $c_1 = 1500$ m/s, and according to (2.2.14) $V \approx 1$. Hence, the amplitude of the pressure of the wave in water will be twice of that in the incident wave in air. Quite on the contrary, if the wave is incident from the water, we have (taking into account that $\varrho c / \varrho_1 c_1 \approx 2.9 \times 10^{-4} \ll 1$):

$$V = \frac{\varrho c - \varrho_1 c_1}{\varrho c + \varrho_1 c_1} \approx -1 + \frac{2\varrho c}{\varrho_1 c_1} \approx -1 \;\; ;$$

$$W = 1 + V \approx \frac{2\varrho c}{\varrho_1 c_1} \approx 5.8 \times 10^{-4} \;\; .$$

The amplitude of the transmitted wave is very small in this case. We see that there is no symmetry in reflection and transmission of a wave with respect to sound pressure. The same also occurs for the particle velocity v as can be easily shown. Such a symmetry does exist, however, for the normal components of the specific energy flux I_z, except in the case of total reflection. Indeed, we have according to (2.1.14b) $I_z = |p|^2 \cos \theta / 2\varrho c$ and $I_{1z} = |p_1|^2 \cos \theta_1 / 2\varrho_1 c_1$ for incident and refracted waves, respectively. The ratio of these quantities

$$R \equiv \frac{I_{1z}}{I_z} = \frac{\varrho c \cos \theta_1}{\varrho_1 c_1 \cos \theta} |W|^2 \tag{2.2.23}$$

is *the transmission coefficient for sound energy*. Taking into account expression (2.2.18) for W, we obtain from (2.2.23)

$$R = 4 \frac{\cos \theta}{\varrho c} \cdot \frac{\cos \theta_1}{\varrho_1 c_1} \left(\frac{\cos \theta}{\varrho c} + \frac{\cos \theta_1}{\varrho_1 c_1} \right)^{-2} \;\; . \tag{2.2.24}$$

The last expression is symmetric with respect to quantities representing both media, therefore it does not change when the direction of propagation is reversed.

We have obtained the symmetry relation using the quantities V and W for the simplest model of the media (two homogeneous fluid half-spaces in contact). In Chap. 6 we will generalize these relations to the case of arbitrary layered moving media without obtaining explicit expressions for V and W.

2.3 Locally Reacting Surfaces

In some cases the impedance of the boundary may be independent or almost independent of the angle of incidence. Consider, for example, the case where the sound velocity in the upper medium is much greater than in the lower one ($n = c/c_1 \gg 1$). By using the refraction law (2.2.5) we obtain $\cos \theta_1 = (1 - n^2 \sin^2 \theta)^{1/2} \approx 1$ and (2.2.10) yields $Z_1 \approx \varrho_1 c_1$ for any θ.

In such cases, the calculation of the reflection coefficient can be simplified. We do not need to consider the sound field in the lower medium and can use one rather than the two boundary conditions of (2.2.8), namely, the continuity of impedance at the boundary, see (2.2.7):

$$z = 0 \quad , \quad \frac{\partial p}{\partial z} + \gamma p = 0 \quad , \quad \gamma \equiv \frac{i\omega\varrho}{Z_1} \quad . \tag{2.3.1}$$

Relation (2.3.1) is called a boundary condition of the third kind or the *impedance boundary condition* (conditions of the first and the second kinds were discussed in Sect. 1.2).

In the case of the plane-wave reflection considered above the simplification of the problem connected with the use of (2.3.1) is not crucial since (2.2.12) obtained in a simple way is valid without any assumption about Z_1. In more sophisticated cases, however, where the wave fronts or boundaries are not planar, the use of the impedance boundary condition could simplify the problem significantly.

A particularly interesting case exists where Z_1 is not approximately constant but exactly. Let the sound wave be incident on the boundary at $z = 0$ below which a set of narrow channels of depth h terminated by a rigid boundary $z = -h$ (the comblike structure of Fig. 2.3) is located. We shall assume the width of the channels to be small both in comparison to the wavelength and to the depth h. Let us find the impedance Z_1 of this comblike structure in the plane $z = 0$ (Fig. 2.3) which will determine the reflection coefficient by (2.2.12). The incident sound wave will excite plane waves in the channels (as in narrow tubes) which travel in them in both directions. We shall neglect energy losses due to friction from the walls. The sound in each channel can be written as

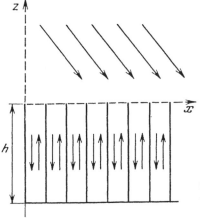

Fig. 2.3. Plane wave incident onto a comblike structure of narrow channels of depth h with a rigid boundary at $z = -h$, exciting plane waves in the channels

$$p_1 = A \exp(ikz) + B \exp(-ikz) \quad . \tag{2.3.2}$$

At the channel bottom $z = -h$ the boundary condition (1.1.24) must be satisfied. Substituting (2.3.2) into this condition yields $A = B \exp(2ikh)$. Taking this relation into account we obtain from (2.2.7) and (2.3.2) for the impedance at $z = 0$

$$Z_1 = i\varrho c \cdot \cot kh \quad , \tag{2.3.3}$$

a quantity that does not depend on θ.

Impedance boundary conditions are widely used in studying the acoustics of buildings. Sound absorbing material with open vertical pores has an angle-independent impedance as in the case of comblike structure. The boundary condition (2.3.1) can also be sometimes used in atmospheric acoustics. In [2.4], for example, the experimental data for the sound field over rather diverse surfaces including fresh snow, asphalt, etc., were satisfactorily explained in terms of angle-independent impedance. The use of such an impedance is generally valid in all cases in which the sound disturbance in the medium is not transmitted along its boundary. Therefore, the normal velocity at each point of the surface will be completely determined by the value of pressure at this point. Such surfaces are known as *locally reacting*.

In a comblike structure the transfer of the disturbance along the boundary is impossible because of the presence of the walls of the tubes. In the case of refraction of the wave at the boundary of two homogeneous half-spaces when $c \gg c_1$, the disturbance is not transferred along the boundary in the lower half-space because the refracted wave propagates almost normally to the boundary.

2.4 Reflection from a Plane Layer

2.4.1 The Input Impedance of a Layer

We shall suppose that a plane acoustic wave is incident on a plane layer of thickness d (Fig. 2.4) at an arbitrary angle of incidence. The numbers 3, 2, and 1 will denote, respectively, the medium through which the incident wave travels, the layer, and the medium into which the wave is transmitted. The angles between the propagation

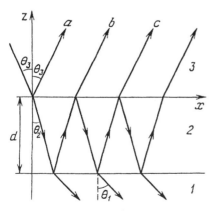

Fig. 2.4. System of waves arising when a plane wave is incident on a layer (2) between two media (1 and 3)

directions in each of the media and the normal to the boundaries of the layer will be denoted by θ_j, $j = 1, 2, 3$. The xz-plane will again be considered as the plane of incidence, as in Sect. 2.2.

According to (2.2.16) the reflection coefficient from a layer is

$$V = \frac{Z_{in} - Z_3}{Z_{in} + Z_3} \quad , \tag{2.4.1}$$

where

$$Z_3 = \frac{\varrho_3 c_3}{\cos \theta_3} \tag{2.4.2}$$

is the impedance of the plane wave in medium 3, Z_{in} is the input impedance of the layer, that is, the impedance at the boundary 2-3, which is to be determined.

It is shown in Sect. 2.2 that the horizontal component of the wave vector is the same all over the layered medium. Therefore, by analogy to (2.2.4) we have

$$k_1 \sin \theta_1 = k_2 \sin \theta_2 = k_3 \sin \theta_3 \quad . \tag{2.4.3}$$

After dropping the factor $\exp[\mathrm{i}(k_2 x \sin \theta_2 - \omega t)]$ which defines the dependence on horizontal coordinates and time, the acoustic field in a layer, see (1.2.10), can be written as

$$p_2 = A \exp(\mathrm{i}k_2 z \cos \theta_2) + B \exp(-\mathrm{i}k_2 z \cos \theta_2) \quad . \tag{2.4.4}$$

Note that impedance Z_1 at $z = 0$ is known through (2.2.10). The relationship between the constants A and B may be found from the continuity of the impedance at the boundary $z = 0$:

$$\left. \frac{-\mathrm{i}\omega \varrho_2 p_2}{\partial p_2/\partial z} \right|_{z=0} = Z_1 \quad , \quad \text{or} \quad \frac{A}{B} = \frac{Z_1 - Z_2}{Z_1 + Z_2} \quad , \tag{2.4.5}$$

where

$$Z_2 = \frac{\varrho_2 c_2}{\cos \theta_2} \tag{2.4.6}$$

is the impedance of the plane wave in medium 2. At $z = d$ the impedance of wave (2.4.4) is equal to the unknown input impedance of the layer. After taking into account (2.4.5) we find the very important formula

$$Z_{in} = Z_2 \frac{Z_1 - \mathrm{i}Z_2 \tan \varphi}{Z_2 - \mathrm{i}Z_1 \tan \varphi} \tag{2.4.7}$$

which expresses the input impedance at the boundary under consideration in terms of that at adjacent boundary. Here

$$\varphi = k_2 d \cos \theta_2 \tag{2.4.8}$$

is the phase advance of the plane wave passing through the layer.

2.4.2 The Reflection and Transmission Coefficients

Substituting (2.4.7) into (2.4.1) we obtain for the reflection coefficient from the layer

$$V = \frac{(Z_1 + Z_2)(Z_2 - Z_3)\exp(-2i\varphi) + (Z_1 - Z_2)(Z_2 + Z_3)}{(Z_1 + Z_2)(Z_2 + Z_3)\exp(-2i\varphi) + (Z_1 - Z_2)(Z_2 - Z_3)} \quad . \qquad (2.4.9)$$

In the particular case when impedances of the surrounding half-spaces are equal $(Z_1 = Z_3)$, Eq. (2.4.9) can be written in the form

$$V = \frac{Z_2^2 - Z_1^2}{Z_1^2 + Z_2^2 + 2iZ_1 Z_2 \cot \varphi} \quad . \qquad (2.4.10)$$

For the case where the layer thickness $d \to 0$ or $d \neq 0$, but $Z_2 \to Z_1$, we obtain $V = (Z_1 - Z_3)/(Z_1 + Z_3)$, from (2.4.9), i.e., the reflection coefficient at the interface of homogeneous half-spaces 1 and 3, cf. (2.2.16).

Now consider the transmitted wave travelling in medium 1. The acoustic pressure will again be given by (2.2.3) where coefficient W needs to be determined. The condition that the field given by (2.4.4) in the layer at $z = 0$ is equal to the pressure of the transmitted wave gives

$$W = A + B \quad . \qquad (2.4.11)$$

On the other hand, writing the incident and the reflected waves in medium 3 as $\exp[-ik_3(z - d)\cos\theta_3]$ and $V\exp[ik_3(z - d)\cos\theta_3]$, we obtain from the pressure equality at both sides of the boundary

$$1 + V = A\exp(i\varphi) + B\exp(-i\varphi) \quad . \qquad (2.4.12)$$

Dividing (2.4.11) by (2.4.12) and using (2.4.5) we find

$$W = \frac{1 + V}{\cos\varphi - iZ_2 \sin\varphi/Z_1} \quad . \qquad (2.4.13)$$

Substituting (2.4.9) for V into (2.4.13) yields

$$W = \frac{4Z_1 Z_2}{(Z_1 - Z_2)(Z_2 - Z_3)\exp(i\varphi) + (Z_1 + Z_2)(Z_2 + Z_3)\exp(-i\varphi)} \quad . $$

$$(2.4.14)$$

For $d \to 0$ we obtain (2.2.9) from (2.4.13) and (2.2.17) from (2.4.14). When $d \neq 0$ and $Z_2 \to Z_1$ we derive from (2.4.13, 14) $W = (1+V)\exp(i\varphi)$ and $W = 2Z_1\exp(i\varphi)/(Z_1 + Z_3)$, respectively. These relations differ from (2.2.9 and 17) deduced in Sect. 2.2 only by the phase factor $\exp(i\varphi)$. The appearance of the latter is caused by displacement of the reference point of the phase of the refracted wave from $z = d$ to $z = 0$.

2.4.3 Another Approach to the Reflection and Transmission Coefficients Calculation

There is another derivation of the expression for the reflection coefficient that is of interest. The resulting wave reflected from the layer may be regarded (Fig. 2.4) as a superposition of: (a) the wave reflected from the front surface of the layer (boundary between media 3 and 2, the first number is taken for the medium from which the wave is indicent); (b) the wave penetrating the front surface of the layer, passing through the layer, reflecting from its back surface (boundary between the media 2 and 1), passing through the layer again and finally leaving it through its upper boundary between the media 2 and 3; (c) the wave penetrating the layer, undergoing two reflections at the back surface and one at the upper surface, passing twice through the layer back and forth, and then again leaving the layer, etc. The complex amplitude change is represented by the factor V_{ij} for each reflection from the boundary i-j and by the factor W_{ij} for the wave penetrating the boundary; a single pass through the layer gives the phase change φ defined by (2.4.8). (The "phase" φ is complex in the case of inhomogeneous plane wave.) The reflection and transmission coefficients of the boundary, (2.2.9, 16), were determined in Sect. 2.2 and are equal to

$$V_{ij} = \frac{Z_j - Z_i}{Z_j + Z_i} \quad , \quad V_{ji} = -V_{ij} \quad , \quad W_{ij} = 1 + V_{ij} \quad ; \quad i, j = 1, 2, 3 \quad .$$

(2.4.15)

We take the amplitude of the incident wave as unity. Summing up all the waves which form the total reflected wave and using the expression for the sum of infinite geometrical progression we obtain

$$
\begin{aligned}
V &= V_{32} + W_{32}V_{21}W_{23} \exp{(2i\varphi)} \sum_{n=0}^{\infty} [V_{23}V_{21} \exp{(2i\varphi)}]^n \\
&= V_{32} + W_{32}V_{21}W_{23} \frac{\exp{(2i\varphi)}}{1 - V_{23}V_{21} \exp{(2i\varphi)}}
\end{aligned}
$$

(2.4.16)

for the amplitude of the reflected wave, i.e., for the reflection coefficient from the layer. After some transformations, by using (2.4.15) we find

$$V = \frac{V_{32} + V_{21} \exp{(2i\varphi)}}{1 + V_{32}V_{21} \exp{(2i\varphi)}} \quad .$$

(2.4.17)

Exactly in the same manner, by summing up all the penetrated waves we find for the transmission coefficient of the layer

$$W = \frac{W_{32}W_{21}}{\exp{(-i\varphi)} + V_{32}V_{21} \exp{(i\varphi)}} \quad .$$

(2.4.18)

When V_{21}, V_{32}, W_{21}, and W_{32} are expressed in terms of impedances according to (2.4.15), equations (2.4.17 and 18) transform into (2.4.9 and 14).

2.4.4 Two Special Cases

Half-wave layer. Let the advance in the phase of the wave over the thickness of the layer be equal to an integral number of half-periods, that is

$$\varphi \equiv k_2 d \cos{\theta_2} = l\pi \quad , \quad l = 1, 2, \ldots \quad .$$

(2.4.19)

For normal incidence this means that $d = l\lambda_2/2$, where $\lambda_2 = 2\pi/k_2$ is the wavelength in the layer. Substituting (2.4.19) into (2.4.7) we obtain for the input impedance

$Z_{in} = Z_1$. Hence, the reflection coefficient is $V = (Z_1 - Z_3)/(Z_1 + Z_3)$, according to (2.4.1). Thus, the half-wave layer has no effect on the incident wave: the reflection coefficient is just the same as if media 3 and 1 were in direct contact with one another. In particular, if media 3 and 1 have the same impedances ($Z_1 = Z_3$), the reflection coefficient is zero.

This property of a half-wavelength layer can be used for the construction of frequency or direction filters. In the first case when the frequency of an incident plane nonmonochromatic wave obeys condition (2.4.19) (with some l) it will pass through the filter. In the second case an ensemble of waves of the same frequency but different incident directions impinges on the filter. The wave (or waves) that are allowed to pass are those with directions θ_2 satisfying the same condition (2.4.19).

Quarter-wave transmission layer. Now let

$$\varphi \equiv k_2 d \cos \theta_2 = \tfrac{\pi}{2} + \pi l \quad , \quad l = 0, 1, 2, \ldots \quad . \tag{2.4.20}$$

For normal incidence and $l = 0$ this means that $d = \lambda_2/4$. Equation (2.4.7) yields $Z_{in} = Z_2^2/Z_1$ for the input impedance. Thus, the reflection coefficient is

$$V = \frac{Z_2^2 - Z_1 Z_3}{Z_2^2 + Z_1 Z_3} \quad . \tag{2.4.21}$$

According to (2.4.1) complete transmission takes place for

$$Z_2 = \sqrt{Z_1 Z_3} \quad . \tag{2.4.22}$$

Thus, the reflection of a monochromatic wave at the boundary of two arbitrary media can be completely eliminated by placing a quarter-wave layer between them which has an impedance equal to the geometric mean of the impedances of the two media.

Systems of more than one layer have numerous applications as filters, sound insultors, and anti-reflection coatings [Ref. 2.3; Chap. 2].

2.4.5 Penetration of a Wave Through a Layer

We assume that the sound velocity in the layer is greater than that in the medium from which the wave is incident. For an infinitely thick layer, total internal reflection would occur beyond some critical angle of incidence. However, in the case of a layer of finite thickness the wave will partially penetrate the layer. This effect is analogous to the phenomenon of penetration of particles through a potential barrier in quantum mechanics.

From equation (2.4.3) we have $\sin \theta_2 = k_3 k_2^{-1} \sin \theta_3 = c_2 c_3^{-1} \sin \theta_3$. When $\theta_3 > \arcsin (c_3/c_2)$ (the angle of incidence is greater than the critical angle for total internal reflection) we obtain $\sin \theta_2 > 1$, i.e., θ_2 is a complex angle and the plane waves in the layer are inhomogneous. Here $\cos \theta_2 = \pm i (c_2^2 c_3^{-2} \sin^2 \theta_3 - 1)^{1/2}$. According to the formula analogous to (2.2.15) the plus sign must be chosen in front of i. (Note that when changing sign (2.4.8,9) do not yield the correct limit $\lim_{d \to \infty} V = V_{32}$.) Thus, the impedance Z_2 (2.4.6) and the quantity φ (2.4.8) are purely imaginary:

$$Z_2 = -i|Z_2| \quad , \quad |Z_2| = \varrho_2 c_2 \frac{1}{\sqrt{c_2^2 c_3^{-2} \sin^2 \theta_3 - 1}} \quad ,$$

$$\varphi = i|\varphi| \quad , \quad |\varphi| = k_2 d \sqrt{c_2^2 c_3^{-2} \sin^2 \theta_3 - 1} \quad .$$

Consider in detail the case in which the media on both sides of the layer are identical. Then $Z_3 = Z_1$ are real, and (2.4.10) gives

$$V \equiv |V| \exp(i\beta) = \frac{Z_1^2 + |Z_2|^2}{|Z_2|^2 - Z_1^2 + 2iZ_1|Z_2| \coth |\varphi|} \quad . \tag{2.4.23}$$

The modulus and phase of the reflection coefficient are then easily found to be

$$|V| = (Z_1^2 + |Z_2|^2)[(Z_1^2 - |Z_2|^2)^2 + 4Z_1^2|Z_2|^2 \coth^2 |\varphi|]^{-1/2} \quad , \tag{2.4.24a}$$

$$\beta = \arctan[2Z_1|Z_2|(Z_1^2 - |Z_2|^2)^{-1} \coth |\varphi|] \quad . \tag{2.4.24b}$$

We see from (2.4.24a) that $|V| < 1$, i.e., there is always partial penetration of the wave through the layer. As the thickness of the layer increases, the reflection coefficient increases and $|V| \to 1$ for $d \to \infty$.

2.5 Reflection and Transmission Coefficients for an Arbitrary Number of Layers

Let us suppose that there are $n-1$ homogeneous layers numbered $2, 3, \ldots, n$ between two semi-infinite media, which we denote by 1 and $n+1$ (Fig. 2.5). Let a plane wave be incident on the boundary of the latter at an arbitrary angle θ_{n+1}. Our objective is to determine the amplitudes of the reflected and transmitted waves in the media $n+1$ and 1, respectively.

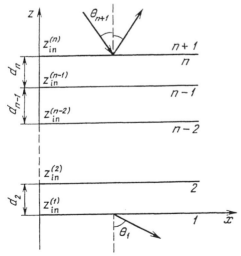

Fig. 2.5. Wave vectors of incident, reflected, and refracted plane waves at a system of $n-1$ layers, each with input impedance $Z_{in}^{(i)}$ and thickness d_i

The input impedance $Z_{in}^{(n)}$ of the entire system of layers can be calculated by means of $(n-1)$-fold application of (2.4.7). Indeed, setting $Z_{in}^{(1)} = Z_1$, $d = d_2$ we obtain the input impedance $Z_{in}^{(2)}$ at the upper boundary of the lowest (second) layer

$$Z_{in}^{(2)} = Z_2(Z_{in}^{(1)} - iZ_2 s_2)/(Z_2 - iZ_{in}^{(1)} s_2) \quad , \tag{2.5.1}$$

where the following notations are used

$$s_j = \tan \varphi_j \quad , \quad \varphi_j = k_j d_j \cos \theta_j \quad , \quad \sin \theta_j = \frac{k_{n+1}}{k_j} \sin \theta_{n+1} \quad ,$$

$$Z_j = \frac{\varrho_j c_j}{\cos \theta_j} \quad ; \quad j = 1, 2, \ldots, n + 1 \quad . \tag{2.5.2}$$

Then, after the change $Z_{in}^{(1)} \to Z_{in}^{(2)}$, $Z_2 \to Z_3$, $s_2 \to s_3$ on the right side of (2.5.1) we obtain the expression for $Z_{in}^{(3)}$, that is, the input impedance of the layer with index 3, etc. After $Z_{in}^{(n-1)}$ has been found, the required input impedance of the system of layers is determined by

$$Z_{in}^{(n)} = Z_n \frac{Z_{in}^{(n-1)} - iZ_n s_n}{Z_n - iZ_{in}^{(n-1)} s_n} \tag{2.5.3}$$

and the reflection coefficient is given by the expression similar to (2.2.16):

$$V = \frac{Z_{in}^{(n)} - Z_{n+1}}{Z_{in}^{(n)} + Z_{n+1}} \quad . \tag{2.5.4}$$

It can be easily seen that if the advance of the phase in some layer j equals an integral number of half-waves, such a layer has no effect on the reflection, regardless of the value of the impedance Z_j. Now we write the input impedance for the system of two layers ($n = 3$) in the explicit form

$$Z_{in}^{(3)} = Z_3 \frac{Z_1(Z_2 - Z_3 s_2 s_3) - iZ_2(Z_2 s_2 + Z_3 s_3)}{Z_2(Z_3 - Z_2 s_2 s_3) - iZ_1(Z_3 s_2 + Z_2 s_3)} \tag{2.5.5}$$

and for the system of three layers ($n = 4$)

$$Z_{in}^{(4)} = M Z_4/N \quad , \tag{2.5.6}$$

where

$$M \equiv Z_1(Z_2 Z_3 - Z_3^2 s_2 s_3 - Z_3 Z_4 s_2 s_4 - Z_2 Z_4 s_3 s_4)$$
$$- iZ_2(Z_2 Z_3 s_2 + Z_3^2 s_3 + Z_3 Z_4 s_4 - Z_2 Z_4 s_2 s_3 s_4) \quad ,$$

$$N \equiv Z_2(Z_3 Z_4 - Z_2 Z_4 s_2 s_3 - Z_2 Z_3 s_2 s_4 - Z_3^2 s_3 s_4)$$
$$- iZ_1(Z_3 Z_4 s_2 + Z_2 Z_4 s_3 + Z_2 Z_3 s_4 - Z_3^2 s_2 s_3 s_4) \quad .$$

It can be easily verified that in the case when the impedances of some neighboring layers are equal (2.5.5 and 6) are reduced to the corresponding relations for smaller n.

Our goal now is to determine the transmission coefficient for an arbitrary number of layers. We denote the coordinate of the upper boundary of the layer j by z_j. Then the acoustic pressure in each of the media can be written (the factor $\exp[i(k_{n+1} x \sin \theta_{n+1} - \omega t)]$ is omitted) as

$$p_j = A_j \exp[ik_j(z - z_{j-1}) \cos \theta_j] + B_j \exp[-ik_j(z - z_{j-1}) \cos \theta_j] \quad ,$$
$$z_{j-1} \le z \le z_j \quad , \quad j = 2, 3, \ldots, n+1 \quad , \quad z_{n+1} \equiv +\infty \quad ;$$
$$p_1 = B_1 \exp[-ik_1(z - z_1) \cos \theta_1] \quad , \quad z < z_1 \quad . \tag{2.5.7}$$

The pressure, as well as the impedance, should be continuous at the boundary. Thus,

$$z = z_j \quad , \quad p_j = p_{j+1} \quad ;$$
$$-i\omega \varrho_j p_j \frac{1}{\partial p_j / \partial z} = -i\omega \varrho_{j+1} p_{j+1} \frac{1}{\partial p_{j+1} / \partial z} = Z_{\text{in}}^{(j)} \quad . \tag{2.5.8}$$

Substitution of p_j and p_{j+1} from (2.5.7) into (2.5.8) and taking into account $z_{j+1} - z_j = d_{j+1}$ yields three equations for the amplitudes A_j, B_j, A_{j+1} and B_{j+1}:

$$A_j \exp(i\varphi_j) + B_j \exp(-i\varphi_j) = A_{j+1} + B_{j+1} \quad ,$$

$$\frac{A_j \exp(i\varphi_j) + B_j \exp(-i\varphi_j)}{A_j \exp(i\varphi_j) - B_j \exp(-i\varphi_j)} = -\frac{Z_{\text{in}}^{(j)}}{Z_j} \quad ,$$

$$\frac{A_{j+1} + B_{j+1}}{A_{j+1} - B_{j+1}} = -\frac{Z_{\text{in}}^{(j)}}{Z_{j+1}} \quad . \tag{2.5.9}$$

Here $A_1 = \varphi_1 = 0$. Solving the last two equations with respect to A_j/B_j and A_{j+1}/B_{j+1} and substituting the results into the first equation of (2.5.9) we find

$$\frac{B_j}{B_{j+1}} = \exp(i\varphi_j) \frac{Z_{\text{in}}^{(j)} + Z_j}{Z_{\text{in}}^{(j)} + Z_{j+1}} \quad , \tag{2.5.10a}$$

$$\frac{A_j}{B_j} = \exp(-2i\varphi_j) \frac{Z_{\text{in}}^{(j)} - Z_j}{Z_{\text{in}}^{(j)} + Z_j} \quad . \tag{2.5.10b}$$

By substituting into (2.5.10a) the values $j = 1, 2, \ldots, n$ successively and multiplying each resulting equation by the successive one we finally get for the transmission coefficient, that is for the ratio of the pressure amplitudes in the transmitted and in the incident wave

$$W \equiv \frac{B_1}{B_{n+1}} = \prod_{j=1}^{n} \exp(i\varphi_j) \frac{Z_{\text{in}}^{(j)} + Z_j}{Z_{\text{in}}^{(j)} + Z_{j+1}} \quad . \tag{2.5.11}$$

In any layer B_j can be analogously expressed in terms of B_{n+1}. Then A_j can be found with the aid of (2.5.10b). Thus, the acoustic field can be determined in the entire discrete layered medium.

One can also approach this problem by another method. To completely determine the acoustic field in a medium with $n - 1$ homogeneous layers between two homogeneous half-spaces it is sufficient, [as follows from (2.5.7)], to find $2n$ amplitude coefficients: A_j, B_j ($j = 2, 3, \ldots, n$), B_1 and A_{n+1}. (The amplitude B_{n+1} of the incident wave in the upper half-space is assumed to be known). The conditions of continuity of the pressure and the normal component of particle velocity at the boundaries between media yield linear equations for determining the unknown A_j, B_j:

$$A_j \exp{(\mathrm{i}\varphi_j)} + B_j \exp{(-\mathrm{i}\varphi_j)} = A_{j+1} + B_{j+1} \quad ,$$
$$j = 1, 2, \ldots, n \quad ; \quad A_1 = 0 \quad ,$$

$$\frac{k_j}{\varrho_j} \cos \theta_j [A_j \exp{(\mathrm{i}\varphi_j)} - B_j \exp{(-\mathrm{i}\varphi_j)}]$$

$$= \frac{k_{j+1}}{\varrho_{j+1}} \cos \theta_{j+1} (A_{j+1} - B_{j+1}) \quad . \tag{2.5.12}$$

The solution of this system of equations can be found with the well-known methods of linear algebra. However, the concept of impedance is useful since the solution can then be obtained with fewer calculations and has a clearer physical meaning.

In practical applications to real layered systems it is of interest not only to calculate the reflection and transmission coefficients of a given system, but also to solve an inverse problem, i.e., to determine parameters that give the desired acoustic properties of a system (for example, low reflectivity at some angle and frequency range). For more information on this subject see [Ref. 2.3; Chap. 2; 2.18, 2.19]. References to more recent papers can be found in [2.5, 2.20].

2.6 Moving Layers. Impedance of Harmonic Waves in Moving Media

The problem of plane-wave reflection from moving discretely inhomogeneous layered media includes much more interesting physical phenomena than the case of media at rest. In particular, the reflection and transmission coefficients depend not only on the incidence angle but also on the orientation of the incidence plane relative to the flow direction. Moreover, in some cases as we shall see below, the reflection coefficient may be greater than unity.

Reflection of waves from discretely layered moving media has been previously considered only for some simple cases. Some work appeared to be erroneous due to wrong formulation of boundary conditions (Sect. 1.2.2). In the following discussion we will generalize the results of the previous section for an arbitrary number of moving layers. A key point will be the generalization of the impedance concept to the case of moving layered media.

We will use models where the flow velocity may be discontinuous at the interfaces of some layer. Strictly speaking, in hydrodynamics, such a discontinuous flow is unstable. In real systems thin transition layers occur instead of discontinuities due to fluid viscosity. If the thickness of such layers is small compared to the sound wavelength, however, in acoustical problems they can be approximated as discontinuities. This was shown in Sect. 2.4 for a homogeneous layer at rest. Similar estimates for arbitrary stratification of the sound velocity, density, and flow velocity will be given in Chap. 10. In the following discussion we also neglect sound energy dissipation.

2.6.1 Sound Wave Impedance in a Moving Medium

Our point of departure is the equation for sound pressure in the medium with piecewise-continuous parameters (1.2.25)

$$\frac{\partial^2 p}{\partial \zeta^2} + (k^2\beta^2 - \xi^2)\left(\frac{\varrho_0}{\varrho\beta^2}\right)^2 p = 0 \quad , \quad \beta = 1 - \frac{\boldsymbol{\xi} \cdot \boldsymbol{v}_0}{\omega} \quad . \tag{2.6.1}$$

Here we assume a harmonic dependence of p on time and horizontal coordinates. The coordinate $\zeta(z)$ is defined by (1.2.24). Boundary conditions (1.2.26) at planes $z = $ const can be written in the form analogous to (2.2.8), i.e.,

$$[p]_S = 0 \quad , \quad [Z]_S = 0 \quad , \tag{2.6.2}$$

where Z is defined by

$$Z = \text{const} \frac{p}{\partial p/\partial \zeta} \quad . \tag{2.6.3}$$

This quantity has all the properties of the impedance discussed in previous sections for media at rest. It does not depend on horizontal coordinates, time, or the amplitude of the wave. Furthermore it is continuous across the interfaces between layers. The constant in (2.6.3) can be determined by the condition that in the absence of flow ($v_0 = 0$, $\beta = 1$) (2.6.3) must coincide with (2.2.7). Then taking into account (1.2.24) we obtain

$$Z = -i\omega\varrho_0 \frac{p}{\partial p/\partial \zeta} \quad , \tag{2.6.4a}$$

or in ordinary coordinates

$$Z = -i\omega\varrho\beta^2 \frac{p}{\partial p/\partial z} \quad . \tag{2.6.4b}$$

Note that the definition of impedance given by (2.6.3) is valid for waves of any physical nature provided that the corresponding wave equation is written as a one dimensional Helmholtz equation which does not contain the derivatives of the media parameters in the coefficients.

The impedance of a harmonic wave for moving media has a clear physical meaning. We have noted, while obtaining (1.2.8), that vertical particle displacement in the wave η is related to the vertical velocity component w and sound pressure p by

$$\eta = \frac{1}{-i\omega\beta} w = \frac{1}{\varrho\omega^2\beta^2} \frac{\partial p}{\partial z} \quad .$$

Hence, impedance is the ratio of the sound pressure to the vertical particle displacement, times a constant factor. Apparently, the impedance was introduced independently in the acoustics of moving media in [2.6, 7]. In the latter the significance of the concept of impedance for the use of many results of ordinary acoustics in acoustics of moving media was emphasized.

2.6.2 Plane Wave Reflection from Discretely Layered Moving Media

We keep the geometry of the problem and numbering of the layers as in the previous section (Fig. 2.5). Again the plane of incidence is the xz-plane. In layer number j the wave equation has the general solution, see (1.2.10, 14)

$$p_j = A_j \exp\left[ik_j(z - z_{j-1})(1 + M_j \sin \theta_j)^{-1} \cos \theta_j\right]$$
$$+ B_j \exp\left[-ik_j(z - z_{j-1})(1 + M_j \sin \theta_j)^{-1} \cos \theta_j\right] \quad, \tag{2.6.5}$$

where θ_j is the incidence angle as in (2.5.7) and

$$M_j = \frac{v_{0xj}}{c_j} \quad. \tag{2.6.6}$$

The factor $\exp\left[ik_j x(1 + M_j \sin \theta_j)^{-1} \sin \theta_j - i\omega t\right]$ common to all the layers is omitted. One also has, in this notation,

$$\beta_j \equiv 1 - \frac{\xi \cdot v_{0j}}{\omega} = \frac{1}{1 + M_j \sin \theta_j} \quad. \tag{2.6.7}$$

We see that the flow influences the sound propagation only through the projection of the vector $v_0(z)$ upon the phase propagation direction of the wave. The angles θ_j in different layers are related to each other by the condition that the phase velocities of waves propagation along all the horizontal boundaries be equal. Such a relation is analogous to (2.2.6):

$$\frac{\sin \theta_j}{c_j(1 + M_j \sin \theta_j)} = \frac{\sin \theta_{n+1}}{c_{n+1}(1 + M_{n+1} \sin \theta_{n+1})} \quad. \tag{2.6.8}$$

The general solution given by (2.6.5) is a superposition of two plane waves propagating in directions symmetric with respect to the horizontal plane. We denote Z_j to be the impedance of the plane wave, of which the z-component of the group velocity (1.2.17) is negative. (This wave results from (2.6.5), when $A_j = 0$). Then (2.6.4b and 7) yield

$$Z_j = \frac{\varrho_j c_j}{(1 + M_j \sin \theta_j) \cos \theta_j} \quad. \tag{2.6.9}$$

With the help of (2.6.8) this impedance may be rewritten in a form where the flow velocity appears only implicitly via θ_j:

$$Z_j = \frac{[2 \sin \theta_{n+1}/c_{n+1}(1 + M_{n+1} \sin \theta_{n+1})]\varrho_j c_j^2}{\sin 2\theta_j} \quad. \tag{2.6.10}$$

The input impedance of the lower half-space is $Z_{\text{in}}^{(1)} = Z_1$. The input impedances of the higher layers are obtained by the successive use of (2.4.7) which yields, cf. (2.5.3),

$$Z_{in}^{(j+1)} = Z_{j+1} \frac{Z_{in}^{(j)} - iZ_{j+1}s_{j+1}}{Z_{j+1} - iZ_{in}^{(j)}s_{j+1}} \quad , \tag{2.6.11}$$

where $s_j = \tan \varphi_j$. The phase advance in the layer j is, according to (2.6.5)

$$\varphi_j = k_j d_j \frac{1}{1 + M_j \sin \theta_j} \cos \theta_j$$

$$= k_{n+1} \sin \theta_{n+1} \frac{1}{1 + M_{n+1} \sin \theta_{n+1}} d_j \cot \theta_j \quad . \tag{2.6.12}$$

When the input impedance is known, the reflection and transmission coefficients are determined from (2.5.4 and 11), respectively.

Input impedances of one, two, and three layers are given by (2.4.7), (2.5.5), and (2.5.6), respectively. The results of the above discussion on different special cases (half-wave and quarter-wave layers, tunneling of the wave through the layer, etc.) are still valid. It is important to note, however, that the above expressions now have a new and deeper meaning. For example, the phase advance in the layer now also depends, according to (2.6.12), on the angle between the vectors ξ and v_0. Hence the maximum transparency of the half-wave layer occurs when

$$\varphi_2 = k_3(1 + M_3 \sin \theta_3)^{-1} \sin \theta_3 \cdot d_2 \cdot \cot \theta_2 = l\pi \quad ,$$

$$l = 1, 2, \ldots \quad . \tag{2.6.13}$$

The angle of maximum penetration now depends on the angle between the flow direction and the x-axis.

2.6.3 Reflection at a Single Interface

Consider more closely the reflection of a plane monochromatic wave from a plane interface between moving homogeneous media. (References [2.8, 9] are devoted to this problem.) We have from (2.2.16) and (2.6.10):

$$V = \frac{\varrho_2 c_2^2 \sin 2\theta_1 - \varrho_1 c_1^2 \sin 2\theta_2}{\varrho_2 c_2^2 \sin 2\theta_1 + \varrho_1 c_1^2 \sin 2\theta_2} \quad . \tag{2.6.14}$$

Here, the flow influences V only via the incidence and refraction angles. Note that for gaseous media $\varrho_2 c_2^2/\varrho_1 c_1^2 = \gamma_2/\gamma_1$ is true, where γ is the ratio of the specific heats at constant pressure and constant volume. According to the relativity principle, as for the V value, the velocity difference of the media at the boundary is important but not the individual velocities of each medium. The reflection coefficient given by (2.6.14) obeys this requirement. Indeed, taking into account (2.6.8) we can rewrite (2.6.14) in the form

$$V = \frac{[1 - (v_2 - v_1)c_1^{-1} \sin \theta_1] - \varrho_1 c_1 \cos \theta_2/\varrho_2 c_2 \cos \theta_1}{[1 - (v_2 - v_1)c_1^{-1} \sin \theta_1] + \varrho_1 c_1 \cos \theta_2/\varrho_2 c_2 \cos \theta_1} \quad , \tag{2.6.15a}$$

where

$$\sin \theta_2 = \frac{(c_2/c_1) \sin \theta_1}{1 - (v_2 - v_1)c_1^{-1} \sin \theta_1} \quad . \tag{2.6.15b}$$

For brevity, we omit the indices 0 and x in the flow velocity ($v_{0xj} \equiv v_j$). V does not change if the signs of $\sin \theta_1$ and ($v_2 - v_1$) are changed simultaneously. We assume, therefore, that $\sin \theta_1 \geq 0$. In the reference system moving together with upper medium this means that the wave's system propagates in the direction of positive x. The reflection coefficient V can be obtained for any incidence angle (under the assumption $\theta_1 \geq 0$) by proper choosing the sign of the flow velocity difference $\pm |v_2 - v_1|$.

When the media are at rest there exist two possible types of reflection: $|V| < 1$ (partial reflection) and $|V| = 1$ (total reflection). It follows from (2.6.15) that the necessary and sufficient condition for total reflection is $|\sin \theta_2| \geq 1$. Hence, total reflection occurs if the incidence angle obeys the condition

$$\frac{v_2 - v_1 - c_2}{c_1} \leq \frac{1}{\sin \theta_1} \leq \frac{v_2 - v_1 + c_2}{c_1} \quad . \tag{2.6.16}$$

When the media are moving the third case is possible, when $|V| > 1$, hence, we have *reflection with amplification*. One can see from (2.6.15) that this occurs when $\cos \theta_2$ is real, i.e., $|\sin \theta_2| < 1$, and the quantity in square brackets in (2.6.15a) is negative. From these considerations it follows that $|V| > 1$ for the incidence angles satisfying the inequality

$$\sin \theta_1 > \frac{c_1}{v_2 - v_1 - c_2} > 0 \quad . \tag{2.6.17}$$

Since $\sin \theta_1 \leq 1$, the amplification is possible only when the relative velocity of the media is sufficiently large

$$v_2 - v_1 > c_1 + c_2 \quad . \tag{2.6.18}$$

The conditions under which (2.6.17) is valid can be stated as follows:

a) There exists an ordinary (homogeneous) plane wave in the lower medium (energy flux in z direction is not zero).

b) The projection v_2 of the flow velocity in the lower medium upon the ξ direction exceeds the wave velocity along the boundary $c_1/\sin \theta_1 + v_1$. In the case where the upper medium is at rest and the flow in the lower one is along ξ, reflection with amplification is possible when the fluid flow outruns the wave trail at the boundary.

A special case is that of *resonance reflection,* where V becomes infinite. The necessary condition for this is satisfying

$$\left(\frac{\varrho_1 c_1}{\varrho_2 c_2}\right)^2 \left[1 - \left(\frac{c_2 \sin \theta_1}{c_1 - (v_2 - v_1)\sin \theta_1}\right)^2\right]$$

$$= (1 - \sin^2 \theta_1)\left(1 - \frac{v_2 - v_1}{c_1}\sin \theta_1\right)^2 \quad . \tag{2.6.19}$$

In this case the denominator in (2.6.15) becomes zero. The condition of (2.6.17) must also be satisfied of course.

Equation (2.6.19) is algebraic and to the sixth order. It determines the resonance incidence angles θ_1. When the parameters of the media are the same ($\varrho_1 = \varrho_2$, $c_1 = c_2$) these angles can be found explicitly. Indeed, let us introduce two new unknown quantities $a = \sin^{-1} \theta_1$ and $b = M - \sin^{-1} \theta_1$, where $M \equiv (v_2 - v_1)/c_1$. Equation (2.6.19) is now equivalent to the system of equations

$$a + b = M \quad , \quad (a^2 - b^2)[(ab)^2 + 2ab - M^2] = 0 \quad , \tag{2.6.20}$$

the solutions of which can be found elementarily. The only resonance angle θ_1 is possible when $2 \leq M \leq 2^{3/2}$. For this angle we have

$$\sin \theta_1 = 2/M \quad . \tag{2.6.21}$$

Two more resonance angles appear when $M > 2^{3/2}$ and for them we have:

$$\sin \theta_1 = \frac{2}{M \pm \sqrt{M^2 + 4 - 4(1 + M^2)^{1/2}}} \quad . \tag{2.6.22}$$

In an analogous way the resonance angles can be found for a more general case, where $\varrho_1 c_1^2 = \varrho_2 c_2^2$ and ϱ_2/ϱ_1 is arbitrary. For gases, this means $\gamma_1 = \gamma_2$.

Wave amplification by reflection does not contradict the energy conservation law: the reflected wave takes some amount of energy from the flow. In other words, the energy of the refracted wave appears to be negative [2.21]. Of course, our expressions for the reflection coefficient cease to be correct when the incidence angle is resonant or near resonant. First of all, this is because the amplitude of the reflected wave becomes large and the approximation of linear acoustics is no longer reasonable. However, the infinite value of the reflection coefficient suggests that in this case auto-oscillations could arise. Analysis of the reflection coefficient behavior appears to be very useful in the study of such oscillations [2.10].

It follows from (2.6.16 and 17) that ordinary reflection (with $|V| < 1$) occurs when

$$\frac{1}{\sin \theta_1} > \frac{v_2 - v_1 + c_2}{c_1} > 0 \quad . \tag{2.6.23}$$

Obviously, $|V| < 1$ at any positive incidence angle if $v_1 + c_1 > v_2 + c_2$; for neither the conditions of total reflection nor reflection with amplification are met for any θ_1 when $0 \leq \theta_1 \leq \pi/2$.

From (2.6.15) we see that the reflection coefficient is zero at the incidence angles satisfying (2.6.19) and (2.6.23). These angles can be found easily from (2.6.20) if the parameters of the media are identical. It appears that there exist two such angles: $\theta_1 = 0$ and

$$\sin \theta_1 = \frac{2}{M + \sqrt{M^2 + 4 + 4(1 + M^2)^{1/2}}} \tag{2.6.24}$$

for any $M \neq 0$.

Figures 2.6 and 7 illustrate the results obtained above. The former shows $|V|$ as a function of incidence angle in the case of moving media with identical parameters and $M = \pm 3$. The curves $V = $ const are shown in Fig. 2.7 in the plane θ_1, M.

Several cases of wave reflection from moving media where there is a homogeneous layer between the two half-spaces are considered in [2.11–13]. Reflection from a thin plate separating two moving media was discussed in [2.12, 14]. Detailed study of sound waves at plane interface between fluids moving with different velocities and of stability of this interface is presented in [2.22] and in a review paper [2.15].

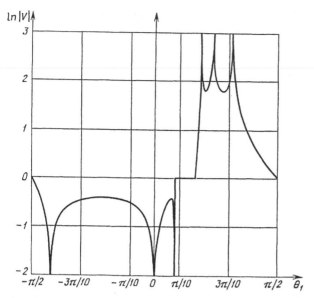

Fig. 2.6. Modulus of the reflection coefficient at a boundary of two identical moving media as a function of incidence angle

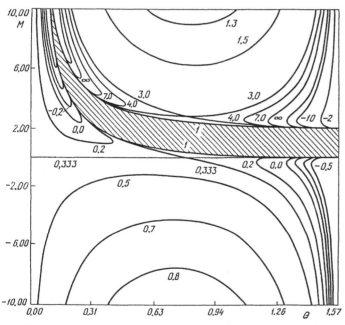

Fig. 2.7. *Curves* of constant reflection coefficient as a function of the incidence angle of the sound wave and relative flow velocity of media with equal sound speeds. The *shaded region* is the region of total reflection, in which V is complex, The shape of the curves is independent of the media densities. Numerical values of V in the figure refer to the case $\varrho_2 = 2\varrho_1$

3. Monochromatic Plane-Wave Reflection from Continuously Layered Media

Stratification of continuously layered media is defined by the dependence of the density $\varrho(z)$, the sound velocity $c(z)$, and in the case of moving media also of the flow velocity $v_0(z)$ on the vertical coordinate z. Exact solutions to the wave equation exist only for a few cases. These cases, however, play an important role as model systems, as a first approximation to real physical problems, in estimation of the range of applicability of various approximation methods, etc. For this reason each new soluble case is of considerable interest.

In this chapter we will consider reflection of a plane sound wave incident from the homogeneous half-space $(z > 0)$ on the continuously stratified fluid half-space $(z < 0)$. In Sects. 3.1–6 the media are assumed to be at rest; moving media are examined in Sect. 3.7. This problem has been widely treated in the mathematical and physical literature, see for example, monographs [3.1–5]. Exact analytic solutions for the field of a point source were considered in [3.37, Chap. 4]. Original works will be cited below.

3.1 General Relations

Let ϱ_1 and c_1 be the density and sound velocity in a homogeneous half-space $(z > 0)$ and $\varrho_2 = \text{const}$, $c(z)$ the same quantities in an inhomogeneous fluid $(z < 0)$. The function $c(z)$ is assumed to be smooth.

The incident wave with unit amplitude is written:

$$p_i = \exp\left[i\left(\xi x - \sqrt{k_1^2 - \xi^2}\, z\right)\right] \quad , \quad k_1 = \frac{\omega}{c_1} \quad ,$$
$$\xi = k_1 \sin \theta \quad , \quad z > 0 \quad , \tag{3.1.1}$$

where θ is the angle of incidence. The time factor $\exp(-i\omega t)$ is omitted. The geometry of the problem is illustrated in Fig. 3.1.

The reflected wave is given by

$$p_r = V \exp\left\{i\left[\xi x + \sqrt{k_1^2 - \xi^2}\, z\right]\right\} \quad , \quad z > 0 \quad . \tag{3.1.2}$$

The sound pressure in the lower half-space can be written as $p(\boldsymbol{r}, \omega) = \Phi(z) \exp(i\xi x)$. Substituting this into (1.2.18) we obtain the equation for $\Phi(z)$

$$\frac{d^2\Phi}{dz^2} + (k^2 - \xi^2)\Phi = 0 \quad , \quad k = k(z) = \frac{\omega}{c(z)} \quad , \quad z < 0 \quad . \tag{3.1.3}$$

We assume that (3.1.3) has at least one bounded solution (only such solutions have

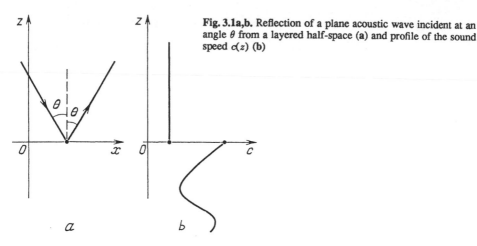

Fig. 3.1a,b. Reflection of a plane acoustic wave incident at an angle θ from a layered half-space (a) and profile of the sound speed $c(z)$ (b)

physical meaning). If the inhomogeneous part of lower half-space is limited or $c(z)$ rapidly approaches a constant value at large negative z (Sect. 3.2), the sound field again becomes the plane wave

$$p = W \exp\left\{i\left[\xi x - \sqrt{k_2^2 - \xi^2}\,z\right]\right\} \quad , \quad k_2 = \frac{\omega}{c_2} \quad ,$$
$$c_2 = \lim_{z \to -\infty} c(z) \quad . \tag{3.1.4}$$

The quantity W is the transmission coefficient and is independent of z. When the solution to (3.1.3) is found the reflection coefficient is determined by [cf. (2.2.7, 12)]

$$V = \frac{Z \cos\theta - \varrho_1 c_1}{Z \cos\theta + \varrho_1 c_1} \quad , \quad Z = -i\omega\varrho_2 \left.\frac{\Phi}{d\Phi/dz}\right|_{z=0} \quad . \tag{3.1.5}$$

By using (3.1.4) and the pressure continuity condition at $z = 0$ one can find for the transmission coefficient

$$W = (1 + V)\Phi^{-1}(0) \lim_{z \to -\infty} \left\{\Phi(z) \exp\left[i\sqrt{k_2^2 - \xi^2}\,z\right]\right\} \quad . \tag{3.1.6}$$

Equation (3.1.3) is a one-dimensional Helmholtz equation and seems to be simple, its solution in quadratures, however, is possible only in exceptional cases. The general method for finding such solutions consists of reducing this equation to some basic soluble one by the substitution of dependent or independent variables. (For methods of reduction of ordinary second-order differential equations see, for example, [Ref. 3.6, Part I, Sect. 25].) Almost all known exact solutions of the Helmholtz equation can be found by this way. It is reasonable to choose as basic equations those of the most general kind, which contains many "free" parameters in its coefficients and, hence, in solutions.

Without loss of generality a linear ordinary second-order differential equation can be written in the form

$$\frac{d^2 W}{d\eta^2} + g(\eta)W = 0 \quad . \tag{3.1.7}$$

Assume that the fundamental system of solutions to this equation that is, two linearly independent solutions, is known. The important question is: for what classes of sound velocity $c(z)$ can the solution of (3.1.3) be expressed in terms of the solution of (3.1.7) by the substitution of independent or dependent variables of the type

$$\Phi(z) = Q(z)W(\eta) \quad , \quad \eta = \eta(z) \quad . \tag{3.1.8}$$

We assume that the complex valued function $\eta(z)$ is sufficiently smooth (the existence and continuity of the first three derivatives are required) and that $d\eta/dz \neq 0$. Substitution of (3.1.8) into (3.1.7) gives after some cumbersome but simple calculations the equation for the function Φ:

$$\Phi'' + \{(0.5 \ln \eta')'' - [(0.5 \ln \eta')']^2 + (\eta')^2 g(\eta)\}\Phi = 0 \quad . \tag{3.1.9}$$

The prime denotes the derivative of the function with respect to its argument (here z).

The function $Q(z)$ is chosen to be

$$Q = (\eta')^{-1/2} \quad . \tag{3.1.10}$$

which allows eilimination of the Φ' term in (3.1.9).

Comparing (3.1.3 and 9) we see that a solution to (3.1.3) can be expressed in terms of the known functions $W(\eta)$ if the squared wave number can be represented as

$$k^2(z) = \xi^2 + (0.5 \ln \eta')'' - [(0.5 \ln \eta')']^2 + (\eta')^2 g(\eta) \quad , \tag{3.1.11}$$

with proper choice of the function $\eta(z)$. If this is the case, then our only problem is to find a particular solution to (3.1.7) which provides the sound field satisfying conditions at infinity by (3.1.8).

The left-hand side of (3.1.11) does not depend on the incident angle of the wave. Hence, in this method oblique incidence with arbitrary ξ can be treated only if an arbitrary additive constant is present in the right-hand side. For nondispersive media one also has $k \sim \omega$. Hence, the case of nonmonochromatic wave with fixed incidence angle can be treated only if the miltiplicative arbitrary constant is present in the right-hand side of (3.1.11). Thus, we see that not every choice for the functions $\eta(z)$ and $g(\eta)$ is suitable. For brevity, those functions $k(z)$ (wave number "profiles") which allow exact solutions to the problem of the reflection of a plane wave are called "solvable profiles".

3.2 Solvable Profiles $k(z)$ from the Confluent Hypergeometric Equation

In this section, we will consider the case [3.7, 8] where (3.1.3) may be reduced to the confluent hypergeometric equation, which we take in the form of the Whittaker equation:

$$\frac{d^2W}{d\eta^2} + \left(-\frac{1}{4} + \frac{l}{\eta} + \frac{1 - 4m^2}{4\eta^2}\right) W = 0 \quad . \tag{3.2.1}$$

Hence, we assume in (3.1.7):

$$g(\eta) = -\frac{1}{4} + \frac{l}{\eta} + \frac{1 - 4m^2}{4\eta^2} \quad , \tag{3.2.2}$$

where l and m are arbitrary complex parameters. Solutions to (3.2.1) can be expressed in terms of the confluent hypergeometric functions, in particular Whittaker functions $W_{l,m}(\eta)$ and $M_{l,m}(\eta)$. The properties of these functions are well known [3.9–11]. The confluent hypergeometric functions form a very broad class of two-parameter functions including the widely used cylindrical functions as well as Lagger and Weber functions.

Let us consider the following class of independent-variable substitutions:

$$\eta(z) = qf(z) \quad , \quad q \neq 0 \quad , \tag{3.2.3}$$

where q is a complex number and $f(z)$ is a real-valued function. It follows from the conditions imposed on $\eta(z)$ in obtaining (3.1.11) that f is a smooth monotonic function of z. For the $k(z)$ profiles specified by (3.1.11), (3.2.2, 3) the reflection coefficient can be expressed in terms of $f(z)$ without a concrete definition of this function.

Three cases are possible depending on the behavior of $f(z)$ at $z \to -\infty$:

1) $\displaystyle\lim_{z \to -\infty} \eta(z) \equiv \eta(-\infty) = \eta_0 \quad , \quad 0 < |\eta_0| < +\infty \quad .$

Since f is monotonic function, in this case we have $\displaystyle\lim_{z \to -\infty} f' = 0$. Because of (3.1.10), the bounded solution $\Phi(z)$ can be obtained only if $W(\eta_0) = 0$, hence

$$\Phi(z) = \mathrm{const}(f')^{-1/2}[W_{-l,m}(-\eta_0)W_{l,m}(\eta) - W_{l,m}(\eta_0)W_{-l,m}(-\eta)] \quad , \tag{3.2.4}$$

where $W_{\pm l,m}(\pm \eta)$ are Whittaker functions. They form the fundamental system of solutions for (3.2.1) [3.11]. The solution (3.2.4) is not trivial. Indeed, because the point η_0 is not singular in (3.2.1), $W_{l,m}(\eta)$ and $W_{-l,m}(\eta)$, which comprise the fundamental system can not be zero simultaneously.

2) $\eta(-\infty) = 0 \quad .$

If $2m$ is not an integer, it is convenient to choose the $M_{l,\pm m}(\eta)$ functions as fundamental solutions of (3.2.1). These functions behave differently in the neighborhood of the point $\eta = 0$ [3.11]

$$M_{l,m}(\eta) = \eta^{1/2+m}[1 + O(\eta)] \quad . \tag{3.2.5}$$

If $2m$ is an integer, then the function $M_{l,|m|}$ remains a solution to (3.2.1) but the function $M_{l,-|m|}$ is not determined. In this case we obtain the second solution for fundamental system by using the first in the usual way [Ref. 3.6, Part I, Sect. 17]:

$$M(\eta) = M_{l,|m|}(\eta) \int^{\eta} M_{l,|m|}^{-2}(u)du \quad . \tag{3.2.6}$$

The asymptotics of this function at $\eta \to 0$ can be easily found from (3.2.6) after accounting for (3.2.5):

$$M \simeq \text{const} \cdot \eta^{1/2-|m|} \quad , \quad \text{if} \quad 2|m| = 1, 2, \dots \quad ;$$

$$M \simeq \text{const} \cdot \eta^{1/2} \ln \eta \quad , \quad \text{if} \quad m = 0 \quad . \tag{3.2.7}$$

When $|\text{Re}\,\{m\}| \geq 0.5$, only one of the solutions $M_{l,|m|}$ and M is zero at the point $\eta = \eta(-\infty)$. The choice of the solution is then made after accounting for finiteness in $\Phi(z)$ in the same fashion as in the case $0 < |\eta(-\infty)| < +\infty$. In contrast, any solution to (3.2.1) becomes zero at $\eta = 0$ if $|\text{Re}\,\{m\}| < 0.5$, and additional considerations must be made to find suitable physical solutions. It follows from (3.1.11) and (3.2.2) that the replacement $m^2 - i\delta|m|^2$ for m^2, $0 < \delta \ll 1$ means the addition of a small imaginary positive part to $k^2(z)$ causing, as we will see in Chap. 7, the appearance of dissipation in the medium. At large $-z$ the amplitude of the wave travelling from above decreases due to this dissipation while the amplitudes of waves generated by sources at $z \to -\infty$ will increase. These considerations allow to separate the suitable physical solution. This is

$$\Phi(z) = \text{const}(f')^{-1/2} M_{l,m}(\eta) \quad ,$$

$$\arg m \in \begin{cases} (-\frac{\pi}{2}, \frac{\pi}{2}] & , \quad |\text{Re}\,\{m\}| \geq 0.5 \quad , \\ (-\pi, 0] & , \quad |\text{Re}\,\{m\}| < 0.5 \quad . \end{cases} \tag{3.2.8}$$

3) $\eta(-\infty) = \infty$

In (3.2.3) we assume that $\lim\limits_{z \to -\infty} f(z) = +\infty$ whereas the sign of q is assumed to be not fixed. As fundamental system of solutions of (3.2.1) we choose $W_{l,m}(\eta)$ and $W_{-l,m}(-\eta)$. As $\eta \to \infty$ these functions behave as (see [3.11])

$$W_{l,m}(\eta) = \eta^l \exp(-\eta/2)[1 + O(\eta^{-1})] \quad ,$$

$$|\arg \eta| \leq \pi - \delta \quad , \quad \delta > 0 \quad . \tag{3.2.9}$$

This formula gives the asymptotics of only one of the linear independent solutions if q is real. In this case the asymptotics of the Whittaker function with a negative argument can be obtained from the asymptotics (3.2.9) of the other linear independent solution in the same way as (3.2.7) was obtained.

The choice of a physically suitable solution to the wave equation can be easily accomplished, if $\text{Re}\,\{q\} \neq 0$, by using the condition of finiteness of the function Φ in (3.1.8). The case where $\text{Re}\,\{q\} = 0$ differs only in technical details from the case $|\text{Re}\,\{m\}| < 0.5$ with $\eta(-\infty) = 0$, considered above. Therefore, we will write out just the results of the calculations

$$\Phi(z) = \text{const}(f')^{-1/2}$$

$$\times \begin{cases} W_{l,m}(\eta) & , \quad \arg q \in [-\frac{\pi}{2}, \frac{\pi}{2}) \quad , \\ W_{-l,m}(-\eta) & , \quad \arg q \in (-\pi, -\frac{\pi}{2}) \cup [\frac{\pi}{2}, \pi] \quad . \end{cases} \tag{3.2.10}$$

In total, (3.2.4, 8, and 10) describe the sound field in the lower (stratified) medium in the case of any substitution of the kind given in (3.2.3). The reflection coefficient of the plane monochromatic wave (for given ω and ξ) can be then calculated according to (3.1.5) for any profile (3.1.11) with $g(\eta)$ given by (3.2.2).

We now prove that *any* smooth real profile $k^2(z)$ can be treated in this way by suitable choice of function f and parameters l, m, and q. To this end, we must show that (3.1.11), considered as the differential equation for $f(z) = \eta(z)/q$, has a solution in the case of any (smooth) profile $k^2(z)$, at least for some l, m, and q. Supposing $l = 0$, $m = \frac{1}{2}$, $q = 1$, we obtain a simpler equation for f from (3.1.11):

$$(0.5 \ln f')'' - [(0.5 \ln f')']^2 - (f'/2)^2 = k^2 - \xi^2 \quad . \tag{3.2.11}$$

The substitution

$$f(z) = \ln\left[1 + \int\limits_0^z v^{-2}(u)du\right] \tag{3.2.12}$$

reduces (3.2.11) to an equation similar to (3.1.3):

$$v''(z) + (k^2 - \xi^2)v(z) = 0 \quad . \tag{3.2.13}$$

Solutions to (3.1.3) must be real and smooth due to the intrinsic nature of the problem. The smoothness and monotony of $f(z)$ follow from (3.2.12, 13). Hence, for any sufficiently smooth function $k^2(z)$ there exists an $f(z)$ satisfying all the conditions imposed in the derivation of (3.1.11) and (3.2.4, 8, 10).

This result is mostly formal, however: generally speaking, for the functions $\eta(z)$ or $f(z)$ (3.1.11) is not simpler than the initial equation (3.1.3). However, there does exist a possibility of finding some profiles $k^2(z)$ which contain arbitrary parameters and correspond to some comparatively simple function f. For such profiles exact solutions for wave fields are known. The choice of the concrete values for the parameters can be used to make the best approximation of real profile of interest.

We turn now to examples of profiles given by (3.1.11) which allow exact solutions in terms of the confluent hypergeometric functions and consider three types of substitutions for $f(z)$:

Example A. Let

$$-\tfrac{\pi}{2} \le \arg q < \tfrac{\pi}{2} \quad , \qquad f(z) = (|z| + z_1)^b \quad , \qquad \text{where} \quad b \neq 0 \quad .$$

Equations (3.1.11) and (3.2.2) yield:

$$k^2(z) = \xi^2 + 0.25b^2(|z| + z_1)^{-2}[-q^2(|z| + z_1)^{2b} + 4lq(|z| + z_1)^b$$
$$+ b^{-2} - 4m^2] \quad . \tag{3.2.14}$$

If $b > 0$, the input impedance of the lower half-space is according to (3.1.5) and (3.2.10)

$$Z = 2i\omega\varrho_2 z_1\{1 - b + 2bqz_1^b[\ln W_{l,m}(qz_1^b)]'\}^{-1} \quad . \tag{3.2.15}$$

(In expressions for the input impedance primes denote the derivative *with respect to the argument of special function*, here qz_1^b.) In the case of $b < 0$ impedance Z is obtained from (3.2.15) by substitution of $M_{l,m}$ for $W_{l,m}$. The possibility of using the Whittaker functions for describing the field (at normal incidence) for profiles (3.2.14) if $b = 1$ and $b = 2$ was pointed out by *Westcott* [3.12]. We shall assume that $z_1 > 0$. Formally our solutions are also true if $z_1 < 0$, but in this case k^2 becomes infinite at finite z, which is impossible in acoustics.

We have from (3.2.14) if $b = 1$

$$k^2(z) = k_2^2 + \alpha_1(|z/z_1| + 1)^{-1} + \alpha_2(|z/z_1| + 1)^{-2} \quad , \tag{3.2.16}$$

where k_2^2, α_1, and α_2 are arbitrary constants. They are related to the parameters l, m, q in (3.2.15) by

$$q = -2i\sqrt{k_2^2 - \xi^2} \quad , \quad \mathrm{Im}\left\{\sqrt{k^2 - \xi^2}\right\} \geq 0 \quad , \tag{3.2.17a}$$

$$l = \alpha_1 z_1/q \quad , \quad m = \sqrt{1/4 - \alpha_2 z_1^2} \quad . \tag{3.2.17b}$$

Here k_2 is the wave number at $z = -\infty$ and z_1 is the vertical scale of inhomogeneities in the lower medium. We see that reflection of a wave with any ξ can be treated due to the arbitrariness of q. The frequency may also be arbitrary depending on the choice of l and m. Hence, the profile is specified by four parameters. Figure 3.2 shows several typical $k^2(z)$ for various combinations of the parameters as a function of the dimensionless coordinate z/z_1. The type of the profile depends on the quantities α_1 and α_2. In particular, a maximum or minimum in the sound velocity can occur in the medium. In these cases, waveguide (in the vicinity of the minimum) or antiwaveguide (in the vicinity of the maximum) conditions for the propagation of waves can arise [3.1].

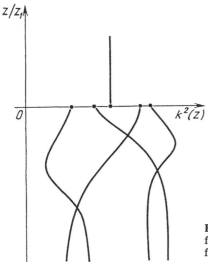

Fig. 3.2. Typical wave number squared profiles $k^2(z)$ as a function of z/z_1, corresponding to different choices of the free parameters values in expression (3.2.16)

A special case for the profile given by (3.2.16) arises when $\xi^2 = k_2^2$. Now (3.2.15 and 17) can not be used directly because we have assumed $q \neq 0$ while obtaining (3.2.15). The input impedance can be easily obtained if $\alpha_1 = 0$ by proceeding towards the limit $q \rightarrow 0$. When $\alpha_1 \neq 0$, however, this procedure is hindered because the parameter l in the Whittaker function tends to infinity. It is instructive to note, however, that we again obtain the profile given by (3.2.16) if we assume in (3.2.14), when $\xi = \pm k_2$, that

$$b = \tfrac{1}{2} \quad , \quad l = 0 \quad , \quad m = \sqrt{1 - 4\alpha_2 z_1^2} \quad , \quad \text{and}$$

$$q = 4\sqrt{-\alpha_1 z_1} \quad . \tag{3.2.18}$$

The input impedance of the lower half-space will again be given by (3.2.15) where the values of the parameters in (3.2.18) must be used. It can be shown that the same result can be obtained, if we proceed to the limit $q \to 0$ in the case $q \neq 0$. Note the simple expression for the reflection coefficient when $\xi = \pm k_2$, $\alpha_1 = 0$, $\varrho_1 = \varrho_2$:

$$V = \frac{i}{2} \left[k_1 z_1 + i \sqrt{1/4 - k_1^2 z_1^2} \right]^{-1} \quad . \tag{3.2.19}$$

Consider the behavior of the wave in the lower half-space at $z \to -\infty$. By using (3.2.9, 10, 17) and retaining only the main terms in the expansion in a power series of $|z_1/z|$, we obtain if $\xi \neq \pm k_2$,

$$\Phi(z) \approx \mathrm{const} \cdot \exp\left[i\sqrt{k_2^2 - \xi^2}(z_1 - z) + \frac{i}{2}\alpha_1 z_1 (k_2^2 - \xi^2)^{-1/2} \ln|z| \right] \quad . \tag{3.2.20}$$

Analogously, by using (3.2.18) we find if $\xi = \pm k_2$,

$$\Phi(z) \approx \mathrm{const} \cdot \exp\left[-2\sqrt{|\alpha_1 z_1 z|} \right] \quad , \quad \alpha_1 < 0 \quad ; \tag{3.2.21a}$$

$$\Phi(z) \approx \mathrm{const} \cdot \exp\left[2i\sqrt{|\alpha_1 z_1 z|} \right] \quad , \quad \alpha_1 > 0 \quad . \tag{3.2.21b}$$

If $\xi = \pm k_2$ and $\alpha_1 = 0$ simultaneously, then it follows from (3.2.20, 21) that with the error $O(|z|^{-1})$, $p \approx \mathrm{const} \cdot \exp(i\xi x)$ in the lower half-space at large $|z|$ and does not depend on z.

Note that the parameter α_2 is absent in the main terms of the above asymptotic expansions. This is because at large $|z|$ the term $\alpha_2(|z/z_1| + 1)^{-2}$ in (3.2.16) is small compared to k_2^2. The term $\alpha_1(|z/z_1| + 1)^{-1}$ is also small, but has cumulative and therefore increasing effects during the propagation of the wave. We see that only if $\alpha_1 = 0$ does the sound field become a plane wave as $z \to -\infty$. This means that the approach of $k^2(z)$ towards k_2^2 as $k^2(z) = k_2^2 + O(|z|^{-1})$ when $z \to -\infty$ is not fast enough for introducing the concept of the transmission coefficient according to relation (3.1.4).

The sound field decreases exponentially at large $|z|$ in the lower medium if $\xi^2 > k_2^2$. The critical incidence angle is $\theta = \delta \equiv \arcsin(k_2/k_1)$. In the opposite case, when $\xi^2 < k_2^2$, the transmitted wave carries away part of the energy of the incident wave. In this case, besides the usual phase of the wave given by $(k_2^2 - \xi^2)^{1/2}(z_1 - z) + \xi x$, an additional term $0.5\alpha_1 z_1 (k_2^2 - \xi^2)^{-1/2} \ln|z|$ is present at large $|z|$, which varies slowly when $|z|$ varies. At the critical incidence angle (when $\xi = k_2$) and large $|z|$, the field in the lower medium is either a plane wave with amplitude independent of z (if $\alpha_1 > 0$) or a wave with exponentially decreasing amplitude (if $\alpha_1 < 0$).

If the incidence angle exceeds the critical angle δ, total reflection takes place. Indeed, in this case according to (3.2.17a), q becomes real and positive. The Whittaker function in (3.2.15) is real if its argument is real (m also assumed to be real). Then it follows from (3.2.15) and (3.1.5) that $\mathrm{Re}\{Z\} = 0$, $|V| = 1$. It is easy to show using (2.1.11), (3.2.20), and (3.2.21) that at $\theta \geq \delta$ (for any m) the time averaged z-component of the energy flux at $z = -\infty$ is zero, as it must be in the case of the total reflection.

Harmonic waves in media with a special kind of the profile (3.2.16) were considered by *Rytov* and *Yudkevich* [3.13], and also by *Wait* [3.5]. The case where $\alpha_1 = 0$ was considered by these authors when the solution can be expressed in terms of cylindrical funcitons. The latter are related to Whittaker functions by the transformation:

$$H_{\nu}^{(1)}(r) = \sqrt{2/\pi r}\,\exp\left[\,-\,i\pi(2\nu+1)/4\right]W_{0,\nu}(-2ir) \quad , \qquad [3.10]$$
$$J_{\nu}(r) = [2^{4\nu+1}i^{2\nu+1}\,\Gamma^2(\nu+1)r]^{-1/2}M_{0,\nu}(2ir) \quad , \qquad [3.11] \quad . \qquad (3.2.22)$$

By using these expressions and noting that the time factor chosen in [3.5] is $\exp(i\omega t)$ rather than $\exp(-i\omega t)$ as used here, one can readily find identity of the result obtained by (3.2.15) to the result given in [Ref. 3.5, Chap. 3, Sect. 3].

By assuming $b = \frac{3}{2}$ we obtain from (3.2.14):

$$k^2(z) = \xi^2 - \tfrac{9}{16}q^2(|z| + z_1) - \tfrac{9}{4}ql(|z| + z_1)^{-1/2}$$
$$+ \tfrac{1}{4}(1 - 9m^2)(|z| + z_1)^{-2} \quad . \qquad (3.2.23)$$

In this case, exact solutions for *arbitrary* ξ are possible only if $l = 0$ and $m = \frac{1}{3}$, when k^2 depends linearly on z. The Whittaker function $W_{0,1/3}$ in the expression for the input impedance (3.2.15) can be expressed in terms of the well-known Airy function [3.10]

$$W_{0,1/3}(4r^{3/2}/3) = 2\sqrt{\pi}r^{1/4}v(r) \quad . \qquad (3.2.24)$$

A linear profile $k^2(z)$ is used very often to obtain numerical and asymptotic solutions for the wave equation for general types of continuously layered media when exact solutions do not exist. Exact solutions for linear profile will be discussed in Sect. 3.5.

There is one more case where the reflection coefficient can be found for an arbitrary angle of incidence, assuming $b = 2$. We obtain from (3.2.14) the profile

$$k^2(z) = \beta_1 + \beta_2(|z/z_1|^2 + 1)^2 + \beta_3(|z/z_1| + 1)^{-2} \quad , \qquad (3.2.25)$$

where the arbitrary constants β_1, β_2, and β_3 are related to parameters l, m, and q by

$$q = \frac{\sqrt{-\beta_2}}{z_1} \quad , \qquad l = \frac{\beta_1 - \xi^2}{4q} \quad , \qquad m = \frac{\sqrt{1 - 4\beta_3 z_1^2}}{4} \quad . \qquad (3.2.26)$$

Oblique incidence can be treated because of the arbitrariness of l. The arbitrariness of q and m allows us to choose the frequency and vertical scale of the inhomogeneities z_1. The condition $q \neq 0$ excludes the case of $\beta_2 = 0$ in (3.2.25). This case was considered above, however; the corresponding profile is obtained from (3.2.16) if $\alpha_1 = 0$, $k_2^2 = \beta_1$, $\alpha_2 = \beta_3$. Some possible types of profiles for different relationships between β_2 and β_3 in (3.2.25) are shown in Fig. 3.3. The abscissa is the dimensionless quantity $[k^2(z) - \beta_1]/\beta_2$. Note also that the profile given by (3.2.25) at $\beta_3 \neq 0$ can be used for modeling of real systems only if the inhomogeneities at large $|z/z_1|$ do not play an important role in reflection. In this case, in (3.2.25) $k(z)$ becomes infinite (sound velocity: $c \to 0$) when $z \to -\infty$.

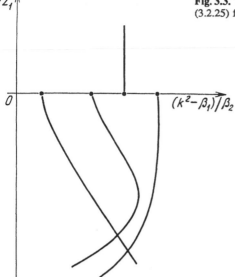

Fig. 3.3. Typical solvable profiles calculated according to (3.2.25) for various values of β_2 and β_3

When $\beta_3 = 0$, there exist among the profiles given by (3.2.25) the parabolic waveguide ($\beta_2 < 0$), considered by *Iamada* [3.14] and many other authors, and the parabolic antiwaveguide ($\beta_2 > 0$), discussed by *Masterov* and *Muromtseva* [3.15]. In these pratical cases the function $W_{l,-1/4}(qz_1^2)$ in (3.2.15) can be expressed in terms of a Weber parabolic cylinder function $D_{2l-1/2}(\sqrt{q}z_1)$ [3.10], which was used in [3.14, 15]. In quantum mechanics the profile $k^2(z)$ of the type described by (3.2.25) with $\beta_3 = 0$ corresponds to the potential of the harmonic oscillator or parabolic potential barrier depending on the sign of β_2 [see, for example, Ref. 3.16, Sects. 23, 50].

Several researchers (see [Ref. 3.1, Sect. 22], [Ref. 3.4, Chap. 7] and references therein) have addressed the problem of normal incidence of a wave upon the medium with the profile

$$k^2(z) = \alpha(|z/z_1| + 1)^\gamma \quad . \tag{3.2.27}$$

The case where $\gamma = -2$ has been discussed above on the basis of the profile of (3.2.16), therefore we shall confine the discussion now to cases where $\gamma \neq -2$. If there are no other limitations on the values of γ, then the choice $f = (|z| + z_1)^b$ permits us to treat the case of normal incidence only. The profile given by (3.2.27) at $\xi = 0$ can be obtained from (3.2.14) with

$$b = (\gamma + 2)/2 \quad , \quad l = 0 \quad , \quad m = \pm(2 + \gamma)^{-1} \quad ,$$
$$q = -4i(2 + \gamma)^{-1} z_1^{-\gamma/2} \alpha^{1/2} \quad . \tag{3.2.28}$$

The arbitrariness of q allows consideration of a wave of any frequency. We find from (3.2.15, 22), if $\gamma > -2$ that:

$$Z = 2i\omega \varrho_2 z_1 \{1 + 4z_1(-\alpha)^{1/2} [\ln H^{(1)}_{(2+\gamma)^{-1}} (2(2 + \gamma)^{-1} \alpha^{1/2} z_1)]'\}^{-1}. \tag{3.2.29}$$

50

If $\gamma < -2$, the substitution of $J_{-(2+\gamma)-1}(-2(2+\gamma)^{-1}\alpha^{1/2}z_1)$ for $H^{(1)}_{(2+\gamma)-1}(2(2+\gamma)^{-1}\alpha^{1/2}z_1)$ in (3.2.29) is necessary. The obtained result, as well as (3.2.29), reproduce the equations that were derived in [Ref. 3.5, Chap. 3, Sect. 9].

Example B. Let

$$f = \exp(-az) \quad , \quad -\tfrac{\pi}{2} \leq \arg q < \tfrac{\pi}{2}$$

in (3.2.3). With the help of (3.1.11) and (3.2.2) we obtain the solvable profile

$$k^2(z) = \alpha_1 + \alpha_2 \exp(-az) + \alpha_3 \exp(-2az) \quad . \tag{3.2.30}$$

For this profile it is possible to examine the case of oblique incidence and arbitrary frequency. Several examples of profiles from the family given by (3.2.30) are shown in Fig. 3.4, plotted in dimensionless coordinates k^2/α_3 and $|a|z$.

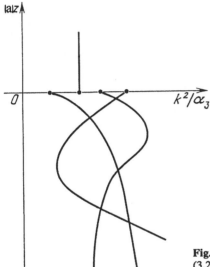

Fig. 3.4. Examples of solvable profiles given by expression (3.2.30) for various values of α_1/α_3 and α_2/α_3

If $a > 0$, at $z \to -\infty$ we have $f(z) \to +\infty$, and if $a < 0$, $f(z) \to 0$. Now with the help of (3.1.5), (3.2.8, 10) we obtain for the input impedance of the lower half-space

$$Z = i\omega\varrho_2 a \cdot \begin{cases} [-\tfrac{1}{2} + q(\ln W_{l,m}(q))']^{-1} \quad , \quad a > 0 \quad , \\ [-\tfrac{1}{2} + q(\ln M_{l,m}(q))']^{-1} \quad , \quad a < 0 \quad . \end{cases} \tag{3.2.31}$$

The parameters in (3.2.31) are related to the constants in (3.2.30) by

$$q = \frac{2\sqrt{-\alpha_3}}{a} \quad , \quad l = \frac{\alpha_2}{2a\sqrt{-\alpha_3}} \quad , \quad m = a^{-1}\sqrt{\xi^2 - \alpha_1} \quad . \tag{3.2.32}$$

The possibility of using the Whittaker functions for the profile given by (3.2.30) was pointed out by different authors [3.6, 17−19]. The particular case of $\alpha_2 = 0$ when the solutions to the wave equation are expressed in terms of cylindrical functions was considered in [Ref. 3.5; Chap. 3] for different relations between α_1 and α_3. Taking into account the relations between the Whittaker and cylindrical functions given by (3.2.22) one can easily prove that the result given by the general formula of (3.2.31) and the results obtained for all the limiting cases considered in [3.5] agree. Exact solutions for the profile given by (3.2.30) with $\alpha_2 = 0$ are widely used for obtaining the numerical solution to (3.1.3) in a medium with arbitrary smooth $k(z)$. The medium is approximated by a system of layers with profiles given by (3.2.30) [3.20].

Example C. Although the solvable profiles discussed above are rather diverse, each one has no more than one extremum, i.e., maximum or minimum, in the sound velocity in the inhomogeneous medium. In practice one has often to consider inhomogeneous media which have many extrema of $c(z)$ and hence, several waveguides or antiwaveguides. We would like, therefore, to point out one more solvable profile. It is generated by the function $f(z)$ given by the parametric relations

$$f = u^2 - 1 \quad , \quad z = -z_1 + 2(u - \operatorname{arccoth} u)/b \quad . \tag{3.2.33}$$

It follows from (3.2.33) that $f'(z) \neq 0$. When $z + z_1$ changes from $-\infty$ to $+\infty$, the function $f(z)$ takes all positive values and increases monotonically, if $b > 0$, and $f(z)$ decreases monotonically if $b < 0$. The profile $k(z)$ is found from (3.1.11), (3.2.2 and 33)

$$k^2(z) = a_1 b^2 - b^2(u^4 + a_3 + 6a_2 u^{-2} - 5a_2 u^{-4})/a_2 \quad , \tag{3.2.34}$$

where a_j, $j = 1, 2, 3$ are arbitrary constants related to the parameters of Whittaker's equation and q. They are given by

$$q = 8a_2^{-1/2} \quad , \quad -\tfrac{\pi}{2} \leq \arg q < \tfrac{\pi}{2} \quad ;$$

$$l = \frac{a_1 - \xi^2 b^{-2}}{q} - q, \quad m = \sqrt{2q^2\left(a_3 - \frac{4l}{q} - 1\right) + \frac{3}{2}} \quad . \tag{3.2.35}$$

The solvable profile of (3.2.34) has five free parameters. As one can see from (3.2.34, 35) on the right-hand side of (3.2.34) there is an additive arbitrary constant, hence, exact solutions can be obtained at oblique incidence. From the other hand, because we have no multiplicative arbitrary constant in (3.2.34) we can not obtain the frequency dependence of the reflection coefficient.

Finding the limit of f at $z \to -\infty$ from (3.2.33) and using (3.1.5), (3.2.8, and 10), we obtain for the input impedance of the lower half-space:

$$Z = -2i\omega\varrho_2 \begin{cases} [f''/f' - 2qf'(\ln M_{l,m}(qf))'] & , & b > 0 \quad ; \\ [f''/f' - 2qf'(\ln W_{l,m}(qf))'] & , & b < 0 \quad . \end{cases} \tag{3.2.36}$$

Here, the values f, f', and f'' are taken at $z = 0$.

Several typical profiles $k^2(z)$ given by (3.2.34) are illustrated in Fig. 3.5. Analysis of (3.2.34) shows that sound velocity could have from zero to three extrema depending on the a_2 and a_3 values. In the case of three extrema, four *turning points* can exist, i.e., points where $k(z) = \xi$ and the vertical component of wave vector becomes zero.

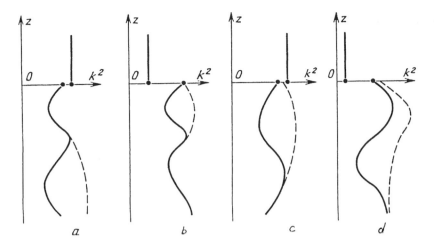

Fig. 3.5a–d. Typical $k^2(z)$ profiles following from (3.2.34). The inhomogeneous half-space can contain two inner waveguides (a, b), an inner and a surface waveguide (c) or an inner waveguide and an antiwaveguide (d). (*Dashed lines*) show profiles obtainable from basic profiles by continuous variation of the parameters in (3.2.34)

In concluding the discussion of profiles which are solvable in terms of confluent hypergeometric functions, it is useful to note the following. When some relationships between the parameters l and m hold, the functions $M_{l,m}$ and $W_{l,m}$ reduce to simpler functions. Although this decreases the number of free parameters in the $k(z)$ profile, it nevertheless can sometimes be worthwhile. Significant simplification occurs when $M_{l,m}$ or $W_{l,m}$ reduce to elementary functions. We point out several such cases following [Ref. 3.10, Chap. 13].

The wave field in the lower half-space is expressed in terms of the function $M_{l,m}(qf)$ if $f(z) \to 0$ at $z \to -\infty$. The function $M_{l,m}$ is elementary for each of the following cases:

1. $l = 0,\ m = \pm(n + \frac{1}{2})$;
2. $l = \frac{1}{2} + m + n,\ m = \pm\frac{1}{4}$;
3. $l = b/2 + n,\ m = (b - 1)/2$;
4. $l = -b/2,\ m = (b - 1)/2$;
5. $l = m + n + \frac{1}{2},\ m > 0$.

Here $n = 0, 1, 2, \ldots$; $b \neq 0, -1, -2, \ldots, 1 - n$. If $f(z) \to +\infty$ at $z \to -\infty$, the field is described in terms of the functions $W_{l,m}$ which reduce to elementary functions in each of the following cases:

1. $l = 0,\ m = n + \frac{1}{2}$;
2. $l = (2n + 1)/4,\ m = \pm\frac{1}{4}$;
3. $l = 2m + n + \frac{1}{2},\ m > 0$,
    where $n = 0, 1, 2, \ldots$ as above.

3.3 Solvable Profiles Obtained from the Hypergeometric Equation

Equation (3.3.1)

$$\frac{d^2 F}{d\eta^2} - \frac{(\alpha + \beta + 1)\eta - \gamma}{\eta(1 - \eta)} \cdot \frac{dF}{d\eta} - \frac{\alpha\beta F}{\eta(1 - \eta)} = 0 \quad . \tag{3.3.1}$$

is called hypergeometric. Here, we review some of its basic properties. Solutions to this equation may be presented as "hypergeometric series", see [Ref. 3.11; Chap. 14] or [Ref. 3.10; Chap. 15],

$$F_1 = F(\alpha, \beta, \gamma, \eta) \equiv 1 + \frac{\alpha\beta}{\gamma}\eta + \frac{\alpha(\alpha + 1)\beta(\beta + 1)}{1 \cdot 2 \cdot \gamma(\gamma + 1)}\eta^2$$

$$+ \frac{\alpha(\alpha + 1)(\alpha + 2)\beta(\beta + 1)(\beta + 2)}{1 \cdot 2 \cdot 3 \cdot \gamma(\gamma + 1)(\gamma + 2)}\eta^3 + \dots \quad . \tag{3.3.2a}$$

This series converges if $|\eta| < 1$. It can also be shown that the second linearly independent solution which converges inside the circle $|\eta| < 1$ is equal to

$$F_2 = \eta^{1-\gamma} F(\alpha - \gamma + 1, \beta - \gamma + 1, 2 - \gamma, \eta) \quad . \tag{3.3.2b}$$

Equation (3.3.1) has three singular points: $\eta = 0, 1, \infty$. Consequently, there exist three pairs of linearly independent solutions; each pair converges near "its own" singular point. Thus, near the point $\eta = 1$ we have the fundamental system of solutions:

$$F_3 = F(\alpha, \beta, \alpha + \beta - \gamma + 1, 1 - \eta) \quad , \tag{3.3.3a}$$

$$F_4 = (1 - \eta)^{\gamma-\alpha-\beta} F(\gamma - \alpha, \gamma - \beta, \gamma - \alpha - \beta + 1, 1 - \eta) \quad , \tag{3.3.3b}$$

and near the point $\eta = \infty$:

$$F_5 = \eta^{-\alpha} F(\alpha, \alpha - \gamma + 1, \alpha - \beta + 1, 1/\eta) \quad , \tag{3.3.4a}$$

$$F_6 = \eta^{-\beta} F(\beta, \beta - \gamma + 1, \beta - \alpha + 1, 1/\eta) \quad . \tag{3.3.4b}$$

Each of these expressions is an analytical function which is a solution to (3.3.1) in the entire convergence region of the corresponding series.

Each of the solutions F_j, $j = 1, 2, 3, \dots, 6$ can be continued beyond the boundary of the convergence region of the corresponding series by the method of analytic continuation. In this way we immediately obtain three solutions in a new region, namely, one continued from the convergence region and two given by (3.3.2–4). Equation (3.3.1) must have exactly two linearly independent solutions in each region. Hence, a linear relation with a constant coefficient must exist among these three solutions. Only one of such relations we shall need below. It appears that analytic continuation of the solution F_5 into the region $|\eta| < 1$ can be expressed in this region in terms of F_1 and F_2 by [3.10, 11]

$$F_5 = (-1)^{-\alpha} \frac{\Gamma(\alpha - \beta + 1)\Gamma(1 - \gamma)}{\Gamma(1 - \beta)\Gamma(1 + \alpha - \gamma)} F_1$$
$$+ (-1)^{\gamma - \alpha - 1} \frac{\Gamma(\alpha - \beta + 1)\Gamma(\gamma - 1)}{\Gamma(\gamma - \beta)\Gamma(\alpha)} F_2 \quad , \tag{3.3.5}$$

Where $\Gamma(\cdot)$ is the gamma function.

For the function

$$W = \eta^{\gamma/2}(1 - \eta)^{(\alpha + \beta - \gamma + 1)/2} F \tag{3.3.6}$$

the hypergeometric equation reduces to (3.1.7) with

$$g(\eta) = -\eta^{-2}[K_1 + K_2\eta(1 - \eta)^{-1} + K_3\eta(1 - \eta)^{-2}] \quad , \tag{3.3.7}$$

where

$$4K_1 = \gamma(\gamma - 2) \quad ,$$
$$4K_2 = 1 - (\alpha - \beta)^2 + \gamma(\gamma - 2) \quad ,$$
$$4K_3 = (\alpha + \beta - \gamma)^2 - 1 \quad . \tag{3.3.8}$$

Expressions (3.1.11) and (3.3.7) determine the types of $k(z)$ profiles for which the exact solution to the wave equation can be expressed in terms of hypergeometric functions. The most interesting profiles are those, as was pointed out above, in which besides other free parameters, additive and muliplicative arbitrary constants are present, too. We will consider two types of the substitutions of $\eta(z)$ where these conditions are met.

Case A. Assume $\eta(z) = -\exp[a(z + z_1)]$, $a \neq 0$. Then expressions (3.1.11) and (3.3.7) yield

$$k^2(z) = k_0^2 \left\{ 1 - N \frac{\exp[a(z + z_1)]}{1 + \exp[a(z + z_1)]} - 4M \frac{\exp[a(z + z_1)]}{\{1 + \exp[a(z + z_1)]\}^2} \right\} \quad , \tag{3.3.9}$$

where k_0, N, and M are related to K_1, K_2, and K_3 from (3.3.8) by

$$K_1 = (\xi^2 - k_0^2)a^{-2} - \tfrac{1}{4} \quad , \quad K_2 = -k_0^2 N a^{-2} \quad , \quad K_3 = -4k_0^2 a^{-2} M \quad . \tag{3.3.10}$$

The reflection of a wave from a medium of the type described by (3.3.9) at normal incidence was first considered by *Epstein* in 1930 [3.21] and the inhomogeneous layer specified by (3.3.9) is therefore called *Epstein's layer*. Exact solutions for any incidence angle and frequency can be obtained for this layer.

Let us express the parameters α, β, and γ of the hypergeometric equation in terms of the layer parameters a, M, N. From (3.3.8, 10) we find

$$\alpha = \tfrac{1}{2} + \tfrac{1}{2}\sqrt{1 - 16Mk_0^2 a^{-2}} + i|a|^{-1}\left(\sqrt{k_0^2 - \xi^2} - \sqrt{k_0^2 - \xi^2 - k_0^2 N}\right) \quad ,$$

$$\beta = \tfrac{1}{2} + \tfrac{1}{2}\sqrt{1 - 16Mk_0^2 a^{-2}} + i|a|^{-1}\left(\sqrt{k_0^2 - \xi^2} + \sqrt{k_0^2 - \xi^2 - k_0^2 N}\right) \quad ,$$

$$\gamma = 1 + 2i|a|^{-1}\sqrt{k_0^2 - \xi^2} \quad . \tag{3.3.11}$$

The signs of square roots are chosen here with the condition that their imaginary parts must be nonnegative. The system of equations made up by (3.3.8, 10) has other solutions besides (3.3.11), but we do not need them for a discussion of the wave equation solutions.

For the $k(z)$ profile given by (3.3.9) the exact solutions in terms of hypergeometric functions can be obtained when z_1, a, M, and N are arbitrary *complex* quantities. From a physical point of view the profiles could be quite different for real and imaginary values of these parameters. Thus, $k(z)$ can be infinite at some z if a is real and z_1 is a complex quantity. The dependence of $k(z)$ on z is periodic (with period $2\pi/|a|$), if a is pure imaginary. We will discuss the most interesting case (from the physical point of view), when a and z_1 are real. Two cases, for positive and negative a, are possible.

1) $a > 0$. Then $\eta \to 0$ at $z \to -\infty$. In accordance with (3.1.8, 10), and (3.3.6) the general solution to (3.1.3) is

$$\Phi = \exp\left[(\gamma - 1)a(z + z_1)/2\right]\{1 + \exp\left[a(z + z_1)\right]\}^{(\alpha+\beta-\gamma+1)/2} F \quad , \tag{3.3.12}$$

where F is the general solution of the hypergeometric equation which at $|\eta| < 1$ is conveniently expressed as

$$F = AF_1 + BF_2 \quad . \tag{3.3.13}$$

We see from (3.3.2) that $F_1 \to 1$ and $F_2 = \eta^{1-\gamma}[1 + O(\eta)]$ when $\eta \to 0$. Hence, we obtain for the sound field in the lower half-space far from the boundary:

$$\Phi \approx A \exp\left[i\sqrt{k_0^2 - \xi^2}(z + z_1)\right] + B \exp\left[-i\sqrt{k_0^2 - \xi^2}(z + z_1)\right] \quad . \tag{3.3.14}$$

where we use (3.3.11) for α, β, and γ.

Note that the first term in (3.3.14) corresponds to the wave coming from infinity ($z = -\infty$) if $\xi^2 < k_0^2$, and to the wave with amplitude tending toward infinity as $z \to -\infty$ if $\xi^2 > k_0^2$. Hence, we must assume that $A = 0$ at any ξ. Using (3.3.11) and (3.3.12) we now obtain for the field in the lower half-space ($z < 0$)

$$\Phi = \text{const} \cdot \exp\left[-i\sqrt{k_0^2 - \xi^2}(z + z_1)\right] [1 + \exp\left(a(z + z_1)\right)]^{1/2 + \frac{1}{2}\sqrt{1 - 16Mk_0^2 a^{-2}}}$$

$$\times F\left(\frac{1}{2} + \frac{1}{2}\sqrt{1 - 16Mk_0^2 a^{-2}} - \frac{i}{a}\left(\sqrt{k_0^2 - \xi^2} + \sqrt{k_0^2 - \xi^2 - k_0^2 N}\right) \quad , \right.$$

$$\frac{1}{2} + \frac{1}{2}\sqrt{1 - 16Mk_0^2 a^{-2}} - \frac{i}{a}\left(\sqrt{k_0^2 - \xi^2} - \sqrt{k_0^2 - \xi^2 - k_0^2 N}\right) \quad ,$$

$$\left. 1 - \frac{2i}{a}\sqrt{k_0^2 - \xi^2}, \ -\exp\left[a(z + z_1)\right]\right) \quad . \tag{3.3.15}$$

Using this expression and (3.1.5, 6) one can calculate the reflection and transmission coefficients.

2) $a < 0$. In this case $\eta \to -\infty$ as $z \to -\infty$. It is convenient now to choose in (3.3.12)

$$F = AF_5 + BF_6 \tag{3.3.16}$$

as a general solution of the hypergeometric equation. Here we have, according to (3.3.4), $F_5 = \eta^{-\alpha}[1+O(\eta^{-1})]$ and $F_6 = \eta^{-\beta}[1+O(\eta^{-1})]$ at $\eta \to -\infty$. Quite similar to the case where $a > 0$, we obtain $B = 0$ from the condition at $z \to -\infty$. As a result, the field in the lower medium ($z < 0$) is then

$$
\Phi = \text{const} \cdot \exp\left[-\left(\frac{a}{2} + \frac{a}{2}\sqrt{1 - 16Mk_0^2 a^{-2}} + i\sqrt{k_0^2 - \xi^2 - k_0^2 N}\right)(z + z_1)\right]
$$

$$
\times \{1 + \exp[a(z + z_1)]\}^{1/2 + 1/2\sqrt{1 - 16Mk_0^2 a^{-2}}}
$$

$$
\times F\left(\frac{1}{2} + \frac{1}{2}\sqrt{1 - 16Mk_0^2 a^{-2}} - ia^{-1}\left(\sqrt{k_0^2 - \xi^2} - \sqrt{k_0^2 - \xi^2 - k_0^2 N}\right)\right. ,
$$

$$
\frac{1}{2} + \frac{1}{2}\sqrt{1 - 16Mk_0^2 a^{-2}} + ia^{-1}\left(\sqrt{k_0^2 - \xi^2} + \sqrt{k_0^2 - \xi^2 - k_0^2 N}\right) ,
$$

$$
\left. 1 + 2ia^{-1}\sqrt{k_0^2 - \xi^2 - k_0^2 N} \quad , \quad -\exp[-a(z + z_1)]\right) . \tag{3.3.17}
$$

The expressions for the reflection and transmission coefficients following from (3.3.15, 17) are rather cumbersome. In Sect. 3.4 we consider plane wave reflection from Epstein's layer by a somewhat simpler approach under assumption that the wave number is given by (3.3.9) throughout the medium, not only for $z < 0$. Such an assumption induces some loss of generality in the solutions. The results in Sect. 3.4 can be obtained from the results of this section by assuming $\varrho_1 = \varrho_2$, $k_1 = k_0$, and $z_1 \to +\infty$. Such a consideration is useful, however, because of the considerable interest in problems relating to wave reflection from Epstein's layer, due especially to the simplicity and tractability of the results.

Case B. Let $\eta(z) = \cosh^{-2} a(z + z_1)$, $a \neq 0$. Equations (3.1.11) and (3.3.7) then yield

$$
k^2(z) = k_0^2[1 + N \sinh^{-2} 2a(z + z_1)
$$
$$
+ M \cosh 2a(z + z_1) \sinh^{-2} 2a(z + z_1)] \quad . \tag{3.3.18}
$$

The parameters k_0, N, and M are related to K_1, K_2, K_3 by

$$
k_0^2 = \xi^2 + a^2(1 - 4K_1) \quad , \quad N = a^2[3 - 8(K_1 - K_2 + K_3)] \quad ,
$$
$$
M = -8a^2(K_2 + K_3 - K_1) \quad . \tag{3.3.19}
$$

Exact solutions, in terms of hypergeometric functions, exist for oblique incidence and arbitrary frequency. The profile given by (3.3.18) was suggested by *Heading* [3.18].

As in the case of the Epstein layer, the profile given by (3.3.18) is quite different when a is real or complex. Thus $k(z)$ is periodic when a is pure imaginary, whereas if a is real, $k(z) \to k_0$ at $z \to -\infty$. A rather complete discussion of the possible profiles following from (3.3.18) for different N, M, and z_1 was given in [Ref. 3.3, Chap. 3].

When a is a real quantity and $z_1 < 0$, $c(z)$ becomes zero at $z = z_1$. The finiteness of k^2 at *any* z will be ensured by complex z_1. Taking, for example, $az_1 = i\pi/4 + az_2$ (z_2 is real) we obtain from (3.3.18)

$$k^2(z) = k_0^2[1 - N \cosh^{-2} 2a(z + z_2)$$
$$- iM \sinh 2a(z + z_2) \cosh^{-2} 2a(z + z_2)] \quad . \tag{3.3.20}$$

Here the parameters N and iM can be made real by suitable choice of the quantities K_2 and K_3. The choice of the physical solutions to (3.1.3) for the profile of (3.3.20) obeying the conditions as $z \to -\infty$ is fulfilled in the same fashion as for the Epstein profile.

In Cases A and B we have considered two kinds of substitutes for the independent variables in the hypergeometric equation. Other functions $\eta(z)$ which also generate solvable profiles $k^2(z)$ for arbitrary incidence angles and frequencies exist. Some of these substitutes were given in, for example, [Ref. 3.3, Chap. 3]. All the substitutes reported in the literature do not offer any *new* solvable profiles, when compared with those obtainable from (3.3.9) and (3.3.18), however.

Note also that the essentially different families of the solvable profiles given by (3.2.14, 30, 34), and (3.3.9, 18) do have some common specific cases, however. The profile $k^2(z) = A + B \cosh^{-2} b(z + z_0)$, for example, can be obtained from (3.3.9) by assuming $k_0^2 = A$, $N = 0$, $M = -B$, $a = 2b$, and $z_1 = z_0$ as well as from (3.3.20) by assuming $k_0^2 = A$, $N = -B$, $M = 0$, $a = b/2$, and $z_2 = z_0$.

3.4 Plane-Wave Reflection from an Epstein Layer

3.4.1 Expressions for Reflection and Transmission Coefficients

Let us consider the $k(z)$ profile given by (3.3.9) for $-\infty < z < +\infty$ and real a. We assume that the density $\varrho(z)$ is constant and also that $z_1 = 0$, $a = -b < 0$. The last two relations do not diminish the generality of the problem in the case of unbounded media and simplify the notation, but do depend on the choice of the origin of z. Equation (3.3.9) now becomes

$$\frac{k^2(z)}{k_0^2} = 1 - N \frac{\exp(-bz)}{1 + \exp(-bz)} - 4M \frac{\exp(-bz)}{[1 + \exp(-bz)]^2} \quad . \tag{3.4.1}$$

Here, we have $k^2 \to k_0^2$ as $z \to +\infty$, and $k^2 \to k_0^2(1 - N)$ as $z \to -\infty$. The sound velocity exponentially approaches constant values with increasing $|z|$. In other words, the medium is essentially inhomogeneous only in some layer around $z = 0$. This permits us to state the problem of plane wave reflection without adding a homogeneous half-space to the inhomogeneous medium.

Let us suppose that the incident plane wave of unit amplitude is given as $z \to +\infty$ by

$$p_i = \exp(i\xi x - ik_0 z \cos \theta_0) \quad , \quad \sin \theta_0 = \xi/k_0 \quad , \tag{3.4.2}$$

where θ_0 is the angle of incidence. Due to the presence of the inhomogeneous layer there also exists a reflected wave as $z \to +\infty$

$$p_r = V \exp(i\xi x + ik_0 z \cos \theta_0) \quad . \tag{3.4.3}$$

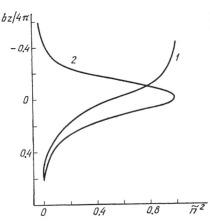

Fig. 3.6. Dependence of squared modified refraction index on z for a symmetric (1) and transition (2) Epstein layer

As $z \to -\infty$ we have the transmitted plane wave

$$p = W \exp{(i\xi x - ik_1 z \cos{\theta_1})} \quad , \quad k_1 = k_0\sqrt{1 - N} \quad , \tag{3.4.4}$$

where θ_1 is the angle of refraction. The angles of incidence and refraction are related to each other by Snell's law: $k_0 \sin{\theta_0} = k_1 \sin{\theta_1}$. The problem is to determine V and W.

Figure 3.6 illustrates dependence of the squared refraction index $n^2 = k^2(z)/k_0^2$ on z for two cases: $M = 0$, $N \neq 0$ and $M \neq 0$, $N = 0$. The quantity $\tilde{n}^2$, which equals $(1 - n^2)/N$ and $(1 - n^2)/M$ respectively is plotted along the abscissa. We see that in the first case (3.4.1) describes a transition layer where the refraction index changes smoothly from $n = 1$ at large positive z to $n = n_\infty \equiv \sqrt{1 - N}$ at large negative z. In the second case we have a symmetrical layer where the refraction index is an even function of z and n approaches unity at large $|z|$ and has maximum deviation from unity at $z = 0$, where $n^2 = 1 - M$. The ratio $z/S\lambda_0 = bz/4\pi$ is plotted along the ordinate, where the dimensionless quantity

$$S \equiv 2k_0/b \tag{3.4.5}$$

is usually referred to as the relative thickness of the layer. Simple calculations show that for the symmetrical layer the effective thickness, defined as the distance along the z-axis between points on both sides of the middle of the layer at which $(1 - n^2)/M$ is equal to one-half of its maximum value at $z = 0$, is

$$l = 0.28\lambda_0 S \quad . \tag{3.4.6}$$

For a transition layer, l is an interval between the points along the z-axis at which $1 - n^2 = qN$ and $1 - n^2 = q^{-1}N$, where $q = 0.85$.

Equations (3.3.11) relate the parameters of the hypergeometric equation to that of the inhomogeneous layer and incident wave. In the notations of present section these formulas become

$$\alpha = \tfrac{1}{2} + d_2 + id_1 + (iS/2)(\cos{\theta_0} - \sqrt{1 - N} \cos{\theta_1}) \quad ,$$
$$\beta = \tfrac{1}{2} + d_2 + id_1 + (iS/2)(\cos{\theta_0} + \sqrt{1 - N} \cos{\theta_1}) \quad ,$$
$$\gamma = 1 + iS \cos{\theta_0} \quad . \tag{3.4.7}$$

59

Here the real numbers d_1 and d_2 are determined by

$$2(d_2 + id_1) \equiv \sqrt{1 - 4S^2 M} \quad , \quad d_1 \geq 0 \quad . \tag{3.4.8}$$

As we have seen in Sect. 3.3 (see (3.3.16, 17)), to satisfy physical requirements we should use the function F_5 determined by (3.3.4a) as a solution of the hypergeometric equation. Then after accounting for (3.4.7) we have for large negative z:

$$p = (-1)^{-\alpha} A \exp(i\xi x - ik_1 z \cos \theta_1) \quad , \quad A = \text{const} \quad . \tag{3.4.9}$$

At positive z, when $|\eta| < 1$, the solution transforms into a linear combination of F_1 and F_2 with the coefficients given by (3.3.5). This enables us to analyze the asymptotic behavior of the solution of the wave equation as $z \to +\infty$. It turns out, (cf. (3.3.13, 14)), that for large positive z F_1 corresponds to the plane wave propagating in the direction of positive z, whereas F_2 corresponds to the wave propagating in the direction of negative z. Using (3.3.12–14) and (3.3.5) we obtain for the sound field as $z \to +\infty$

$$p = (-1)^{-\alpha} A \exp(i\xi x) \left\{ \frac{\Gamma(\alpha - \beta + 1)\Gamma(1 - \gamma)}{\Gamma(1 - \beta)\Gamma(1 + \alpha - \gamma)} \exp(-ik_0 z \cos \theta_0) \right.$$
$$\left. + \frac{\Gamma(\alpha - \beta + 1)\Gamma(\gamma - 1)}{\Gamma(\gamma - \beta)\Gamma(\alpha)} \exp(ik_0 z \cos \theta_0) \right\} \quad . \tag{3.4.10}$$

The latter expression can be treated as the superposition of incident (3.4.2) and reflected (3.4.3) waves. Comparing the coefficients before the exponential factors, we obtain for the reflection coefficient

$$V = \frac{\Gamma(\gamma - 1)\Gamma(1 - \beta)\Gamma(1 + \alpha - \gamma)}{\Gamma(1 - \gamma)\Gamma(\gamma - \beta)\Gamma(\alpha)} \quad , \tag{3.4.11}$$

and by (3.4.9, 10) for the transmission coefficient of the layer

$$W = \frac{\Gamma(1 - \beta)\Gamma(1 + \alpha - \gamma)}{\Gamma(\alpha - \beta + 1)\Gamma(1 - \gamma)} \quad . \tag{3.4.12}$$

Thus, we have had to use only the asymptotic values of the hypergeometric series. We see that the reflection and transmission coefficients for an isolated Epstein layer are expressed through Γ functions, which are functions of a single variable. Recall that in the previous section for the case of a boundary between an Epstein layer and a homogeneous medium we had much more cumbersome results, which contained the hypergeometric function of four variables.

We now analyze (3.4.11) for the transition and symmetrical Epstein layers.

3.4.2 Transition Epstein Layer

In this case, that is, when $M = 0$ and $N \neq 0$, we obtain from (3.4.7, 8)

$$d_1 = 0 \quad , \quad d_2 = \tfrac{1}{2} \quad , \quad \alpha = 1 + (iS/2)(\cos \theta_0 - \sqrt{1 - N} \cos \theta_1) \quad ,$$
$$\beta = 1 + (iS/2)(\cos \theta_0 + \sqrt{1 - N} \cos \theta_1) \quad , \quad \gamma = 1 + iS \cos \theta_0 \quad . \tag{3.4.13}$$

Then (3.4.11) takes the form

$$V = \frac{\Gamma(iS\cos\theta_0)\Gamma[\frac{-i}{2}S(\cos\theta_0 + \sqrt{1-N}\cos\theta_1)]\Gamma[1-\frac{i}{2}S(\cos\theta_0 + \sqrt{1-N}\cos\theta_1)]}{\Gamma(-iS\cos\theta_0)\Gamma[\frac{i}{2}S(\cos\theta_0 - \sqrt{1-N}\cos\theta_1)]\Gamma[1+\frac{i}{2}S(\cos\theta_0 - \sqrt{1-N}\cos\theta_1)]} \cdot$$

(3.4.14)

A simpler relation results for the modulus of the reflection coefficient $\varrho \equiv |V|$. Here we shall use the well known relations [Ref. 3.10, Chap. 6]:

$$\Gamma(w^*) = [\Gamma(w)]^* \quad , \tag{3.4.15}$$

$$\Gamma(w)\Gamma(1-w) = \pi/\sin\pi w \quad , \tag{3.4.16}$$

$$\Gamma(1+w) = w\Gamma(w) \quad . \tag{3.4.17}$$

The asterisks denote the complex conjugate; w is an arbitrary complex number. For real a the relation

$$|\Gamma(ia)\Gamma(1+ia)| = |\Gamma(-ia)\Gamma(1+ia)| = \pi/\sinh\pi|a| \tag{3.4.18}$$

follows from (3.4.15) and (3.4.16).

It is necessary to distinguish the two cases: $\sin\theta_0 \geq 1-N$ and $\sin\theta_0 < 1-N$. (For negative N the second case always takes place). In the first case we have $\cos\theta_1 = i|\cos\theta_1|$. Then the numerator and denominator in the right-hand side of (3.4.14) are, by virtue of (3.4.15), complex conjugates. Therefore, $\varrho = 1$, i.e., we have total reflection. In the second case $\cos\theta_1$ is a real number. Using the identity in (3.4.18) we obtain

$$\varrho = \frac{\sinh\left[\frac{\pi}{2}S|\cos\theta_0 - \sqrt{1-N}\cos\theta_1|\right]}{\sinh\left[\frac{\pi}{2}S(\cos\theta_0 + \sqrt{1-N}\cos\theta_1)\right]} \cdot \tag{3.4.19}$$

Here, if $\cos\theta_1 \neq 0$, then $\varrho < 1$ is always true.

It was shown above that $1 - N = n_\infty^2$, where n_∞ is the refraction index of the medium far from the layer, on the side opposite the incident wave, i.e. for $z \to -\infty$. Therefore, the condition of total reflection can be written in the form

$$\sin\theta_0 \geq n_\infty \quad . \tag{3.4.20}$$

Consider the reflection coefficient (3.4.14) when the thickness S of the transition layer tends to zero. At $w \to 0$ we have $\Gamma(1+w) \to \Gamma(1) = 1$ and by virtue of (3.4.17) we also have $\Gamma(w) = \Gamma(1+w)/w = w^{-1}[1+O(w)]$. Using these relations, we obtain from (3.4.14) as $S \to 0$

$$V = \frac{\cos\theta_0 - \sqrt{1-N}\cos\theta_1}{\cos\theta_0 + \sqrt{1-N}\cos\theta_1} \cdot$$

This is the Fresnel formula for the reflection coefficient at the boundary of homogeneous media with the relative index of refraction given by $\sqrt{1-N} = n_\infty$. (Recall that in this discussion the density of the layered medium is assumed constant.)

The modulus of the reflection coefficient ϱ from a transition layer is shown in Fig. 3.7 as a function of the angle of incidence θ_0 for $n_\infty = 1.1$ and $n_\infty = 0.9$ and different values of the ratio l/λ_0, where l is the effective thickness of the layer, as given by (3.4.6). As above the wave is assumed to be incident from $z = +\infty$, where

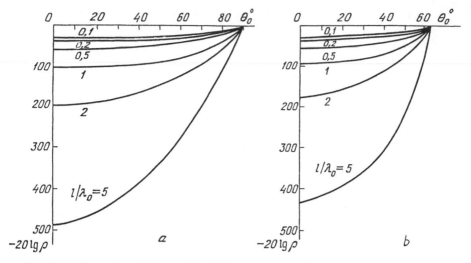

Fig. 3.7a,b. Dependence of the modulus of the sound wave reflection coefficient from a transition layer on the angle of incidence for various values of the layer thickness (a) at $n_\infty = 1.1$, (b) at $n_\infty = 0.9$

$n = 1$ and λ_0 is the wavelength. The abscissa and the ordinate represent the angle of incidence and the modulus of the reflection coefficient expressed in decibels. We see that as the thickness of the layer is increased the reflection coefficient falls off rapidly. Thus, in the case of normal incidence, an increase in the thickness of layer l from $0.1\lambda_0$ to $5\lambda_0$ with $n_\infty = 1.1$ causes the amplitude of the reflected wave to decrease from -27 db ($\varrho = 4 \times 10^{-2}$) to -486 db ($\varrho = 5 \times 10^{-25}$). The greater the thickness of the layer, the more sharply does the reflection coefficient depend on the angle of incidence. At $\theta_0 \to \frac{\pi}{2}$, we have $\varrho \to 1$ ($\ln \varrho \to 0$). When $n_\infty = 0.9$, total reflection occurs for $\theta_0 > 1.120$ (that is, for $\theta_0 > 64°10'$).

3.4.3 Symmetrical Epstein Layer

We now analyze (3.4.11) for the reflection coefficient in the case of a symmetrical layer given by (3.4.1) with $N = 0$ and $M \neq 0$. The index of refraction approaches unity at large $|z|$ at both sides of the layer, and deviates most strongly from unity at the center of the layer: $n^2(0) \equiv n_0^2 = 1 - M$. Setting $\cos \theta_1 = \cos \theta_0$ in (3.4.7), substituting the resulting values of α, β, and γ into (3.4.11), and using (3.4.16), we obtain for the reflection coefficient

$$V = \frac{\Gamma(\mathrm{i}S\cos\theta_0)}{\pi\,\Gamma(-\mathrm{i}S\cos\theta_0)}\cos\,(\pi d_2 + \mathrm{i}\pi d_1)\Gamma(\tfrac{1}{2} - d_2 - \mathrm{i}d_1 - \mathrm{i}S\cos\theta_0)$$

$$\times\ \Gamma(\tfrac{1}{2} + d_2 + \mathrm{i}d_1 - \mathrm{i}S\cos\theta_0)\ . \tag{3.4.21}$$

Note that for a symmetrical layer with parameters $M = M_0$ and $S = S_0$ the reflection coefficient at normal incidence is equal to that for the incidence angle θ_0 and the layer with the parameter values $M = M_0 \cos^2 \theta_0$, $S = S_0/\cos \theta_0$. Indeed, according to (3.4.8), such a substitution of M and S leaves the parameters d_1, d_2, and $S \cos \theta_0$, which determine ϱ by (3.4.21), unchanged.

Let us consider this formula for two cases: $4S^2 M \geq 1$ and $4S^2 M < 1$. In the first case we have, according to (3.4.8), $d_2 = 0$ and $d_1 = (S^2 M - \frac{1}{4})^{1/2}$. From (3.4.15) and (3.4.16) one can obtain

$$|\Gamma(1 - b + ia)\Gamma(b + ia)| = \pi |\sin(\pi b + i\pi a)|^{-1}$$
$$= \pi [\cosh^2 \pi a - \cos^2 \pi b]^{-1/2} \quad , \tag{3.4.22}$$

where a and b are real numbers. Now we have from (3.4.21, 22) the expression for the square of the modulus of the reflection coefficient in elementary functions:

$$\varrho^2 = \frac{\cosh^2 \pi d_1}{\cosh(\pi d_1 + \pi S \cos \theta_0)\cosh(\pi d_1 - \pi S \cos \theta_0)} \quad . \tag{3.4.23}$$

In the second case we have, according to (3.4.8), $d_1 = 0$ and $d_2 = (\frac{1}{4} - S^2 M)^{1/2}$. Then it follows from (3.4.21, 22)

$$\varrho^2 = \frac{\cos^2 \pi d_2}{\cosh^2(\pi S \cos \theta_0) - \cos^2 \pi d_2} \quad . \tag{3.4.24}$$

As $M \to 0$, the layer vanishes; the medium becomes homogeneous. Under these conditions, we have $d_2 \to \frac{1}{2}$ and according to (3.4.24), $\varrho \to 0$, as should be expected. Note also that if the sound velocity in the layer is less than that in the media surrounding it, i.e., $n_0 > 1$, then M is negative and the modulus of the reflection coefficient will be given by (3.4.24) for any layer thickness S. As the latter approaches infinity, this formula yields $\varrho \to 0$, i.e., reflection disappears, which is natural since the gradient of the index of refraction approaches zero.

The situation in the limit $S \to \infty$ is somewhat more complicated when $n_0 < 1$, i.e., when the velocity in the layer is greater than that in the adjoining media. In this case $M > 0$, and for sufficiently large S we have to use (3.4.23). Two cases should be considered:

1) $d_1 > S \cos \theta_0$,

2) $d_1 < S \cos \theta_0$. $\tag{3.4.25}$

In the first case, assuming M to be fixed, we obtain for $S \to +\infty$

$$d_1 \approx S\sqrt{M} = S\sqrt{1 - n_0^2} \quad , \tag{3.4.26}$$

$$\cosh(\pi d_1 \pm \pi S \cos \theta_0) \approx \frac{1}{2}\exp(\pi d_1 \pm \pi S \cos \theta_0) \quad ;$$
$$\cosh \pi d_1 \approx \frac{1}{2}\exp \pi d_1 \quad . \tag{3.4.27}$$

As a result, (3.4.23) yields $\varrho \to 1$, i.e., we have total reflection. In the second case we have $\cosh(\pi d_1 - \pi S \cos \theta_0) \approx \frac{1}{2}\exp(\pi S \cos \theta_0 - d_1)$. As a result, as $S \to +\infty$, (3.4.23) reveals exponential decay of the reflection coefficient as the layer thickness or the wave frequency increase

$$\varrho \approx \exp\left[-2\pi S\left(\cos \theta_0 - \sqrt{1 - n_0^2}\right)\right] \quad . \tag{3.4.28}$$

If we take into account (3.4.26), the conditions of (3.4.25) can be written in the form: 1) $\sin \theta_0 > n_0$; 2) $\sin \theta_0 < n_0$. The first means that the ray corresponding to a plane incident wave turns in the layer while refracting and goes back into the medium from which it came (Chap. 8). The penetration of the wave through the layer occurs because of the tunneling effect, similar to particles penetration through potential barriers known in quantum mechanics. With increase in layer thickness such penetration is reduced and we obtain total reflection.

It is worth noting that when $\sin \theta_0 = n_0$ and the ray turns at $z = 0$ ("barrier" thickness becomes zero), we have from (3.4.23) that $\varrho^2 \to \frac{1}{2}$ as $S \to +\infty$ [since $\cosh(\pi d_1 - \pi S \cos \theta_0) \to 1$], i.e., half of the incident energy is reflected.

Figure 3.8 illustrates the dependence of $-20\log \varrho$ on the incidence angle θ_0 for $n_0 = 0.9$ and $n_0 = 1.1$ for various values of the effective layer thickness l defined by (3.4.6).

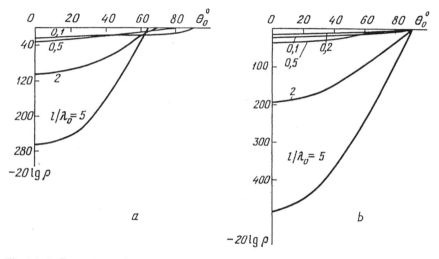

Fig. 3.8a,b. Dependence of the modulus of the sound wave reflection coefficient from a symmetric layer on the angle of incidence for various values of the layer thickness (a) at $n_0 = 0.9$ and (b) at $n_0 = 1.1$

We have not yet analyzed the transmission coefficient W of the layer. Usually quantity $|W|^2$ is of interest alone, and it can be obtained from the law of energy conservation after determination of ϱ. Thus, for the symmetrical layer $|W|^2 = 1 - \varrho^2$.

Nevertheless, one case deserves special attention: that of a symmetrical layer with $M > 0$, i.e., when the sound velocity in the layer is greater than that in the adjoining media. When $4S^2 M \geq 1$ we have, according to (3.4.8), $d_2 = 0$, and $2d_1 = (4S^2 M - 1)^{1/2}$. Then (3.4.7) and (3.4.12) give

$$W = \frac{\Gamma(0.5 - iS\cos\theta_0 - id_1)\Gamma(0.5 - iS\cos\theta_0 + id_1)}{\Gamma(1 - iS\cos\theta_0)\Gamma(-iS\cos\theta_0)} . \qquad (3.4.29)$$

Now using (3.4.22), we obtain a simple expression for the squared modulus of the transmission coefficient:

$$|W|^2 = \frac{\sinh^2(\pi S \cos\theta_0)}{\cosh(\pi d_1 + \pi S \cos\theta_0)\cosh(\pi d_1 - \pi S \cos\theta_0)} \quad . \tag{3.4.30}$$

It can be easily verified that the sum of ϱ^2 from (3.4.23) and $|W|^2$ is equal to unity, as could be expected. At large layer thickness ($S \to \infty$), using approximate expressions similar to (3.4.27) we obtain

$$|W|^2 \approx 1 \quad , \quad \sin\theta_0 < n_0 \quad ,$$
$$|W|^2 \approx \exp\left[-2\pi S(\sqrt{1 - n_0^2} - \cos\theta_0)\right] \quad , \quad \sin\theta_0 > n_0 \quad ,$$
$$|W|^2 \approx \tfrac{1}{2} \quad , \quad \sin\theta_0 = n_0 \quad . \tag{3.4.31}$$

We see that when the ray corresponding to the plane wave does not penetrate into the region of negative z, the transmission coefficient decays *exponentially* with the layer thickness or wave frequency increasing.

3.5 Reflection of a Plane Wave from a Half-Space with a Linear Law for the Squared Refraction Index

It was shown in Sect. 3.2 that when k^2 is linearly dependent on z exact solutions of the wave equation in terms of confluent hypergeometric functions can be found. In this section we shall consider this case in more detail. We begin with a summary of the main properties of the special kind of confluent hypergeometric functions, namely, the *Airy functions*. Airy functions are widely applied in diffraction and wave-propagation theory, and are used repeatedly in this book.

3.5.1 The Airy Functions

In the half-space $z < 0$ with the squared wave number

$$k^2(z) \equiv k_0^2 n^2(z) \quad , \quad n^2 = 1 \pm az \quad , \quad a > 0 \tag{3.5.1}$$

(3.1.3), for the vertical dependence Φ of sound pressure, is written as

$$\frac{d^2\Phi}{dz^2} + (k_0^2 - \xi^2 \pm k_0^2 az)\Phi = 0 \quad . \tag{3.5.2}$$

Let us introduce a new independent variable t in place of z:

$$t = t_0 \mp z/H \quad , \quad t_0 \equiv (\xi^2 - k_0^2)H^2 \quad , \quad H \equiv (ak_0^2)^{-1/3} \quad . \tag{3.5.3}$$

Then we obtain instead of (3.5.2)

$$\Phi''(t) = t\Phi(t) \quad . \tag{3.5.4}$$

This is the Airy equation, solved by Airy functions, which have been thoroughly investigated and tabulated ([Ref. 3.10, Chap. 10] and [Ref. 3.22]).

Two linearly independent solutions of (3.5.4) $u(t)$ and $v(t)$ can be represented by the real and imaginary parts of the integral

$$\Phi(t) = \pi^{-1/2} \int_\Gamma \exp(ts - s^3/3)ds \quad , \tag{3.5.5}$$

where the contour Γ in the complex plane s runs along the ray $\arg s = -2\pi/3$ from infinity to zero, and along the real axis from zero to plus infinity. The integral in (3.5.5) converges for all complex values of t and is an entire transcendental function of t. It is easy to prove that the function determined by this integral satisfies (3.5.4). For $t = 0$ we have from (3.5.5)

$$\Phi(0) = 2\sqrt{\pi}\exp{(i\pi/6)/3^{2/3}}\,\Gamma(2/3) \approx 1.089929 + i0.629271 \quad,$$
$$\Phi'(0) = 2\sqrt{\pi}\exp{(-i\pi/6)/3^{4/3}}\,\Gamma(4/3) \approx 0.794570 - i0.458745 \quad. \tag{3.5.6}$$

Being an entire transcendental function, $\Phi(t)$ can be expanded in a power series that converges for any t. This series has the form

$$\Phi(t) = \Phi(0)\sum_{n=0}^{\infty}\frac{t^{3n}}{\prod\limits_{m=1}^{n}(3m-1)3m} + t\Phi'(0)\sum_{n=0}^{\infty}\frac{t^{3n}}{\prod\limits_{m=1}^{n}3m(3m+1)} \quad. \tag{3.5.7}$$

In the case of real t we denote

$$\Phi(t) = u(t) + iv(t) \quad, \tag{3.5.8}$$

where $u(t)$ and $v(t)$ are real-valued functions which are two linearly independent solutions of (3.5.4). Their series expansions (which we shall not write down), are immediately obtained from (3.5.7). From (3.5.6) we have for the Wronskian

$$u'(t)v(t) - u(t)v'(t) = 1 \quad. \tag{3.5.9}$$

The integral representations of the functions $u(t)$ and $v(t)$ are readily obtained from (3.5.5). In particular, $v(t)$ is expressed as the Airy integral

$$v(t) = \pi^{-1/2}\int_{0}^{\infty}\cos{(st + s^3/3)}ds \quad. \tag{3.5.10}$$

The functions $u(t)$ and $v(t)$ are also defined for complex t and are entire transcendental functions. The following relations hold

$$\Phi(t) = u(t) + iv(t) \quad, \quad \Phi[t\exp{(i\pi/3)}] = 2\exp{(i\pi/6)}v(-t) \quad,$$
$$\Phi[t\exp{(2i\pi/3)}] = \exp{(i\pi/3)}[u(t) - iv(t)] \quad, \quad \Phi[t\exp{(i\pi)}] = u(-t) + iv(-t) \quad,$$
$$\Phi[t\exp{(4\pi i/3)}] = 2\exp{(i\pi/6)}v(t) \quad,$$
$$\Phi[t\exp{(5i\pi/3)}] = \exp{(i\pi/3)}[u(-t) - iv(-t)] \quad. \tag{3.5.11}$$

In particular, these relations give the expressions for $\Phi(t)$ on the six rays $\arg t = n\pi/3$ ($n = 0, 1, 2, 3, 4, 5$) in terms of the real functions $u(t)$ and $v(t)$.

Let us now write the asymptotic expansions for the Airy functions and their derivatives. For this we introduce a system of coefficients:

$$a_1 = \frac{5}{72} \quad, \quad a_2 = (5\cdot 11)\cdot\frac{7}{2\cdot(72)^2} \quad,\dots \quad,$$
$$a_n = \prod_{m=1}^{n}(6m-1)(6m-5)/n!(72)^n \quad;$$

$$b_1 = \frac{7}{72} \quad , \quad b_2 = (7 \cdot 13) \cdot \frac{5}{2 \cdot (72)^2} \quad , \ldots \quad ,$$

$$b_n = - \prod_{m=1}^{n} (6m+1)(6m-7)/n!(72)^n \quad . \tag{3.5.12}$$

We then have

$\boxed{t > 0, \ w \equiv (2/3)t^{3/2}}$

$$u(t) = t^{-1/4} \exp w \left(1 + \sum_{n=1}^{\infty} a_n w^{-n} \right) \quad ,$$

$$u'(t) = t^{1/4} \exp w \left(1 - \sum_{n=1}^{\infty} b_n w^{-n} \right) \quad ;$$

$$v(t) = 0.5 t^{-1/4} \exp(-w) \left[1 + \sum_{n=1}^{\infty} a_n(-w)^{-n} \right] \quad ,$$

$$v'(t) = -0.5 t^{1/4} \exp(-w) \left[1 - \sum_{n=1}^{\infty} b_n(-w)^{-n} \right] \quad . \tag{3.5.13}$$

$\boxed{t < 0, \ w \equiv (2/3)(-t)^{3/2}}$

$$u(t) = (-t)^{-1/4} \left\{ \cos\left(w + \frac{\pi}{4} \right) \left[1 + \sum_{n=1}^{\infty} (-1)^n a_{2n} w^{-2n} \right] \right.$$

$$\left. + \sin\left(w + \frac{\pi}{4} \right) \sum_{n=1}^{\infty} (-1)^{n+1} a_{2n-1} w^{1-2n} \right\} \quad ,$$

$$u'(t) = (-t)^{1/4} \left\{ \sin\left(w + \frac{\pi}{4} \right) \left[1 + \sum_{n=1}^{\infty} (-1)^{n+1} b_{2n} w^{-2n} \right] \right.$$

$$\left. + \cos\left(w + \frac{\pi}{4} \right) \sum_{n=1}^{\infty} (-1)^{n+1} b_{2n-1} w^{1-2n} \right\} \quad ,$$

$$v(t) = (-t)^{-1/4} \left\{ \sin\left(w + \frac{\pi}{4} \right) \left[1 + \sum_{n=1}^{\infty} (-1)^n a_{2n} w^{-2n} \right] \right.$$

$$\left. + \cos\left(w + \frac{\pi}{4} \right) \sum_{n=1}^{\infty} (-1)^n a_{2n-1} w^{1-2n} \right\} \quad ,$$

$$v'(t) = (-t)^{1/4} \left\{ \cos\left(w + \frac{\pi}{4} \right) \left[-1 + \sum_{n=1}^{\infty} (-1)^n b_{2n} w^{-2n} \right] \right.$$

$$\left. + \sin\left(w + \frac{\pi}{4} \right) \sum_{n=1}^{\infty} (-1)^{n+1} b_{2n-1} w^{1-2n} \right\} \quad . \tag{3.5.14}$$

The Airy functions are expressed in terms of cylindrical functions of order $\frac{1}{3}$ in the following way:

67

$$\boxed{t>0,\ w \equiv (2/3)t^{3/2}}$$

$$u(t) = \sqrt{\pi t/3}\,[I_{-1/3}(w) + I_{1/3}(w)] \quad,$$
$$v(t) = 3^{-1}\sqrt{\pi t}\,[I_{-1/3}(w) - I_{1/3}(w)] \quad ; \tag{3.5.15}$$

$$\boxed{t<0,\ w \equiv (2/3)(-t)^{3/2}}\,,$$

$$u(t) = \sqrt{-\pi t/3}\,[J_{-1/3}(w) - J_{1/3}(w)] \quad,$$
$$v(t) = 3^{-1}\sqrt{-\pi t}\,[J_{-1/3}(w) + J_{1/3}(w)] \quad,$$
$$\Phi(t) = \sqrt{\pi t/3}\,\exp(2\pi i/3)H^{(1)}_{1/3}(w) \quad. \tag{3.5.16}$$

The function $v(t)$ is more frequently encountered in applications than $u(t)$. Figure 3.9 shows a plot of the function $v(t)/v(0)$, where $v(0) \approx 0.6293$, see (3.5.6). This function oscillates when $t < 0$ and decays to zero rapidly and monotonically when $t > 0$.

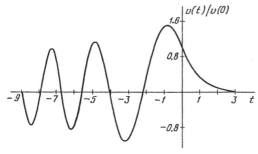

Fig. 3.9. Airy function $v(t)$

It is also useful to write the values of the zeroes of $v(t)$ and $v'(t)$ which all lie in $t < 0$. We designate $v(-y_l) = 0$, $v'(-x_l) = 0$, $l = 1, 2, \ldots$, and arrange x_l and y_l in the order of increasing values. Table 3.1 lists values of the zeros of $v(t)$ and $v'(t)$ as a function of l.

Table 3.1. Zeros of $v(t)$ and $v'(t)$

l	1	2	3	4	5
y_l	2.33811	4.08795	5.52056	6.78671	7.94417
x_l	1.01879	3.24820	4.82010	6.16331	7.37218

Let us denote

$$y_l = (1.5w_l)^{2/3} \quad, \qquad x_l = (1.5w_l')^{2/3} \quad. \tag{3.5.17}$$

The values w_l and w_l' are given approximately (although very accurately even for small l) by the expressions

$$w_l = (l - \tfrac{1}{4})\pi + 0.0884194(4l - 1)^{-1}$$
$$- 0.08328(4l - 1)^{-3} + 0.4065(4l - 1)^{-5} - \ldots \quad ;$$

$$w'_l = (l - \tfrac{3}{4})\pi - 0.1237872(4l - 3)^{-1}$$
$$+ 0.07758(4l - 3)^{-3} - 0.3890(4l - 3)^{-5} + \dots \quad . \tag{3.5.18}$$

As can be readily seen from the second equation (3.5.11), the function $\Phi(t)$ has zeroes t_l, $l = 1, 2, 3, \dots$, lying on the ray $\arg t = \pi/3$ and expressing in terms of y_l by the formula

$$t_l = y_l \exp(i\pi/3) \quad . \tag{3.5.19}$$

Similarly, the function $\Phi'(t)$ becomes zero on the same ray at the points

$$t_l = x_l \exp(i\pi/3) \quad . \tag{3.5.20}$$

The zeros of the function $u(t)$ and that of its derivative lie on the negative part of the real axis t and in the sectors $\frac{\pi}{3} < \arg t < \frac{\pi}{2}$ and $-\frac{\pi}{2} < \arg t < -\frac{\pi}{3}$.

Detailed tables of the Airy functions and values related to them can be found in [3.10, 22, 23]. In [3.10] as well as in some other works the symbols

$$\mathrm{Ai}(t) \equiv \pi^{-1/2} v(t) \quad , \quad \mathrm{Bi}(t) \equiv \pi^{-1/2} u(t) \tag{3.5.21}$$

are used in place of the notations u and v introduced by *Fock*.

3.5.2 The Case when dn^2/dz is Positive

We now return to the problem of plane wave reflection from a layered half-space with a wave number given by (3.5.1). First we assume that the index of refraction is decreasing with increasing distance from the boundary [upper signs in (3.5.1, 3)]. The general solution of (3.5.2) has the form $\Phi = Au(t) + Bv(t)$. By requiring finiteness of the field at $z \to -\infty$ and taking into account (3.5.13), we obtain $A = 0$. Then (3.1.5) gives for the reflection coefficient

$$V = -\frac{v'(t_0) - igv(t_0)}{v'(t_0) + igv(t_0)} \quad , \tag{3.5.22}$$

where

$$g = \frac{\varrho_2}{\varrho_1} k_1 H \cos\theta \tag{3.5.23}$$

is a real number and θ is the incidence angle of a plane wave in the homogeneous half-space. The same result follows from (3.2.15, 24), where q and z_1 are connected with H and t_0 by

$$q = \tfrac{4}{3} H^{-3/2} \quad , \quad z_1 = H t_0 \quad . \tag{3.5.24}$$

For the case when the densities of both half-spaces are equal and the sound velocity is continuous at the boundary, i.e., $k_1 = k_0$, we have from (3.5.3):

$$t_0 = -k_0^2 H^2 \cos^2\theta < 0 \quad , \quad g = \sqrt{-t_0} \quad . \tag{3.5.25}$$

Taking into account that the Airy function $v(t)$ is real at real values of the argument, we can rewrite (3.5.22) for V in the form

$$V = \exp(i\varphi) \quad , \quad \varphi = -\pi - 2\arctan\frac{gv(t_0)}{v'(t_0)} \quad . \tag{3.5.26}$$

Hence, we have total reflection from the half-space, and φ is the phase of the reflection coefficient. It will be shown in Chap. 9, that the phase φ can also be obtained in the geometrical-optics approximation for $-t_0 \gg 1$, if we only take into account the additional phase loss of the wave, $\frac{\pi}{2}$, at the turning point $z = z_m$, corresponding to

$$t = 0 \quad , \quad k_0^2 a z_m = \xi^2 - k_0^2 \quad , \quad z_m < 0 \quad . \tag{3.5.27}$$

At the turning point, the ray corresponding to our wave is directed horizontally (parallel to the boundary $z = 0$).

The characters of the field before and after the turning plane are completely different. In the case where $z_m < z < 0$, the dependence of the field on z is oscillating (the region $t < 0$ in Fig. 3.9). The amplitude of the oscillations increases, as z approaches the turning plane. For $z < z_m$ the field decreases monotonically. The maximum amplitude of the sound pressure occurs at $z > z_m$ at the point where $t = -x_1$.

3.5.3 The Case when dn^2/dz is Negative

We now consider another case where the index of refraction increases with increasing distance from the boundary. Here the lower signs in (3.5.1,3) should be taken. To find the proper solution of the Airy equation which satisfies the conditions at infinity, we use the principle of limiting absorption. It follows from (3.5.14) that when a small absorption is introduced [k_0 is replaced by $k_0(1 + i\eta)$, $0 < \eta \ll 1$], only the solution $\Phi = A[u(t) + iv(t)]$ goes to zero as $z \to -\infty$. Then, using (3.1.5), we find the reflection coefficient

$$V = \frac{g(u + iv) - i(u' + iv')}{g(u + iv) + i(u' + iv')} \quad . \tag{3.5.28}$$

Here, the values of the Airy functions and their derivatives are taken at the point $t = t_0$. Taking into account the fifth relation in (3.5.11) we also obtain the same result from (3.2.15, 24) where in the given situation

$$q = -4i/3H^{3/2} \quad , \quad z_1 = -Ht_0 \quad . \tag{3.5.29}$$

Note that at grazing incidence ($\theta \to \frac{\pi}{2}$) we have $g \to 0$ and $V \to -1$.

It is useful to consider (3.5.28) in the three limiting cases:

a). $|t_0| \ll 1$. This case, as is seen from (3.5.3), occurs for large values of the sound velocity gradient ($a/k_0 \gg 1$) or for angles of incidence θ close to θ_1, where $\theta_1 \equiv \arcsin(k_0/k_1)$. Here it is convenient to use expansions of the Airy functions in powers of the argument. In accord with (3.5.7), we have

$$u(t_0) + iv(t_0) \equiv \Phi(t_0) = \Phi(0) + t_0\Phi'(0) + O(t_0^2) \quad ;$$
$$u'(t_0) + iv'(t_0) = \Phi'(0) + O(t_0^2) \quad .$$

Taking the values of $\Phi(0)$, $\Phi'(0)$, and t_0 from (3.5.3, 6) we obtain from (3.5.28)

$$V = \frac{1 + B}{1 - B}\left[1 - 2i\frac{\xi^2 - k_0^2}{(ak_0^2)^{2/3}} \cdot \frac{B^2}{1 + B^2}\right] \quad , \tag{3.5.30a}$$

$$B = \frac{\exp\,(5i\pi/6)}{3^{1/3}g} \cdot \frac{\Gamma(1/3)}{\Gamma(2/3)} \quad , \tag{3.5.30b}$$

(with an accuracy to terms of the order of t_0^2). For constant density and continuous sound velocity in the entire medium, for $|t_0| \ll 1$ we always have $|g| \ll 1$ and the limiting expression (3.5.30) for the reflection coefficient can be further simplified to

$$V = -1 + 2\left(\frac{k_0}{3a}\right)^{1/3} \cos\theta \frac{\exp\,(-i\pi/6)\Gamma(1/3)}{\Gamma(2/3)} + O\left[\left(\frac{k_0}{a}\right)^{2/3} \cos^2\theta\right] \quad . \tag{3.5.31}$$

b). $t_0 \to +\infty$. This case occurs only under the condition that $k_1 \geq \xi > k_0$, i.e., when the sound velocity is greater in the upper part of the layered half-space than in the upper (homogeneous) medium. In addition, the gradient of the index of refraction must be small (i.e., $a/k_0 \ll 1$). The angle of incidence is assumed to be not very close to θ_1.

By using the asymptotic expansions for the Airy function and their derivatives (3.5.13), we find that, as $t_0 \to +\infty$, the reflection coefficient tends to

$$V = \frac{g - \sqrt{-t_0}}{g + \sqrt{-t_0}} = \frac{(\varrho_2/\varrho_1)\cos\theta - \sqrt{\sin^2\theta_1 - \sin^2\theta}}{(\varrho_2/\varrho_1)\cos\theta + \sqrt{\sin^2\theta_1 - \sin^2\theta}} \quad , \tag{3.5.32}$$

i.e., to the Fresnel coefficient at the interface of homogeneous media with parameters k_1, ϱ_1 and k_0, ϱ_2. This should be expected because in this case the wave is inhomogeneous in the lower half-space and does not penetrate into those layers where the wave number is markedly different from k_0.

c). $t_0 \to -\infty$. This case occurs for a slowly changing index of refraction (a/k_0 is small) and angles of incidence $\theta < \theta_1$. We again assume that θ is not very close to θ_1. Retaining in the asymptotic expressions for the Airy functions (3.5.14) the principal terms and corrections of the order w_0^{-1}, where $w_0 = 2(-t_0)^{3/2}/3$ we get

$$V \approx \left[g\left(1 - \frac{5i}{72w_0}\right) - \sqrt{-t_0}\left(1 + \frac{7i}{72w_0}\right)\right]$$
$$\times \left[g\left(1 - \frac{5i}{72w_0}\right) + \sqrt{-t_0}\left(1 + \frac{7i}{72w_0}\right)\right]^{-1} \quad . \tag{3.5.33}$$

Terms of the order w_0^{-1} are small corrections, if $g \neq (-t_0)^{1/2}$, and may be neglected. We then again obtain the Fresnel reflection coefficient (3.5.32). However, if $g = (-t_0)^{1/2}$, when the Fresnel coefficient becomes zero, the principal terms in the numerator of (3.5.33) are mutually canceled and we find

$$V \approx \frac{-i}{12w_0} = -iak_0^2\left(2k_1\cos\theta \cdot \frac{\varrho_2}{\varrho_1}\right)^{-3} \quad . \tag{3.5.34}$$

Thus, a phase shift of $\frac{\pi}{2}$ takes place in the process of reflection. The Fresnel reflection coefficient becomes zero only at a certain angle of incidence, if $k_1 \neq k_0$ or $\varrho_2 \neq \varrho_1$ (Sect. 2.2). In the case when parameters of the medium are everywhere continuous ($k_1 = k_0$, $\varrho_2 = \varrho_1$) (3.5.34) gives the reflection coefficient at all angles of incidence except the grazing angles (the condition $a/k_0 \cos^3\theta \ll 1$ must be satisfied since we require $-t_0 \gg 1$). In this case the modulus of the reflection coefficient can be written in the form

$$|V| = \lambda |dn/dz|_{z=0} |/8\pi \cos^3 \theta \quad , \tag{3.5.35}$$

where λ is the wavelength in the upper medium.

We can consider (3.5.33) as the high-frequency limit ($\omega \to \infty$) of the reflection coefficient from the medium with a given sound velocity profile ($a = \text{const}$). We see that when the *turning planes are absent*, the high-frequency wave is markedly reflected ($|V| \simeq 1$) only from the *discontinuities* of sound velocity or density. In absence of such discontinuities reflection is caused by discontinuities in their first derivatives. In this case, however, the reflection coefficient is small compared to unity and decreases as ω^{-1} when the frequency increases. Those parts of the layered medium where parameters change smoothly contribute little into the reflected sound field. The example of the Epstein layer considered in Sect. 3.4 showed that in the case where $c(z)$ is infinitely differentiable and in the absence of the turning planes, the reflection coefficients exponentially tend to zero when ω increases.

We note that in accord with (2.2.14), the reflection coefficient from the interface of two homogeneous media with the same density and with nearly equal values of the sound velocity c and c_1 is equal to

$$V = \frac{\cos\theta - \sqrt{\cos^2\theta + c_1^2/c^2 - 1}}{\cos\theta + \sqrt{\cos^2\theta + c_1^2/c^2 - 1}} \approx \frac{c^{-1}\Delta c}{2\cos^2\theta} \quad , \tag{3.5.36}$$

where $c^{-1}\Delta c \equiv (c - c_1)/c$ is the relative drop in the sound velocity. The angle of incidence θ is assumed to be not very close to $\frac{\pi}{2}$ ($\cos^2\theta \gg |c_1^2/c^2 - 1|$). The modulus of the reflection coefficient (3.5.36) will be identical with (3.5.35) if we set $c^{-1}\Delta c = -\lambda(dn/dz)_{z=0}/4\pi \cos\theta$, or, since $n = c_1/c(z)$, $c \approx c_1$,

$$\Delta c = \frac{\lambda}{4\pi \cos\theta}\left(\frac{dc}{dz}\right)_{z=0} \quad . \tag{3.5.37}$$

This drop is equal to the drop in the velocity in the inhomogeneous medium at a distance $\Delta z = \tilde{\lambda}/4\pi$, where $\tilde{\lambda} = \lambda/\cos\theta$ is the vertical scale of the sound-field variability. One can say that the wave neglects the details of the profile $c(z)$ with vertical dimension much less than $\tilde{\lambda}$.

It will be seen in Chap. 10 that the properties of plane-wave reflection discussed here for the case of special media also have a more general significance.

3.6 Other Cases with Exact Solutions for Normal Incidence

3.6.1 Smooth $k(z)$ Profiles

The solvable profiles considered in the previous sections practically cover all the cases of nonperiodic infinitely differentiable functions $c(z)$ for which the exact solutions of the wave equation are known. The exact solutions for many periodic functions $c(z)$ can be obtained from the profiles considered above (Sects. 3.2, 3). Other profiles can be similarly obtained [3.38] by use of the differential Mathieu equation [3.6] with

$$g(\eta) = a\cos^2 \eta + b \quad , \tag{3.6.1}$$

in (3.1.7).

We would like to emphasize, however, that all that has been said above concerns the problem of plane-wave reflection for arbitrary angles of incidence θ and wave frequency ω. On the other hand, we have already noted that solutions for the fixed values of θ and ω can be readily obtained for an infinite number of profiles with the use of (3.1.11). Construction of profiles solvable for an arbitrary wave frequency but fixed angle of incidence is a problem of intermediate complexity. The simple and effective method of its solution was suggested by *Abraham* and *Moses* [3.24]. Under our approach this method consists in prescribing some relationship between the functions $g(\eta)$ and $\eta(z)$ in (3.1.11), which were previously chosen independently.

Let the fundamental system of solutions to the basic equation

$$\frac{\partial^2}{\partial \eta^2} W(\eta, q) + [q^2 + g_1(\eta)]W(\eta, q) = 0 \tag{3.6.2}$$

be known for some function $g_1(\eta)$ and arbitrary value of q. We denote some nontrivial real solution of (3.6.2) at $q = 0$ by $w(\eta)$. To construct the solvable [in terms of functions $W(\eta, q)$] profiles we replace the variable as follows

$$z(\eta) = -\int_{\eta_0}^{\eta} w^{-2}(\eta')d\eta' \quad , \quad \eta_0 = \text{const} \quad . \tag{3.6.3}$$

It is clear that $z(\eta)$ is a monotonically decreasing function. That is why an inverse function $\eta(z)$, needed for constructing the solutions of wave equation (3.1.3) in terms of $W(\eta, q)$, also exists. It follows from (3.6.3), that

$$\frac{d}{dz} = \frac{d\eta}{dz}\frac{d}{d\eta} = w^{-2}(\eta)\frac{d}{d\eta} \quad . \tag{3.6.4}$$

By calculating derivatives with the help of (3.6.4) in (3.1.11), we find the solvable profile

$$k^2(z) = w^3(\eta)w''(\eta) + w^4(\eta)[q^2 + g_1(\eta)] + \xi^2 \quad . \tag{3.6.5}$$

For the case of normal incidence ($\xi = 0$), taking into account that $w(\eta)$ satisfies (3.6.2) at $q = 0$, we finally obtain

$$k(z) = qw^2(\eta) \quad . \tag{3.6.6}$$

The profile given by (3.6.6) has a multiplicative constant q, which enables us to consider the reflection of waves with arbitrary frequency. In fact the dependence of the wave number on the vertical coordinate z is given here in parametrical form by (3.6.3, 6). Only $g_1(\eta)$ for which $w(\eta)$ are elementary functions are of interest. As an example we consider

$$g_1(\eta) = -b^2 \quad , \quad b = \text{const} > 0 \quad . \tag{3.6.7}$$

In this case the solutions $W(\eta, q)$ are linear combinations of the exponents $\exp\left[\pm(b^2 - q^2)^{1/2}\eta\right]$. Function w we take in the form

$$w = A\exp(-b\eta) + B\exp(b\eta) \quad . \tag{3.6.8}$$

By integrating in (3.6.3), we find

$$z + z_1 = \frac{1 - \exp(2b\eta)}{2b(A+B)[A+B\exp(2b\eta)]} \quad , \quad z_1 = \text{const} < 0 \quad . \tag{3.6.9}$$

The solvable profile is given by (3.6.6, 8), whereas (3.6.9) enables us to express η and then k *explicitly* in terms of z:

$$k = -4qAB\left\{\left[2ABb(z+z_1) + \frac{A-B}{A+B}\right]^2 - 1\right\}^{-1} \quad , \quad z < 0 \quad . \tag{3.6.10}$$

Note that $k(z) \to 0$ as $z \to -\infty$ and that for $A > -B > 0$ the wave number is bounded in the entire lower medium.

To determine the reflection coefficient for a plane wave normally incident on the layered half-space with the wave number of (3.6.10), we need the solution $W(\eta, q)$ to the basic equation (3.6.2) which ensures appropriate behavior of the sound field as $z \to -\infty$, i.e., see (3.6.9), as $\eta \to \eta_0$, where

$$\eta_0 = (2b)^{-1} \ln(-A/B) \quad . \tag{3.6.11}$$

The dependence of the sound field on the vertical coordinate in terms of the solution of the basic equation is according to (3.1.8, 10) and (3.6.3)

$$\Phi(z) = w^{-1}(\eta)W(\eta, q) \quad . \tag{3.6.12}$$

Since $w(\eta_0) = 0$, the condition of finiteness of the sound field yields $W(\eta_0, q) = 0$. This gives the solution to the reference equations (3.6.2, 7) to an accuracy up to a multiplicative constant. We obtain after simple transformation

$$\Phi(z) = \text{const} \cdot \frac{\sinh\left[\sqrt{b^2 - q^2}(\eta - \eta_0)\right]}{\sinh\left[b(\eta - \eta_0)\right]} \quad . \tag{3.6.13}$$

Then, by using (3.6.4, 9) which relate the variables η, z and the derivatives with respect to these variables, one can easily get the reflection coefficient V from (3.1.5). We shall not write the formula for V.

Two other examples of the profiles which permit exact solutions for a normally incident wave with an arbitrary frequency are given in [3.24] in parametrical form. In the first case

$$k(\eta) = q\coth^2\alpha \cdot \tanh^2(a\eta + \alpha) \quad ,$$
$$z(\eta) = -\eta\tanh^2\alpha + a^{-1}\tanh^2\alpha \cdot [\coth(a\eta + \alpha) - \coth\alpha] \quad , \tag{3.6.14a}$$

where a and α are positive constants and the parameter η takes all nonnegative values. In the second case, in the same notation:

$$k(\eta) = q\tanh^2\alpha \cdot \coth^2(a\eta + \alpha) \quad ,$$
$$z(\eta) = -\eta\coth^2\alpha + a^{-1}\coth^2\alpha \cdot [\tanh(a\eta + \alpha) - \tanh\alpha] \quad . \tag{3.6.14b}$$

The profile given by (3.6.14a) corresponds to the sound velocity monotonically increasing from the limiting value of $\omega q^{-1} \tanh^2 \alpha$ as $z \to -\infty$ to ωq^{-1} at $z = 0$. In the case of (3.6.14b) the sound velocity monotonically decreases from $\omega q^{-1} \coth^2 \alpha$ as $z \to -\infty$ to ωq^{-1} at $z = 0$. For the field in the layered half-spaces specified by (3.6.14) and for the corresponding reflection coefficients, cumbersome relations, which do, however, contain only elementary functions are obtained in [3.24]. In the same manner as for (3.6.14) more general $k(z)$ profiles can be considered. The origin of their z coordinate is displaced with respet to the origin of (3.6.14), which leads to a change in sound velocity at the boundary $z = 0$. Thus, instead of $z(\eta)$ in (3.6.14a) we can take the following:

$$z(\eta) = -z_1 - \eta \tanh^2 \alpha$$
$$+ a^{-1} \tanh^2 \alpha \cdot [\coth (a\eta + \alpha) - \coth \alpha] \quad , \quad z_1 \geq 0 \quad . \tag{3.6.15}$$

In recent years analytical methods of solving one-dimensional inverse problems for the Helmholtz equation have been developed that is, the problem of the reconstruction of the $k(z)$ profile according to the reflection coefficient or other field parameters [3.25, 26], When the solution for $k(z)$ is obtained in closed form, the *method of the inverse problem in the scattering theory* yields new solvable profiles [3.27–29]. Although most of results have been formulated for the Schrödinger equation, they can be easily transferred to the Helmholtz equation. We should also note the interesting generalizations of the Epstein profile suggested by *Rawer* [3.30] which allow exact solutions at normal wave incidence.

3.6.2 "Compound" Layered Media

It is possible to obtain exact solutions for the sound field in a continuously layered medium of very general type by dividing the medium into a number of layers in which the wave number $k(z)$ is defined by one of the equations given in Sects. 3.2–5, or when $k(z)$ is constant. This approach is used by many authors. Combinations of layers with different types of $k(z)$ or with different parameter values (which is more frequently the case), are constructed. In such a fashion we can get a satisfactory approximation to practically any real profile $k(z)$, but it is still impossible to construct an infinitely differentiable profile. Moreover, when the number of layers is large, the results become cumbersome and can be used only for numerical computations.

Rayleigh [Ref. 3.31, Sect. 148b] has considered an interesting case of such a "compound" layered medium for normal incidence of a sound wave. The problem concerned reflection from the $-L \leq z \leq 0$ layer for linear dependence of the sound velocity

$$c(z) = c_1(1 - az) \quad , \quad a = \text{const} \quad . \tag{3.6.16}$$

The layer contacts a homogeneous half-space with sound velocity c_1 at $z = 0$ and another homogeneous half-space, where $c = c_2 \equiv c_1(1 + aL)$, at $z = -L$. Thus, the sound velocity in the entire medium changes continuously. The density is assumed to be constant. At normal incidence the acoustic pressure $p(\mathbf{r}) = \Phi(z)$ in the layer obeys the equation

$$\Phi''(z) + k_1^2(1 - az)^{-2}\Phi(z) = 0 \quad . \tag{3.6.17}$$

The general solution of this equation can be readily found

$$\Phi(z) = A_1(1 - az)^{m_1} + A_2(1 - az)^{m_2} \quad , \quad -L \le z \le 0 \quad , \tag{3.6.18}$$

where

$$m_{1,2} = \tfrac{1}{2} \pm \sqrt{\tfrac{1}{4} - k_1^2 a^{-2}} \quad , \quad \mathrm{Im}\left\{\sqrt{\tfrac{1}{4} - k_1^2 a^{-2}}\right\} \ge 0 \quad . \tag{3.6.19}$$

The sound field is a plane wave departing from the interface in the half-space $z < -L$. Therefore,

$$\Phi = W \exp(-i\omega z/c_2) \quad , \quad z \le -L \quad . \tag{3.6.20}$$

Equating the impedances of waves (3.6.18 and 20) at $z = -L$ we find the ratio of the coefficients A_1 and A_2:

$$\frac{A_1}{A_2} = \frac{(m_2 - ik_1a^{-1})\alpha}{ik_1a^{-1} - m_1} \quad , \tag{3.6.21}$$

where

$$\alpha \equiv \left(\frac{c_2}{c_1}\right)^{m_1 - m_2} = \exp\left[2\sqrt{\tfrac{1}{4} - k_1^2 a^{-2}}\ln\left(\frac{c_2}{c_1}\right)\right] \quad . \tag{3.6.22}$$

Then, as usual, we determine the reflection coefficient for a plane wave by (3.1.5):

$$V = \frac{i}{2}\left[\frac{k_1}{a} + i(\alpha + 1)\frac{\sqrt{\tfrac{1}{4} - k_1^2 a^{-2}}}{\alpha - 1}\right]^{-1} \quad . \tag{3.6.23}$$

For a medium containing more layers, one should use the condition of continuity of the impedance to calculate the sound field successively from the lowest layer to the uppermost one and to find the reflection coefficient. This procedure parallels that considered in Sect. 2.5 for a discretely layered medium.

It is of interest to compare the reflection coefficient given by (3.6.23) and the results of Sects. 3.2 and 4. First, consider the limiting case where the layer thickness L tends to infinity. If we assume $a > 0$, i.e., the sound velocity increases with $|z|$, then the wave number will not become infinite at limited z and the problem is physically meaningful. Then as $L \to \infty$, $\ln(c_2/c_1) \to +\infty$. If

$$k_1 < |a|/2 \quad , \tag{3.6.24}$$

then $\mu \equiv \sqrt{\tfrac{1}{4} - k_1^2 a^{-2}}$ is a real positive number and $\alpha \to +\infty$. The reflection coefficient (3.6.23) assumes the limiting value

$$V = \frac{i}{2}\left[\frac{k_1}{a} + i\sqrt{\tfrac{1}{4} - k_1^2 a^{-2}}\right]^{-1} \quad , \tag{3.6.25}$$

which is also suitable at $k_1 = a/2$. When the opposite inequality

$$k_1 > |a|/2 \tag{3.6.26}$$

is taken, μ is purely imaginary, and the reflection coefficient changes periodically with increasing layer thickness. In actuality, this does not take place, as at sufficiently large L even small sound absorption in the medium is significant. As we shall see in Chap. 7, to take into account the weak absorption it is sufficient to give k_1^2 an additional small positive imaginary term. Hence, it turns out that $\mathrm{Re}\{\mu\} > 0$ and from (3.6.22,23) we again obtain the reflection coefficient (3.6.25).

As $L \to +\infty$, reflection from an inhomogeneous layer transforms into reflection from a half-space. The corresponding profile of the wave number is obtained from (3.2.16) of Sect. 3.2 at $k_2 = 0$, $\alpha_1 = 0$, $\alpha_2 = k_1^2$, $z_1 = 1/a$, and $\varrho_2 = \varrho_1$. Note that the limiting value of the coefficient of reflection from the layer (3.6.25) coincides with the reflection coefficient for the layered half-space (3.2.19), obtained as a special case of more general result, valid for arbitrary angle of incidence.

For finite L (3.6.23) gives the reflection coefficient from the transient layer in which the index of refraction changes continuously from its value in the upper to that in the lower half-space. Under condition (3.6.24) when μ and α are real, we obtain for the modulus of the reflection coefficient ϱ

$$\varrho = \{1 + 4\mu^2 \sinh^{-2}[\mu \ln(c_2/c_1)]\}^{-1/2} \ . \tag{3.6.27a}$$

In another case, when (3.6.26) is satisfied, μ is purely imaginary and (3.6.23) gives

$$\varrho = \{1 + 4|\mu|^2 \sin^{-2}[|\mu| \ln(c_2/c_1)]\}^{-1/2}, \quad |\mu| = (k_1^2 a^{-2} - \tfrac{1}{4})^{1/2} \ . \tag{3.6.27b}$$

For the exact equality $k_1 = |a|/2$, the value of ϱ can be obtained by proceeding to the limit in (3.6.27a) or in (3.6.27b).

We assume the sound velocity to be increasing with $|z|$ (i.e., $a > 0$) and compare reflection from the layer given by (3.6.16) and from the transition Epstein layer. In the latter the modulus of the reflection coefficient is given by (3.4.19), which for a normal incident wave ($\theta_0 = \theta_1 = 0$) takes the form

$$\varrho = \frac{\sinh[\pi S(1 - n_\infty)/2]}{\sinh[\pi S(1 + n_\infty)/2]} \ , \tag{3.6.28}$$

where $n_\infty = (1 - N)^{1/2} < 1$ is the index of refraction as $z \to -\infty$.

Figure 3.10a shows the reflection coefficients given by (3.6.28) (curve 1) and (3.6.27) (curve 2). The modulus of the reflection coefficient ϱ is plotted along the ordinate, and $k_1 L$ is plotted along the abscissa. [The parameter k_1/a in the formulas for ϱ is expressed in terms of the layer thickness L by $k_1/a = k_1 L c_1 (c_2 - c_1)^{-1}$]. For both layers the refraction index of the medium on the side of the layer opposite to an incident wave [that is c_1/c_2 in (3.6.27) and n_∞ in (3.6.28)] is taken to be 0.8. The relation between the layers' thickness is chosen so that $S = k_1 L/3$, where S characterizes the Epstein layer. Figure 3.10b shows the dependence of $n(z)$ on z in these two cases. The curve $n(z)$ for a linear layer is shifted by $L/2$ to aid visual comparison.

We see that for zero thickness of the layers the reflection coefficients are identical and equal to the Fresnel coefficient of reflection from the boundary of homogeneous media. Here the general behavior occurs: the wave "ignores" the details of the media's parameters distribution when they are much smaller than the wavelength. For the smoothing transition layer (curve 1 in Fig. 3.10a) according to (3.6.28) the reflection coefficient decreases monotonically with increasing layer thickness. For

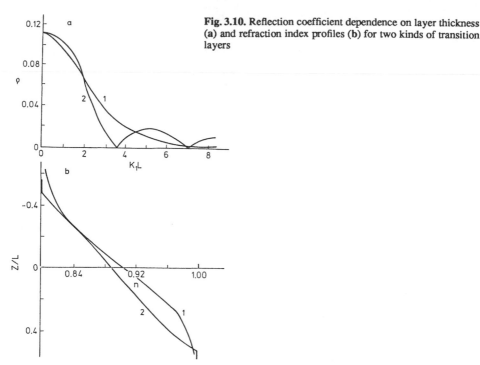

Fig. 3.10. Reflection coefficient dependence on layer thickness (a) and refraction index profiles (b) for two kinds of transition layers

the limited thickness transition layer (curve 2 in Fig. 3.10a) according to (3.6.27b) the reflection coefficient vanishes at the points satisfying $|\mu| \ln (c_2/c_1) = m\pi$, where m is a nonzero integer, and has maxima approximately halfway between these points. Obviously, this is due to the interference of waves reflected from the planes of discontinuity in the derivative dn/dz. At rather large layer thickness (3.6.27b, 28) give, respectively,

$$\pi S/2 \gg 1 \quad , \quad \varrho \approx \exp(-\pi S n_\infty) \quad , \tag{3.6.29}$$

$$k_1 L(c_2/c_1 - 1)^{-1} \gg 1 \quad , \quad \varrho \approx |\sin [|\mu| \ln (c_2/c_1)]|/2|\mu| \quad . \tag{3.6.30}$$

Thus, the reflection coefficient decays exponentially with increasing layer thickness in one case, and is inversely proportional to the layer thickness in the other case (cf. (3.5.35) for the reflection coefficient from the half-space with a discontinuity in dn/dz at the interface).

3.7 Exact Solutions for Media with Continuous Stratification of Sound Velocity, Density, and Flow Velocity

Up to now we have considered media with constant density at rest. In this section we dispense with this limitation and assume that the density is specified by the function $\varrho = \varrho(z)$ and the flow velocity by $v_0 = v_0(z)$. In Sects. 3.1–6 the determination of the reflection coefficient for a known set of solutions of the differential equation fo the sound field was illustrated. Therefore, in this treatment we shall limit ourselves to finding linearly independent solutions without rederiving the formulas for the reflection coefficients, for the solvable profiles.

3.7.1 Motionless Fluid with Density Stratification

It is convenient to take the wave equation for the inhomogeneous medium in the form given by (1.2.28 or 30). Then for the vertical dependence of the acoustic pressure, $\Phi(z)$, we have the ordinary differential equations

$$(\varrho^{-1/2}\Phi)'' + [k^2 - \xi^2 + \varrho''/2\varrho - 3(\varrho'/2\varrho)^2]\varrho^{-1/2}\Phi = 0 \quad , \tag{3.7.1}$$

$$d^2\Phi/d\zeta^2 + (\varrho_0/\varrho)^2(k^2 - \xi^2)\Phi = 0 \quad , \quad \zeta(z) = \varrho_0^{-1}\int_{z_0}^{z}\varrho(z_1)dz_1 \quad , \tag{3.7.2}$$

where primes denote differentiation with respect to z. Equations (3.7.1, 2) are equivalent for media with smooth z-dependence in the parameters. In the various cases studied here we shall use whichever form of the wave equation that is more convenient.

Equations (3.7.1, 2), like (3.1.3), are one-dimensional Helmholtz equations. Their exact solutions can be constructed in terms of solutions of the basic equation in (3.1.7) by choosing suitable substitution of the variables $\eta = \eta(z)$. For an arbitrary density stratification the solution of (3.7.1) can be expressed in terms of the solution of the basic equation for media with a wave number such as

$$k^2(z) = \xi^2 + 3(\varrho'/2\varrho)^2 - \varrho''/2\varrho + (0.5\ln\eta')''$$
$$- [(0.5\ln\eta')']^2 + (\eta')^2 g(\eta) \quad . \tag{3.7.3}$$

The derivation of this expression is quite similar to that for (3.1.11) in Sect. 3.1. In the case of any density stratification each (smooth) function $\eta(z)$ generates the solvable profile $k(z)$. However, in this method the reflection of waves of arbitrary frequency and incident angles can be considered only when the right-hand side of (3.7.3) contains arbitrary additive and multiplicative constants.

We see that when one accounts for density inhomogeneity, *constructing the examples* of media which allow exact solutions becomes the *simpler* problem. Indeed, at any given density stratification and fixed ω and ξ there exists a function $c(z)$ such that the effective wave number in (3.7.1) is constant or depends on z in a given arbitrary way. However, it is more complicated to pick up such pairs of functions $k(z)$ and $\varrho(z)$ which lead to exact solutions of the wave equation at any ω and ξ and at the same time satisfy the given requirements [for instance, $c(z)$ and $\varrho(z)$ are close to some experimentally measured profiles or $c(z)$ is fixed]. In general this problem cannot be solved in closed form just as it cannot be for $\varrho = $ const. Let us examine for what density stratification (except the trivial case of $\varrho = $ const) we can still find exact solutions for the $c(z)$ profiles obtained in Sects. 3.2, 3.

The additional term in (3.7.1) for the effective wave number squared caused by density stratification, can be written as

$$\varrho''/2\varrho - 3(\varrho'/2\varrho)^2 = \varphi' - \varphi^2 \quad , \tag{3.7.4}$$

where

$$\varphi = 0.5(\ln \varrho)' \quad . \tag{3.7.5}$$

If the exact solutions of the wave equation for any profile $k(z)$ and $\varrho = \varrho_0$ at arbitrary incidence angle are known, this equation can be also solved for $\varrho(z)$ satisfying

$$\varphi' - \varphi^2 = -a^2 \quad , \quad a = \text{const} \quad . \tag{3.7.6}$$

Indeed, in this case we have from (3.7.1)

$$\Phi(z, \xi)|_{\varrho = \varrho(z)} = \sqrt{\varrho(z)/\varrho_0} \Phi\left(z, \sqrt{\xi^2 + a^2}\right)\Bigg|_{\varrho \equiv \varrho_0} \quad . \tag{3.7.7}$$

The nonlinear ordinary differential equation (3.7.6) for the function φ is a particular case of the Riccati equation [Ref. 3.6, Part I, Sect. 4]. In (3.7.6) the variables are separable, which permits one to solve this equation easily and to find the corresponding functions $\varrho(z)$ by using (3.7.5). After simple transformations we obtain three functions for the dependence of density on z for which (3.7.7) is valid:

$$\varrho = \varrho_0 \exp\left[\pm 2a(z + z_3)\right] \quad , \tag{3.7.8a}$$

$$\varrho = \varrho_0 \sinh^{-2} a(z + z_3) \quad , \quad z < 0 \quad , \tag{3.7.8b}$$

$$\varrho = \varrho_0 \cosh^{-2} a(z + z_3) \quad . \tag{3.7.8c}$$

Here ϱ_0, a, and z_3 are arbitrary constants. The constant z_3 has dimension of length and is independent of the values z_1 in (3.2.14, 33) and (3.3.9) and of z_2 in (3.3.20). In (3.7.8a, c) the medium density in the lower half-space has no singularities for finite z. The same occurs in (3.7.8b) if $z_3 < 0$. The results of Sects. 3.2–5 can be immediately applied to media with density stratification of the form given by (3.7.8).

Consider now, departing from (3.7.1), density variations given by

$$\varrho(z) = \varrho_0(|z/z_1| + 1)^\beta \quad . \tag{3.7.9}$$

A number of researchers [3.4, 32 and references therein] treated the propagation of elastic waves in fluids and solids with such a density stratification. In this case the additional term to the effective wave number squared in (3.7.1) is $-\beta(\beta + 2)(|z/z_1| + 1)^{-2}/4z_1^2$. It is important to note that it vanishes not only at $\beta = 0$, i.e., at $\varrho \equiv \varrho_0$, but at $\beta = -2$ as well. In the latter case the influence of density stratification for the case of any $c(z)$ is simply reduced to multiplication of $\Phi(z)$ by $[\varrho(z)/\varrho_0]^{1/2} = 1 + |z/z_1|$ [see (3.7.7) where now we should set $a = 0$]. This case is also one of the limiting cases contained in (3.7.8b) (as $a \to 0$, $\varrho_0 \to 0$, $\varrho_0 a^{-2} \to \text{const} < \infty$).

For the $k(z)$ profiles from the family of (3.2.14) obtained in Sect. 3.2 which are solvable in terms of Whittaker functions, one can easily take into account the additional term to the square of the effective wave number. This can be attained by changing the parameter m in the Whittaker equation. Therefore the results obtained for the profiles given by (3.2.14) can be applied to media with the density changing according to (3.7.9). For instance, in the case of the profile in (3.2.16), the effect of the density stratification, besides multiplication of $\Phi(z)$ by $[\varrho(z)/\varrho_0]^{1/2}$, is replacement of α_2 by $\alpha_2 - \beta(\beta + 1)/4z_1^2$ in formulas for the field in an inhomogeneous medium.

Other $\varrho(z)$ profiles leading to exact solutions of the wave equation can be conveniently obtained by using (3.7.2). The general relation for the profiles of the wave number which enables us to express exact solutions of the wave equation through the solution of the basic equation (3.1.7), is deduced in a manner similar to that used in deriving (3.1.11) and has the form

$$k^2[z(\zeta)] = \xi^2 + (\varrho/\varrho_0)^2\{(0.5 \ln \eta')'' - [(0.5 \ln \eta')']^2 + (\eta')^2 g(\eta)\} \quad . \tag{3.7.10}$$

Here $\eta = \eta(\zeta)$, since it is ζ that is the independent variable in (3.7.2). At $\varrho(z) \not\equiv \varrho_0$ (3.7.10) gives one more family of solvable profiles $k(z)$ which do not coincide with (3.7.3). Taking, for example, the profiles of (3.3.9) and (3.3.20) in Sect. 3.3, we shall show how (3.7.10) enables us to find, under given conditions, allowed $\varrho(z)$ by using the solvable (at $\varrho \equiv \varrho_0$) profiles $c(z)$. By allowed $\varrho(z)$ we mean such functions for which the wave equation may be solved at any ω and ξ, as for the case of constant density.

With the same basic equation and the same function η that have led to the solvable profile of (3.3.18), (3.7.10) takes the form

$$k^2[z(\zeta)] = (\varrho/\varrho_0)^2 k_0^2[1 + N \sinh^{-2} b\zeta + M \cosh b\zeta \sinh^{-2} b\zeta] + \xi^2 \quad .$$
$$\tag{3.7.11}$$

This expression for the square of the wave number has additive and multiplicative arbitrary constants not only for $\varrho \equiv \varrho_0$, but also for

$$\varrho/\varrho_0 = -\sinh b\zeta \quad , \tag{3.7.12}$$

and for some other $\varrho(z)$ as well. Consider the case of (3.7.12). The definition of $\zeta(z)$ in (3.7.2) and (3.7.12) give the simple differential equation relating ζ to z

$$d\zeta/dz = -\sinh b\zeta \quad , \tag{3.7.13}$$

from which it follows that

$$\tanh(b\zeta/2) = \exp[-b(z + z_3)] \quad , \quad z_3 = \text{const} \quad , \tag{3.7.14}$$

$$\varrho(z)/\varrho_0 = \sinh^{-1} b_1(z + z_3) \quad , \quad b_1 = -b \quad . \tag{3.7.15}$$

The square of the wave number as a function of z is given by the relation resulting from (3.7.11, 14):

$$k^2(z) = k_1^2[1 + N_1 \sinh^{-2} b_1(z + z_3) + M_1 \coth b_1(z + z_3)] \quad , \tag{3.7.16}$$

where k_1, N_1, M_1 are the arbitrary constants related to the parameters k_0, N, M (and through the latter related to the parameters of the hypergeometric equation) as follows

$$k_1^2 = k_0^2 N + \xi^2 \quad , \quad N_1 = \frac{k_0^2}{k_1^2} \quad , \quad M_1 = \frac{k_0^2 M}{k_1^2} \quad . \tag{3.7.17}$$

Thus, one can solve the wave equation exactly in terms of the hypergeometric functions for a layered medium with the profiles of the wave number given by (3.7.16) and of the density given by (3.7.15), which include six arbitrary constants. Note that at $M = M_1 = 0$ the profile of (3.7.16) coincides with the profile (3.3.20), which is solvable at $\varrho = \text{const}$.

Similarly, a number of density stratifications for which the wave equation is solved exactly for $k(z)$ given by *Epstein* can be suggested. When the same (hypergeometric) basic equation and the same change of variable η that lead to the Epstein layer (3.3.9) are chosen, (3.7.10) yields

$$k^2[z(\zeta)] = (\varrho/\varrho_0)^2 k_0^2 \{1 - N \exp(a\zeta)[1 + \exp(a\zeta)]^{-1} - 4M \exp(a\zeta)$$
$$\times [1 + \exp(a\zeta)]^{-2}\} + \xi^2 \quad . \tag{3.7.18}$$

The right-hand side of (3.7.18) acquires the arbitrary additive and multiplicative constants if one takes the vertical dependence of the density in the form

$$\varrho/\varrho_0 = 1 + \exp(a\zeta) \quad . \tag{3.7.19}$$

Reconstructing the z-dependence of ζ with the use of (3.7.2), we obtain

$$\varrho(z) = \varrho_0 \{1 + \exp[a(z + z_1)]\}^{-1} \quad , \quad z_1 = \text{const} \quad , \tag{3.7.20}$$

$$k^2(z) = k_1^2 - k_1^2 N_1 \exp[a(z + z_1)]\{1 + \exp[a(z + z_1)]\}^{-1}$$
$$- 4k_1^2 M_1 \exp[a(z + z_1)]\{1 + \exp[a(z + z_1)]\}^{-2} \quad , \tag{3.7.21}$$

where

$$k_1^2 = k_0^2 + \xi^2 \quad , \quad N_1 = (1 - 4M)k_0^2 k_1^{-2} \quad ,$$
$$M_1 = (1 - N)k_0^2/4k_1^2 \quad , \tag{3.7.22}$$

and where z_1, ϱ_0, and a are arbitrary constants. The reflection and transmission coefficients for a plane wave in a layered medium with the stratification defined by (3.7.20) and (3.7.21) are given by the expressions treated in detail in Sect. 3.4. One can easily verify that also in the case of the density varying according to

$$\varrho(z) = \varrho_0 \{1 + \exp[a(z + z_1)]\} \quad , \tag{3.7.23}$$

the reflection and transmission coefficients for a medium with $k^2(z)$ varying according to *Epstein* are given by the same expressions but with proper redefinition of the parameters.

3.7.2 Moving Layered Medium

For the vertical dependence of the acoustic pressure, $\Phi(z)$, we obtain from (1.2.21)

$$(\Phi/\varrho^{1/2}\beta)'' + \{k^2\beta^2 - \xi^2 + (2\varrho\beta^2)^{-1}(\varrho\beta^2)''$$
$$- 3[(\varrho\beta^2)'/2\varrho\beta^2]^2\}(\Phi/\varrho^{1/2}\beta) = 0 \quad . \tag{3.7.24}$$

Here the quantity

$$\beta = 1 - \boldsymbol{\xi} \cdot \boldsymbol{v}_0(z)/\omega \tag{3.7.25}$$

is the only one which includes the influence of the flow on the sound field. At normal incidence ($\xi = 0$) the flow does not affect the sound propagation. If ω and $\xi \neq 0$ are fixed, then one can choose the sound-velocity stratification such that (3.7.24) has a solution in elementary functions for any functions $v_0(z)$ and $\varrho(z)$, at least at $\beta \neq 0$.

For instance, choosing

$$k^2(z) = \{a^2 + \xi^2 - (2\varrho\beta^2)^{-1}(\varrho\beta^2)''$$
$$+ 3[(\varrho\beta^2)'/2\varrho\beta^2]^2\}\beta^{-2} \quad , \quad a = \text{const} \tag{3.7.26}$$

we have the general solution of (3.7.24)

$$\varPhi(z) = \varrho^{1/2}\beta[A\exp(iaz) + B\exp(-iaz)] \quad . \tag{3.7.27}$$

Assume that $v_0(z)$ and the derivatives of $\varrho(z)$ are equal to zero outside the layer of finite thickness. Then at $|z| \to \infty$, $k = k_0$, $a = k_0 \cos\theta_0$, and $\xi = k_0 \sin\theta_0$, where $k_0 = \omega/c_0$ is the wave number, and θ_0 is the incidence angle at large $|z|$. Setting $A = 0$ in (3.7.27), we obtain the solution in the form of a wave incident from $z = +\infty$ in the absence of a reflected one. Consider the following example: the flow velocity is directed parallel to the x-axis and has the value $v_0(z) = u_0\exp(-z^2/b^2)$ and the medium density is constant. The square of the index of refraction $n^2 = k^2(z)/k_0^2$ which makes the medium nonreflecting is found from (3.7.26):

$$n^2 = \frac{1}{\beta^2}\left\{1 - 2(1 - \beta)(k_0 b\beta)^{-2}\left[\beta\left(1 + \frac{2z^2}{b^2}\right) - \frac{4z^2}{b^2}\right]\right\} \quad ,$$

$$\beta = 1 - \frac{u_0}{c_0}\sin\theta_0\exp\left(-\frac{z^2}{b^2}\right) \quad . \tag{3.7.28}$$

The functions $v_0(z)$ and $n^2(z)$ are shown in Fig. 3.11 for the values $z > 0$, $u_0/c_0 = 0.5$, $k_0 b = 1$. As $\beta \to 1$ we have $n^2 \to 1$. Variations of β and the corresponding inhomogeneities of n increase with increasing θ_0. If $k_0 b$ is large, the point $z = 0$ is the only extremum of $n(z)$.

In some applications one may want to know the smooth profiles of the flow velocity (independent of ω and ξ) for which one can find the exact solutions of (3.7.24) at constant sound velocity and density. Such solutions are needed in the investigation of sound propagation in the atmosphere where refraction of sound is frequently due to wind and to a lesser extent due to the medium inhomogeneities.

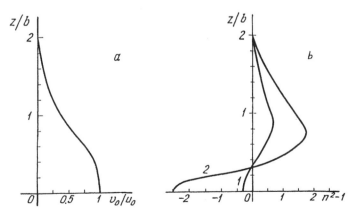

Fig. 3.11a,b. Flow velocity $v_0(z)$ vertical dependence (a) and corresponding refraction index dependence (b) in a nonreflecting medium. Reflection is absent at the angle of incidence $\theta_0 = \pi/6$ (*curve 1*) or $\theta_0 = \pi/3$ (*curve 2*)

Other examples of problems where exact solutions for a moving continuously layered medium are important are sound waves interacting with jets which arise in jet engines or due to bodies motion in fluids, or also in hydrodynamical sound sources.

At present, such exact solutions of (3.7.24) in the closed form are known only for the linear flow profile [3.33, 34]

$$v_0 = b + az \quad .$$
(3.7.29)

In this case

$$\beta = a_1(z_1 - z) \quad , \quad a_1 = \frac{\xi \cdot a}{\omega} \quad , \quad z_1 = z_1(\omega, \xi) = \frac{\omega - \xi \cdot b}{\xi \cdot a}$$
(3.7.30)

and (3.7.24) takes the form

$$[(z - z_1)^{-1}\Phi]'' + [k^2 a_1^2(z - z_1)^2 - \xi^2 - 2(z - z_1)^{-2}][(z - z_1)^{-1}\Phi] = 0 \quad .$$
(3.7.31)

In this equation the effective wave number is a particular case of the family of solvable profiles given by (3.2.25) ($\beta_1 = 0$, $\beta_3 z_1^2 = -2$, $\beta_2 z_1^2 = k^2 a_1^2$). It was shown in Sect. 3.2 that the solutions of wave equations for such profiles are expressed in terms of the Whittaker functions, the index m of the Whittaker functions being $\pm \frac{3}{4}$ for (3.7.31) according to the last expression in (3.2.26). In this particular case the Whittaker functions are reduced to parabolic cylinder functions. Detailed analysis of the solutions of (3.7.31), in particular under the conditions of sound-flow resonant interaction that occurs in the case where $z_1 < 0$, will be presented in Chap. 9.

For the slow flows velocity of which is small compared to the sound velocity ($v_0 \ll c$), (3.7.24) can be simplified by retaining only terms to the first order in v_0. For simplicity the density is assumed to be constant. Then we obtain

$$[(1 - m)^{-1}\Phi]'' + (k^2 - 2k^2 m - m'' - \xi^2)[(1 - m)^{-1}\Phi] = 0 \quad ,$$
(3.7.32)

where

$$m \equiv 1 - \beta = \xi \cdot v_0(z)/\omega \quad .$$
(3.7.33)

Equation (3.7.32) is widely used in the theory of sound propagation in the atmosphere [3.35, 36], since the condition of the smallness of the Mach number v_0/c is usually satisfied by the wind velocity ($v_0/c \lesssim 5 \cdot 10^{-2}$). It is more complicated to find profiles $v_0(z)$ which lead to exact solutions of (3.7.32) at arbitrary ω and ξ, than to find the solvable profiles $k(z)$ for (3.1.3), since (3.7.32) contains $m''(z)$ besides $m(z)$. By testing the functions $m(z) \sim k^2(z)$ with $k^2(z)$ being the solvable profiles for the equation (3.1.3) defined in Sects. 3.2, 3, one can find the exact solutions for two types of dependence of v_0 on z:

$$v_0(z) = a_1(z + z_1)^2 + a_2(z + z_1) + a_3 \quad ,$$
(3.7.34)

$$v_0(z) = a_1 \exp(2\alpha z) + a_2 \exp(\alpha z) + a_3 \quad .$$
(3.7.35)

Here a_j, $j = 1, 2, 3$ are arbitrary horizontal vectors; z_1 and α are scalar constants. Figure 3.12 shows the possible vertical dependence of the projection of the vector v_0 given by (3.7.34) on the arbitrary horizontal direction. The dependence of the

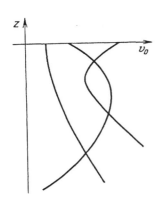

Fig. 3.12. Vertical dependence of projections on coordinate axes of the flow velocity given by (3.7.34) for different values of the free parameters a_1, a_2, a_3 in (3.7.34). For the projections on Ox and Oy axes each of the three curves may be taken independently

projections of v_0 (3.7.35) on z formally coincides with the function $k^2(z)$ in (3.2.30). Thus, Fig. 3.4 illustrates forms of the profiles of $(v_0)_x$ and $(v_0)_y$ corresponding to different values of parameters a_j and α. Solutions of (3.7.32) for the flow profile (3.7.34) are expressed in terms of the parabolic cylinder function if $\boldsymbol{\xi} \cdot a_1 \neq 0$, and in terms of the Airy function otherwise. (Recall that in the latter case the solution of the exact equation (3.7.31) can be found.) The solutions of the wave equation for a homogeneous medium with the flow given by (3.7.35) are expressed in terms of the Whittaker functions $W_{l,\mu}$ and $M_{l,\mu}$. The sound field in the particular case when $\boldsymbol{\xi} \cdot a_2 = 0$ (when the solutions are reduced to Bessel functions) has been treated in [3.36]. A particular case of flow-velocity stratification (3.7.34) has been analyzed in [3.39]. For layered media with slow flows given by (3.7.34, 35), the solutions of the wave equation in terms of known special functions can also be obtained for other stratifications of sound velocity and medium density besides $c(z) = $ const and $\varrho(z) = $ const [3.40, 41]. We shall not consider these cases.

One can also obtain the flow profiles (3.7.34, 35) for which the sound field can be found exactly (under condition of flow slowness) by starting from the wave equation in the form of (1.2.25). Retaining in (1.2.25) only the terms that are linear in m (i.e., in the flow velocity), we obtain

$$\frac{d^2\Phi}{d\zeta^2} + [k^2 - \xi^2 + 2m(k^2 - 2\xi^2)]\Phi = 0 \quad . \tag{3.7.36}$$

For simplicity, we assume that the medium density is constant. The square of the effective wave number in (3.7.36), in contrast to (3.7.32), contains m but not m''. However, the coordinate

$$\zeta = \int\limits_0^z [1 - m(z_1)]^2 dz_1 = z - 2\int\limits_0^z m(z_1)dz_1 + O(m^2) \tag{3.7.37}$$

depends on the horizontal wave vector $\boldsymbol{\xi}$ and flow velocity. This makes it difficult to find the solvable profiles $v_0(z)$ which are suitable at any $\boldsymbol{\xi}$.

The term $2m(k^2 - 2\xi^2)$ in (3.7.36) is small compared to $k^2 - \xi^2$. Moreover, for $m \ll 1$ $\zeta(z)$ differs little from z. Therefore, the replacement of $m[z(\zeta)]$ by $m(\zeta)$ in (3.7.37) does not contribute too large an error; under certain conditions it is of the same order as the omitted terms $O(m^2)$. Then one can transfer *all* the exact solutions, suitable at any ω and ξ for an motionless layered medium with constant density defined in Sects. 3.2, 3, to the case of a moving homogeneous medium. Indeed, let the general solution

$$\Phi = \sum_{j=1}^{2} A_j \Phi_j(b_1, b_2, z) \tag{3.7.38}$$

of the equation

$$\frac{d^2\Phi}{dz^2} + [b_1 g(z) + b_2]\Phi = 0 \tag{3.7.39}$$

be known for any values of b_1 and b_2. Then, for the layered flow given by

$$v_0(z) = a g(z) \quad , \quad a = \text{const} \quad , \tag{3.7.40}$$

in a homogeneous fluid (3.7.36) has the general solution

$$\Phi = \sum_{j=1}^{2} A_j \Phi_j[2\boldsymbol{\xi} \cdot a(k^2 - 2\xi^2)/\omega, \, k^2 - \xi^2, \, \zeta(z)] \quad , \tag{3.7.41}$$

which becomes exact in the limit of small flow velocity. The function $\zeta(z)$, calculated by (3.7.37, 40), must be substituted into (3.7.41).

Now we shall give the conditions under which the replacement $m[z(\zeta)]$ by $m(\zeta)$ in (3.7.36) is possible. We let m_0 be a typical value of $|m(z)|$, and L that of the spatial scale of $v_0(z)$ variation. When $|\zeta - z| \lesssim L$, the replacement of $m[z(\zeta)]$ by $m(\zeta)$ contributes an error of the order of $k^2 m_0^2 |z|/L$ to the coefficient of the equation. It may be negligible compared to $k^2 m$ as long as

$$m_0 |z|/L \ll 1 \quad . \tag{3.7.42}$$

The sound wave "feels" the inhomogeneities of the medium averaging out over the wave's length. Hence, for sharp changes of $m(z)$, when $L \ll k^{-1}$ the condition of (3.7.42) requiring that the perturbation of the effective wave number be small just *at each point* appears to be unjustifiably restrictive and must be changed to

$$k m_0 |z| \ll 1 \quad . \tag{3.7.43}$$

The physical meaning of the requirement established by (3.7.43) is that the change of the phase of the wave must be small over its path $\zeta(z) - z$.

4. Plane-Wave Reflection from the Boundaries of Solids

In this chapter we study the elastic waves behavior in discretely layered solid media. The basic equations and boundary conditions for this problem were discussed in Sect. 1.3. Because the propagation of shear waves of horizontal polarization in a layered solid is independent of that of waves of vertical polarization and is formally analogous to the propagation of sound waves in liquid, in this chapter we shall consider only the case of vertical polarization.

The plane monochromatic elastic wave in a homogeneous solid can be described by two scalar functions $\varphi(x, z)$ and $\psi(x, z)$ as is shown in Sect. 1.3:

$$\varphi = \varphi_1 \exp(i\alpha z) + \varphi_2 \exp(-i\alpha z) \quad,$$
$$\alpha = (k_l^2 - \xi^2)^{1/2} \quad, \quad k_l = \omega/c_l \quad, \quad \text{Im}\{\alpha\} \geq 0 \quad;$$
$$\psi = \psi_1 \exp(i\beta z) + \psi_2 \exp(-i\beta z) \quad,$$
$$\beta = (k_t^2 - \xi^2)^{1/2} \quad, \quad k_t = \omega/c_t \quad, \quad \text{Im}\{\beta\} \geq 0 \quad, \tag{4.0.1}$$

where c_l and c_t are the velocities of the compression and shear wave, respectively. Let the wave vector lie in the xz plane. For brevity, the factor $\exp(i\xi x - i\omega t)$ that is common for all waves will be omitted.

In accord with (1.3.6, 27) for the particles displacement in the wave (4.0.1) we have

$$\boldsymbol{u} = (i\xi\varphi - \partial\psi/\partial z, 0, \partial\varphi/\partial z + i\xi\psi) \quad. \tag{4.0.2}$$

In the discussion below we shall need only two of the components of the stress tensor. Substituting (4.0.2) into the Hooke's law (1.3.1), after simple transformation we find

$$\sigma_{13} = -2\mu\xi(\gamma\psi - i\partial\varphi/\partial z), \quad \sigma_{33} = 2\xi\mu(\gamma\varphi + i\partial\psi/\partial z) \quad, \tag{4.0.3}$$

where

$$\gamma = \xi - k_t^2/2\xi \quad. \tag{4.0.4}$$

4.1 Plane Waves in Elastic Half-Spaces with a Free Boundary

Let an elastic medium fill the region $z > 0$. Then φ_2 and ψ_2 refer to the amplitudes of the longitudinal and transverse waves that are incident upon the boundary $z = 0$. The quantities φ_1 and ψ_1 are the amplitudes of the reflected waves. The components of the stress tensor σ_{3j}, $j = 1, 2, 3$ must vanish at the boundary. Expressing these conditions in terms of the wave amplitudes by the use of (4.0.1, 3) we obtain

$$\beta(\psi_1 - \psi_2) - \gamma(\varphi_1 + \varphi_2) = 0 \quad , \quad \alpha(\varphi_1 - \varphi_2) + \gamma(\psi_1 + \psi_2) = 0 \quad . \qquad (4.1.1)$$

Equation (4.1.1) permits us to find the amplitudes of the reflected waves which are expressed linearly in terms of the amplitudes of the incident waves. It is convenient to write down this linear relationship in the matrix form:

$$\begin{pmatrix} \varphi_1 \\ \psi_1 \end{pmatrix} = [S] \begin{pmatrix} \varphi_2 \\ \psi_2 \end{pmatrix} \quad , \quad \text{where} \quad [S] = \begin{pmatrix} V_{ll} & V_{tl} \\ V_{lt} & V_{tt} \end{pmatrix} \qquad (4.1.2)$$

is called the *scattering matrix*. Its elements have a clear physical meaning. The V_{ll} component is the reflection coefficient for a longitudinal wave, equal to the amplitude φ_1 of the reflected longitudinal wave, when the incident wave is also longitudinal ($\psi_2 = 0$, $\varphi_2 = 1$). The element $V_{lt} = \psi_1/\varphi_2$ at $\psi_2 = 0$. This is the longitudinal-to-transverse wave transformation coefficient. Similarly, V_{tl} has a meaning of the transformation coefficient of the incident transverse wave into the reflected longitudinal and V_{tt} is the reflection coefficient of the transverse wave.

From (4.1.1) eliminating ψ_1, we have

$$(\alpha\beta + \gamma^2)\varphi_1 = (\alpha\beta - \gamma^2)\varphi_2 - 2\gamma\beta\psi_2 \quad .$$

On the other hand, it follows from (4.1.2) that $\varphi_1 = V_{ll}\varphi_2 + V_{tl}\psi_2$. These equalities must be valid at any values of φ_2 and ψ_2. Hence,

$$V_{ll} = (\alpha\beta - \gamma^2)/(\alpha\beta + \gamma^2) \quad , \quad V_{tl} = -2\beta\gamma/(\alpha\beta + \gamma^2) \quad . \qquad (4.1.3)$$

The other two components of the scattering matrix are determined in a similar way. They are equal to

$$V_{tt} = V_{ll} \quad , \quad V_{lt} = \frac{2\alpha\gamma}{\alpha\beta + \gamma^2} \quad . \qquad (4.1.4)$$

This leads to an easily verified relation

$$V_{ll}^2 = V_{ll}V_{tt} = 1 + V_{tl}V_{lt} \quad , \quad \text{or} \qquad (4.1.5)$$

$$\det[S] = V_{ll}V_{tt} - V_{tl}V_{lt} = 1 \quad . \qquad (4.1.6)$$

Let us consider some properties of the coefficients of reflection and transformation. For normal incidence ($\xi = 0$, $\gamma = -\infty$) and also for the grazing incidence ($\alpha = 0$ or $\beta = 0$) we have $V_{ll} = V_{tt} = -1$, $V_{tl} = V_{lt} = 0$, i.e., total reflection of both the transverse and longitudinal waves takes place (with a phase change π) without transformation of one into another. At

$$\alpha\beta = \gamma^2 \qquad (4.1.7)$$

we obtain

$$V_{tt} = V_{ll} = 0 \quad , \quad V_{lt} = \sqrt{\alpha/\beta} \quad , \quad V_{tl} = -\sqrt{\beta/\alpha} \quad , \qquad (4.1.8)$$

that is, reflection is absent. At the boundary, the longitudinal wave transforms entirely into the transverse, and vice versa. For determining $\xi = \xi_0$ corresponding to this exchange of polarization, from (4.1.7) we get

$$\xi_0^2 \sqrt{k_l^2 - \xi_0^2}\sqrt{k_t^2 - \xi_0^2} = (\xi_0^2 - k_t^2/2)^2 \quad . \qquad (4.1.9)$$

We shall see in Sect. 4.2 that this equation has either two real roots or none.

Let a transverse wave satisfying the inequality $k_l < \xi < k_t$ be incident on the boundary. Then the potential φ is an inhomogeneous wave which decays exponentially with increasing distance from the boundary

$$\varphi = \varphi_1 \exp\left[-\sqrt{\xi^2 - k_l^2}\, z\right] \quad . \tag{4.1.10}$$

In this case the reflection and transformation coefficients can be represented in the following form with use of (4.1.3, 4)

$$V_{tt} = -\exp\left(-i\delta\right) \quad , \quad V_{tl} = 2\beta\gamma[\gamma^4 + \beta^2(\xi^2 - k_l^2)]^{-1/2}\exp\left(-\frac{i\delta}{2}\right) \quad ,$$

$$\tan\frac{\delta}{2} = \frac{\beta}{\gamma^2}\sqrt{\xi^2 - k_l^2} \quad . \tag{4.1.11}$$

Reflection is total since $|V_{tt}| = 1$. This case is analogous to the case of total reflection of sound waves considered in Sect. 2.2, with the only difference being that the incident wave is transverse and the longitudinal wave corresponds to a refracted sound wave. Here again (as in the case of sound waves) we have the general rule that if $\xi > k$ (spatial period $2\pi/\xi$ at the boundary is less than the wavelength λ), such a periodicity generates the exponentially decaying (inhomogeneous) waves in the half-space.

The reflection and transformation coefficients can be expressed in terms of the angles θ_l and θ_t which are made by the normals to the fronts of the longitudinal and transverse waves with the z-axis. Here

$$\xi = k_l \sin \theta_l = k_t \sin \theta_t \quad , \tag{4.1.12a}$$

$$\alpha = k_l \cos \theta_l \quad , \quad \beta = k_t \cos \theta_t \quad ,$$

$$\gamma = -k_t \cos 2\theta_t/2 \sin \theta_t \quad . \tag{4.1.12b}$$

Substituting these expressions into (4.1.3), we obtain

$$V_{ll} = V_{tt} = \frac{k_l \cos \theta_l \tan^2 2\theta_t - k_t \cos \theta_t}{k_l \cos \theta_l \tan^2 2\theta_t + k_t \cos \theta_t} \quad . \tag{4.1.13}$$

For the angles of polarization exchange θ_{l0} and θ_{t0} (at $V_{ll} = V_{tt} = 0$), we obtain the equation

$$k_l \cos \theta_{l0} \tan^2 2\theta_{t0} = k_t \cos \theta_{t0} \quad , \tag{4.1.14}$$

where θ_{l0} and θ_{t0} are assumed to be related through (4.1.12a).

Figure 4.1 shows the coefficient V_{ll} [according to Ref. 4.1 by *Arenberg* from whom Figs. 4.2 and 3 are also taken] as a function of the angle of incidence for the longitudinal wave for different values of the Poisson ratio σ. The latter, as is well known [4.2, 3], is connected with th Lamé constants λ and μ and with the ratio of the velocities of the waves c_t/c_l as follows:

$$\frac{c_t^2}{c_l^2} = \frac{\mu}{\lambda + 2\mu} = \frac{1 - 2\sigma}{2(1 - \sigma)} \quad . \tag{4.1.15}$$

As $\sigma \to 0.5$ the problem becomes that for a liquid, for which the coefficient of

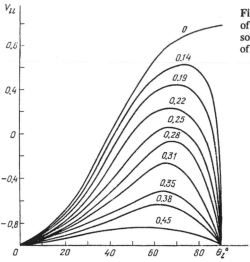

Fig. 4.1. Dependence of the reflection coefficient of longitudinal waves V_{ll} at the free boundary of a solid on the angle of incidence for different values of the Poisson's ratio. From [4.1]

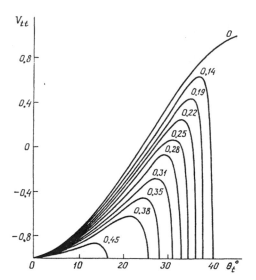

Fig. 4.2. Dependence of the reflection coefficient of transverse waves V_{tt} at the free boundary of a solid on the angle of incidence for different values of the Poisson's ratio. From [4.1]

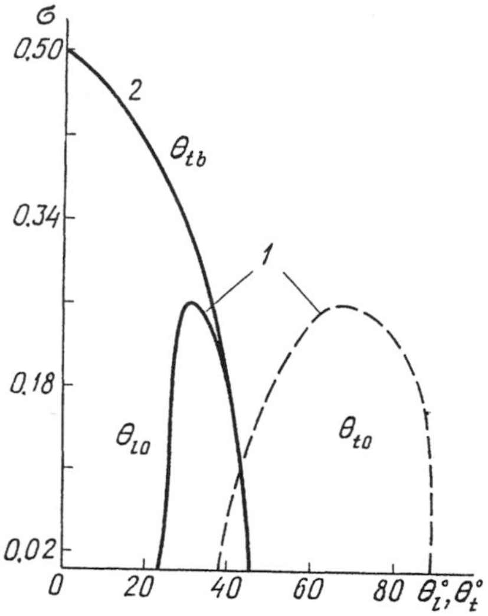

Fig. 4.3. Characteristic angles in reflection of shear waves from the free surface of an elastic half-space. *Curves 1* – angles of polarization exchange; *curve 2* – limiting angle of total reflection of shear waves. From [4.1]

reflection from a free boundary equals -1 for all angles of incidence [(2.2.13) for the case $m = 0$], which we also see in Fig. 4.1. For $\sigma < 0.26$ each curve intersects the line $V_{ll} = 0$ twice. The values of the angle θ_{l0} for the points of intersection can be obtained from (4.1.14).

Figure 4.2 shows the reflection coefficient for transverse waves V_{tt} [equal to V_{ll}, according to (4.1.4)] as a function of the incidence angle of the transverse wave θ_t. In essence, this is the same as the graph in Fig. 4.1 but now θ_t is along the abscissa and θ_t is related to θ_l by (4.1.12a). For $\sin\theta_t > c_t/c_l = k_l/k_t$ we have $\xi > k_l$ from (4.1.12a), that is, the case of total reflection. Since, according to (1.3.14) $c_t/c_l < 2^{-1/2}$, then at $\theta_t > \pi/4$ there will be total reflection for all possible values of σ $(0 \le \sigma \le \frac{1}{2})$.

Figure 4.3 shows the characteristic angles, namely, the limiting angle of total reflection of transverse waves and the angles of incidence for polarization exchange in the longitudinal and transverse waves for different values of σ.

The scattering matrix for the *potentials* φ and ψ has been obtained above. However, in experiment we do not measure the potentials but the displacements u or velocities $v = -i\omega u$. In the plane longitudinal wave $\varphi = \varphi_1 \exp[i(\alpha z + \xi x - \omega t)]$ the amplitude of the displacement is $u_1 = k_l\varphi_1$ according to (4.0.2). Similarly, that for the transverse wave is $u_1 = k_t\psi_1$. Consequently the scattering matrix for the particles' displacements and velocities has the form

$$[S] = \begin{pmatrix} V_{ll} & k_l V_{tl}/k_t \\ k_t V_{lt}/k_l & V_{tt} \end{pmatrix} . \tag{4.1.16}$$

In other words, V_{ll} and V_{tt} are the reflection coefficients not only for the potentials, but also for the displacements and velocities. However, the numerical values of the transformation coefficients change when transforming from potentials to other characteristics of waves. Note that for the matrix $[S]$ given by (4.1.16), (4.1.6) remains valid.

4.2 Reflection from Solid-Solid and Solid-Fluid Interfaces

4.2.1 Two Elastic Half-Spaces in Contact

We now assume that the half-space $z < 0$ is not vacuum, as in the previous section, but rather an elastic medium characterized by quantities with the subscript 1: c_{l1}, c_{t1}, ϱ_1, etc. The potentials of elastic waves in the upper medium are of the same form as in (4.0.1). Similarly, for the lower medium we have

$$\begin{aligned} \tilde{\varphi} &= \tilde{\varphi}_1 \exp(-i\alpha_1 z) + \tilde{\varphi}_2 \exp(i\alpha_1 z) \quad, \\ \tilde{\psi} &= \tilde{\psi}_1 \exp(-i\beta_1 z) + \tilde{\psi}_2 \exp(i\beta_1 z) \quad. \end{aligned} \tag{4.2.1}$$

Figure 4.4 illustrates the assumed notation. According to the principle of superposition, the amplitudes of the waves departing from the interface φ_1, ψ_1, $\tilde{\varphi}_1$, and $\tilde{\psi}_1$

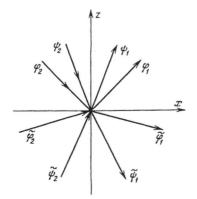

Fig. 4.4. System of waves in reflection at a boundary of two elastic half-spaces

are linearly connected with those of the waves incident on the boundary:

$$
\begin{pmatrix} \varphi_1 \\ \psi_1 \\ \tilde{\varphi}_1 \\ \tilde{\psi}_1 \end{pmatrix} = [S] \cdot \begin{pmatrix} \varphi_2 \\ \psi_2 \\ \tilde{\varphi}_2 \\ \tilde{\psi}_2 \end{pmatrix} , \qquad [S] = \begin{pmatrix} V_{ll} & V_{tl} & \tilde{W}_{ll} & \tilde{W}_{tl} \\ V_{lt} & V_{tt} & \tilde{W}_{lt} & \tilde{W}_{tt} \\ W_{ll} & W_{tl} & \tilde{V}_{ll} & \tilde{V}_{tl} \\ W_{lt} & W_{tt} & \tilde{V}_{lt} & \tilde{V}_{tt} \end{pmatrix} . \tag{4.2.2}
$$

Increasing the number of wave types increases the rank of the scattering matrix $[S]$, see (4.1.2). Its components have a clear physical meaning: V_{ll} is the reflection coefficient of a longitudinal wave incident from the upper medium; V_{lt} is the transformation coefficient of a longitudinal wave into a transverse in the upper medium; W_{tl} is the coefficient of excitation of a longitudinal wave in the lower medium by a transverse wave incident from the upper medium, etc. The tilde denotes wave incidence from the lower medium.

The scattering matrix enables us to find the reflected waves in both the half-spaces for an arbitrary plane incident wave. To calculate $[S]$, by virtue of the symmetry of the problem only 8 of 16 matrix components are required to be determined, the coefficients of the wave incident from the lower medium, i.e., those with tildes are obtained from the corresponding coefficients without tildes by the exchange $c_l \leftrightarrow c_{l1}$, $c_t \leftrightarrow c_{t1}$, $\varrho \leftrightarrow \varrho_1$. The form of the scattering matrix is defined by the boundary conditions given by (1.3.26) of continuity of the displacement vector and of two components of the stress tensor. Let us express the boundary conditions, by using (4.0.2, 3), in terms of the amplitudes of waves incident on the interface and departing from it:

$$
\mu[\beta(\psi_1 - \psi_2) - \gamma(\varphi_1 + \varphi_2)] = \mu_1[\beta_1(\tilde{\psi}_2 - \tilde{\psi}_1) - \gamma_1(\tilde{\varphi}_1 + \tilde{\varphi}_2)] , \tag{4.2.3}
$$

$$
\mu[\alpha(\varphi_1 - \varphi_2) + \gamma(\psi_1 + \psi_2)] = \mu_1[\alpha_1(\tilde{\varphi}_2 - \tilde{\varphi}_1) + \gamma_1(\tilde{\psi}_1 + \tilde{\psi}_2)] , \tag{4.2.4}
$$

$$
\beta(\psi_1 - \psi_2) - \xi(\varphi_1 + \varphi_2) = \beta_1(\tilde{\psi}_2 - \tilde{\psi}_1) - \xi(\tilde{\varphi}_1 + \tilde{\varphi}_2) , \tag{4.2.5}
$$

$$
\alpha(\varphi_1 - \varphi_2) + \xi(\psi_1 + \psi_2) = \alpha_1(\tilde{\varphi}_2 - \tilde{\varphi}_1) + \xi(\tilde{\psi}_1 + \tilde{\psi}_2) . \tag{4.2.6}
$$

Boundary conditions (4.2.3–6) yield a set of four linear equations for determining φ_1, ψ_1, $\tilde{\varphi}_1$, and $\tilde{\psi}_1$. Equation (4.2.2) shows that to determine components of the matrix $[S]$ it is sufficient to solve the set for two cases:

$$
\varphi_2 = 1 , \quad \psi_2 = \tilde{\varphi}_2 = \tilde{\psi}_2 = 0 \quad \text{and} \quad \psi_2 = 1 , \quad \varphi_2 = \tilde{\varphi}_2 = \tilde{\psi}_2 = 0 .
$$

The set of equations (4.2.3–6) may also be written in a matrix form

$$
[N_1] \begin{pmatrix} \varphi_1 \\ \psi_1 \\ \tilde{\varphi}_1 \\ \tilde{\psi}_1 \end{pmatrix} = [N_2] \begin{pmatrix} \varphi_2 \\ \psi_2 \\ \tilde{\varphi}_2 \\ \tilde{\psi}_2 \end{pmatrix} ,
$$

$$
[N_j] = \begin{pmatrix} (-1)^j \mu\gamma & \mu\beta & (-1)^{j+1}\mu_1\gamma_1 & \mu_1\beta_1 \\ \mu\alpha & (-1)^{j+1}\mu\gamma & \mu_1\alpha_1 & (-1)^j\mu_1\gamma_1 \\ (-1)^j \xi & \beta & (-1)^j \xi & \beta_1 \\ \alpha & (-1)^{j+1}\xi & \alpha_1 & (-1)^{j+1}\xi \end{pmatrix} . \tag{4.2.7}
$$

Then $[S] = [N_1]^{-1} \cdot [N_2]$, and the calculation of the scattering matrix is reduced to finding the matrix $[N_1]^{-1}$, the inverse of $[N_1]$, and multiplying two matrices.

After several straightforward but lengthy calculations which are required to determine the S-matrix in both approaches we obtain:

$$V_{ll} = [A_1^2 - \beta A_2^2/\alpha + \alpha_1 \beta_1^{-1}(B_1^2 - \beta B_2^2/\alpha) \\ + m(\beta_1/\beta - \alpha_1/\alpha)k_{t1}^4/4\xi^4]\Delta^{-1} \quad , \tag{4.2.8}$$

$$V_{tt} = [\alpha_1 \beta_1^{-1}(B_1^2 - \beta B_2^2/\alpha) + A_1^2 - \beta A_2^2/\alpha \\ + m(\alpha_1/\alpha - \beta_1/\beta)k_{t1}^4/4\xi^4]\Delta^{-1} \quad , \tag{4.2.9}$$

$$V_{lt} = -2(A_1 A_2 + \alpha_1 \beta_1^{-1} B_1 B_2)\Delta^{-1} \quad , \qquad V_{tl} = -\beta \alpha^{-1} V_{lt} \quad , \tag{4.2.10}$$

$$W_{ll} = k_{t1}^2 \xi^{-2}(A_1 - B_2)\Delta^{-1} \quad , \\ W_{tt} = k_{t1}^2 \xi^{-2}(A_1 - \alpha_1 \beta B_2/\alpha \beta_1)\Delta^{-1} \quad , \tag{4.2.11}$$

$$W_{lt} = k_{t1}^2 \xi^{-2}(A_2 + \alpha_1 \beta_1^{-1} B_1)\Delta^{-1} \quad , \\ W_{tl} = -k_{t1}^2 \xi^{-2}(\beta \alpha^{-1} A_2 + B_1)\Delta^{-1} \quad . \tag{4.2.12}$$

Here we use the following notation

$$A_1 = n^2 - m\gamma_1/\xi \quad , \qquad A_2 = (n^2\gamma - m\gamma_1)/\beta \quad , \\ B_1 = (n^2 - m)\beta_1/\xi \quad , \qquad B_2 = \beta_1 \beta^{-1}(n^2\gamma/\xi - m) \quad , \tag{4.2.13}$$

$$m = \varrho_1/\varrho \quad , \qquad n = c_t/c_{t1} \quad , \tag{4.2.14}$$

$$\Delta = A_1^2 + \beta A_2^2/\alpha + \alpha_1 \beta_1^{-1}(B_1^2 + \beta B_2^2/\alpha) \\ + m(\alpha_1/\alpha + \beta_1/\beta)k_{t1}^4/4\xi^4 \quad . \tag{4.2.15}$$

Detailed tables for values of the coefficients of reflection from an interface of two elastic half-spaces for different incidence angles and ratios of the parameters of the media were given in [4.4]. The reader can find graphs of the angle dependence of the S-matrix components in [4.5], for example.

Due to the symmetry of the problem and thanks to the energy conservation law, several universal properties of the S-matrix can be found. Consider one of these properties. It is known that after multiplying a matrix column or row by a number q, the matrix determinant also becomes q times larger. It can be easily shown by multiplying the first and third columns of the $[N_j]$ matrix in (4.2.7) by $(-1)^j$ and deviding the second and the fourth rows by the same factors $(-1)^j$ that det $[N_1] = $ det $[N_2]$. Hence

$$\det[S] = 1 \quad . \tag{4.2.16}$$

The identity in (4.2.16) is valid for arbitrary parameters of the half-spaces in contact and can be used to check calculations of the reflection and transformation coefficients. It is, indeed, a direct generalization of (4.1.5,6) which were proved for the case of reflection from a free boundary. Other universal properties of the scattering matrix are considered in Chap. 6, see also [4.6,47,48].

4.2.2 Sound Wave Reflection from Solid

Let us consider in more detail the important case of reflection at the boundary of fluid and elastic half-spaces. Assume that the half-space $z > 0$ is filled by fluid in which only sound waves propagate. Comparing (1.3.6) and (1.1.9) yields for the potential φ

$$p = \varrho \omega^2 \varphi \quad , \tag{4.2.17}$$

where p is the acoustic pressure. We shall omit the index l for the quantities referring to the longitudinal waves in fluid.

To obtain the waves' reflection and transmission coefficients at the boundary, we should assume that $\mu \to 0$ in (4.2.8–12). Then $n \to 0$, $k_t \to \infty$, $\gamma \to -\infty$, and $n^2\gamma \to -k_{t1}^2/2\xi$. From (4.2.8 and 15) we find the reflection coefficient for the sound wave incident from the fluid on the boundary of the solid:

$$V = \frac{4m\alpha\xi^2(\alpha_1\beta_1 + \gamma_1^2) - \alpha_1 k_{t1}^4}{4m\alpha\xi^2(\alpha_1\beta_1 + \gamma_1^2) + \alpha_1 k_{t1}^4} \quad . \tag{4.2.18}$$

From (4.2.11, 12) for the coefficients of excitation of longitudinal and transverse waves in the elastic half-space we have

$$W_l = \frac{-4\alpha\gamma_1\xi k_{t1}^2}{4m\alpha\xi^2(\alpha_1\beta_1 + \gamma_1^2) + \alpha_1 k_{t1}^4} \quad , \qquad W_t = \frac{\alpha_1 W_l}{\gamma_1} \quad . \tag{4.2.19}$$

The coefficients V, W_l, and W_t can be expressed in terms of the angle of incidence of the sound wave θ and the refraction angles of the longitudinal and transverse waves θ_l and θ_t in the elastic half-space. Similarly to (4.1.12) we have

$$\xi = k \sin \theta = k_{l1} \sin \theta_l = k_{t1} \sin \theta_t \quad , \tag{4.2.20}$$

and

$$\alpha = k \cos \theta \quad , \qquad \alpha_1 = k_{l1} \cos \theta_l \quad , \qquad \beta_1 = k_{t1} \cos \theta_t \quad ,$$
$$\gamma_1 = -k_{t1} \cos 2\theta_t/2 \sin \theta_t \quad . \tag{4.2.21}$$

Snell's law follows from (4.2.20), cf. (2.2.6),

$$\frac{\sin \theta}{c} = \frac{\sin \theta_l}{c_{l1}} = \frac{\sin \theta_t}{c_{t1}} \quad . \tag{4.2.22}$$

It is also reasonable to introduce the following notation for the impedances:

$$Z = \frac{\varrho c}{\cos \theta} \quad , \qquad Z_l = \frac{\varrho_1 c_{l1}}{\cos \theta_l} \quad , \qquad Z_t = \frac{\varrho_1 c_{t1}}{\cos \theta_t} \quad . \tag{4.2.23}$$

Then (4.2.18, 19) can be written as

$$V = \frac{Z_l \cos^2 2\theta_t + Z_t \sin^2 2\theta_t - Z}{Z_l \cos^2 2\theta_t + Z_t \sin^2 2\theta_t + Z} \quad , \tag{4.2.24}$$

$$W_l = \frac{2\varrho\varrho_1^{-1} Z_l \cos 2\theta_t}{Z_l \cos^2 2\theta_t + Z_t \sin^2 2\theta_t + Z} \quad , \tag{4.2.25}$$

$$W_t = \frac{-2\varrho\varrho_1^{-1} Z_t \sin 2\theta_t}{Z_l \cos^2 2\theta_t + Z_t \sin^2 2\theta_t + Z} \quad . \tag{4.2.26}$$

Let us discuss the obtained formulas. For normal incidence ($\theta = \theta_l = \theta_t = 0$) we have

$$V = \frac{Z_l - Z}{Z_l + Z} \quad , \quad W_l = \frac{2\varrho Z_l}{\varrho_1(Z_l + Z)} \quad , \quad W_t = 0 \quad . \tag{4.2.27}$$

As it was to be expected, in this case the shear waves are not excited. The solid behaves as a liquid with density ϱ_1 and sound velocity c_{l1}. On the other hand, for $\theta = \arcsin(c/\sqrt{2}c_{t1})$ when according to (4.2.22) $\theta_t = \pi/4$, we obtain

$$V = \frac{Z_t - Z}{Z_t + Z} \quad , \quad W_l = 0 \quad , \quad W_t = -\frac{2\varrho Z_t}{\varrho_1(Z_t + Z)} \quad , \tag{4.2.28}$$

i.e., only transverse waves are excited.

In most cases of interest, the sound velocity in the fluid, c, is less than that of the longitudinal waves in the solid, c_{l1}. It can also be less than the velocity of transverse waves, c_{t1}. Let us first consider the case where $c_{t1} < c < c_{l1}$. From (4.2.22) we see that for $\sin \theta > c/c_{l1}$ the angle θ_l is complex while the angle θ_t is real for all θ. Thus, the longitudinal wave in the solid will be an inhomogeneous wave which "glides" along the boundary and decays while departing from it and the transverse wave will be the usual plane wave. Since $\sin \theta_l > 1$, $\cos \theta_l$ and Z_l are purely imaginary quantities. From the requirement of finiteness of the field as $z \to -\infty$ it follows that $\cos \theta_l$ should be a positive imaginary quantity and, consequently, Z_l should be a negative imaginary one.

In this case the reflection coefficient (4.2.24) becomes complex:

$$V = \frac{Z_t \sin^2 2\theta_t - Z - i|Z_l| \cos^2 2\theta_t}{Z_t \sin^2 2\theta_t + Z - i|Z_l| \cos^2 2\theta_t} \quad , \tag{4.2.29}$$

and the square of its modulus is

$$|V|^2 = \frac{(Z_t \sin^2 2\theta_t - Z)^2 + |Z_l|^2 \cos^4 2\theta_t}{(Z_t \sin^2 2\theta_t + Z)^2 + |Z_l|^2 \cos^4 2\theta_t} \quad . \tag{4.2.30}$$

For $\theta_t \neq \frac{\pi}{2}$ the modulus of the reflection coefficient is smaller than unity, which one could expect since part of the energy is carried away from the boundary by the transverse wave. The angle of incidence of the sound wave $\theta = \arcsin(c/c_{l1})$ is critical, in which case $\theta_l = \frac{\pi}{2}$, $Z_l = \infty$, $V = 1$, $W_t = 0$, and $W_l = 2\varrho/\varrho_1(1 - 2c_{t1}^2/c_l^2)$.

Now consider the case $c < c_{t1} < c_{l1}$. As can be seen from (4.2.22), θ_l and θ_t are real angles for $0 \leq \sin \theta < c/c_{l1}$, i.e., we have the case of ordinary reflection at the boundary with the reflection coefficient being real and less than unity. For the critical angle of incidence $\theta = \arcsin(c/c_{l1})$ with respect to longitudinal waves we have $\theta_l = \frac{\pi}{2}$ and, in accord with (4.2.24), $V = 1$. Then $W_t = 0$, and $W_l \neq 0$ that is, only longitudinal waves are excited in the elastic half-space.

At $c/c_{l1} < \sin\theta < c/c_{t1}$ the angle θ_t will be real whereas θ_l will be complex, i.e., we obtain the case considered above, see (4.2.29). The angle $\theta = \arcsin(c/c_{t1})$ is critical with respect to transverse waves. Here $\theta_t = \frac{\pi}{2}$, $Z_t = \infty$, and $V = (Z_l-Z)/(Z_l+Z)$ with $|V| = 1$. For $\theta > \arcsin(c/c_{t1})$ both angles θ_l and θ_t are complex, which means that both longitudinal and transverse waves in the solid are inhomogeneous and propagate along the boundary. The impedances Z_l and Z_t take purely imaginary values. It is seen from (4.2.22) that $\sin^2 2\theta_t$ and $\cos^2 2\theta_t$ are real for any θ. Therefore, the reflection coefficient becomes

$$V = \frac{|Z_l|\cos^2 2\theta_t + |Z_t|\sin^2 2\theta_t - iZ}{|Z_l|\cos^2 2\theta_t + |Z_t|\sin^2 2\theta_t + iZ} \quad , \tag{4.2.31}$$

and $|V| = 1$, i.e., total reflection takes place.

We note, following [4.7], one more circumstance. Equation (4.2.24) for the reflection coefficient can be written in the form

$$V = \frac{Z_b - Z}{Z_b + Z} \quad , \tag{4.2.32}$$

where

$$Z_b \equiv Z_l \cos^2 2\theta_t + Z_t \sin^2 2\theta_t \tag{4.2.33}$$

is the total impedance of the boundary due to the existence of longitudinal and transverse waves in the lower medium. For *real* θ_l and θ_t we have from (4.2.22, 23)

$$\frac{Z_l}{Z_t} = \frac{c_{l1}\cos\theta_t}{c_{t1}\cos\theta_l} \geq \frac{c_{l1}}{c_{t1}} > \sqrt{2} \quad , \tag{4.2.34}$$

and therefore $Z_b/Z_l \leq 1$. Here the equality takes place only at $\theta_t = 0$ and $\theta_t = \frac{\pi}{2}$. Thus, at $0 < \theta_l \leq \frac{\pi}{2}$ the total impedance of the solid boundary is less than the impedance of a liquid with the same ϱ_1 and c_{l1}, i.e., when the excitation of transverse waves in the process of reflection is accounted for, the boundary "softens". Reference [4.7] also shows that with the change of the angle of incidence θ the impedance Z_b changes less than the impedance of the equivalent liquid Z_l, so that in certain cases the reflection from the solid can be approximately considered as the reflection from a medium with an impedance, independent of the angle.

In the above discussion we have written, in terms of potentials, (4.2.25, 26) for the coefficients of excitation by the sound wave of the longitudinal and transverse waves in the lower-space. Now we give similar values for the other characteristics of the elastic waves. According to (4.0.2), the amplitudes of the displacement u_l and u_t in the plane longitudinal and transverse waves are related to the amplitudes of the potentials $\tilde{\varphi}_1$ and $\tilde{\psi}_1$ by $u_l = k_{l1}\tilde{\varphi}_1$ and $u_t = k_{t1}\tilde{\psi}_1$. By using (4.2.17), we find the wave amplitudes in the lower medium

$$u_l = \left(\frac{p_0}{\omega\varrho c_{l1}}\right)W_l \quad , \quad u_t = \left(\frac{p_0}{\omega\varrho c_{t1}}\right)W_t \quad , \tag{4.2.35}$$

where p_0 is the pressure amplitude in the indicent sound wave.

Also of interest are expressions for the normal component of the energy flux density vector of a reflected wave in a fluid I_r and that of longitudinal and transverse waves in a solid I_l and I_t:

$$\frac{I_r}{I} = |V|^2 \quad , \quad \frac{I_l}{I} = \frac{\varrho_1 \tan \theta}{\varrho \tan \theta_l}|W_l|^2 \quad , \quad \frac{I_t}{I} = \frac{\varrho_1 \tan \theta}{\varrho \tan \theta_t}|W_t|^2 \quad . \qquad (4.2.36)$$

Here I is the component of the energy flux density vector in the incident sound wave normal to the boundary. The last two expressions are valid only for the real values of θ_l and θ_t, respectively. When the longitudinal (transverse) wave in the lower medium becomes inhomogeneous, I_l (I_t) vanishes. The reader can, by using (4.2.22–26), prove that the energy conservation law is satisfied: $I = I_r + I_l + I_t$ in all the cases. The set of curves I in Fig. 4.5a shows, according to [4.8], the dependence of the modulus of the reflection coefficient on the angle of incidence for $\varrho_1/\varrho = 3$, $c_{l1}/c = 3$ for three different cases: $c_{l1}/c_{t1} = 1.6$ (curve 1); 1.7 (curve 2); 1.8 (curve 3). Curves II (Fig. 4.5a) and III (Fig. 4.5b) represent, respectively, $(I_l/I)^{1/2}$ and $(I_t/I)^{1/2}$ for the same cases.

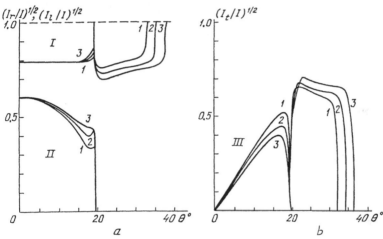

Fig. 4.5. Dependence of the moduli of reflection and transmission coefficients for a plane wave incident from a fluid onto an elastic half-space as a function of incident angle θ, at various speed ratios. *Curves* I: $(I_r/I)^{1/2}$, II: $(I_l/I)^{1/2}$, III: $(I_t/I)^{1/2}$. After [4.8]

4.2.3 Elastic Wave Reflection from Fluid Half-Space

In concluding Sect. 4.2 we also give the expressions for the other components of the scattering matrix at the interface of fluid and solid half-spaces but without a detailed discussion. Since the transverse waves are absent in a fluid, $\psi_1 \equiv \psi_2 \equiv 0$ and, for the case considered, the elements of the second column and the second row in matrix $[S]$ defined in (4.2.2) do not need to be determined. The six coefficients $\tilde{W}_{ll}$, $\tilde{V}_{ll}$, $\tilde{V}_{lt}$, and $\tilde{W}_{tl}$, $\tilde{V}_{tl}$, $\tilde{V}_{tt}$ which characterize the process of reflection of longitudinal and transverse waves incident from the solid on the boundary with the fluid must be found, however. We can obtain the unknown coefficients from (4.2.9–12) by interchanging the parameters of the medium $c_l \leftrightarrow c_{l1}$, $c_t \leftrightarrow c_{t1}$, $\varrho \leftrightarrow \varrho_1$ in the formulas. Proceeding to the limit $\mu \to 0$ is accomplished in the same way as in deriving (4.2.18). As a result we find: for the reflection coefficient of a longitudinal wave

$$\tilde{V}_{ll} = \frac{Z + Z_t \sin^2 2\theta_t - Z_l \cos^2 2\theta_t}{Z + Z_t \sin^2 2\theta_t + Z_l \cos^2 2\theta_t} \quad ; \tag{4.2.37}$$

for the coefficient of transformation of a longitudinal to transverse wave and to a sound wave in a fluid

$$\tilde{V}_{lt} = -\frac{2(1 - \tilde{V}_{ll}) \cot \theta_l \sin^2 \theta_t}{\cos 2\theta_t} \quad , \tag{4.2.38}$$

$$\tilde{W}_{ll} = \frac{(1 - \tilde{V}_{ll}) \tan \theta \cot \theta_l}{\cos 2\theta_t} \quad ; \tag{4.2.39}$$

for the reflection coefficient of a transverse wave

$$\tilde{V}_{tt} = -\frac{Z + Z_l \cos^2 2\theta_t - Z_t \sin^2 2\theta_t}{Z + Z_l \cos^2 2\theta_t + Z_t \sin^2 2\theta_t} \quad ; \tag{4.2.40}$$

for the coefficient of transformation of a transverse wave into a longitudinal wave and into a sound wave in a fluid

$$\tilde{V}_{tl} = \frac{(1 + \tilde{V}_{tt}) \tan \theta_l \cos 2\theta_t}{2 \sin^2 \theta_t} \quad , \tag{4.2.41}$$

$$\tilde{W}_{tl} = \frac{(1 + \tilde{V}_{tt}) \tan \theta}{2 \sin^2 \theta_t} \quad . \tag{4.2.42}$$

The reader can prove that as $Z \to 0$ (4.2.37, 38, 40 and 41) coincide with those obtained in Sect. 4.1 for the reflection from the free boundary of a solid.

Energy relations for the reflection of waves incident from a solid on the interface with a fluid are illustrated in [4.8] by *Ergin*.

4.3 Reflection from a System of Solid Layers

4.3.1 Matrix Propagator

Let us again consider, as in Fig. 2.5, a system of $n - 1$ layers bounded from below by a solid and from above by a fluid. A plane acoustic wave with unit amplitude and angle of incidence θ is incident from the fluid on a set of layers. The sound velocity in the fluid equals c. It is required to determine the amplitude of the reflected wave and the amplitude of the two waves (longitudinal and transverse) in the lower half-space. In each of the layers a pair of longitudinal waves (propagating upwards and downwards symmetrically with respect to the horizontal plane) and a pair of similar transverse waves will appear. All the waves will contain one common factor $\exp(\mathrm{i}\xi x - \mathrm{i}\omega t)$, where

$$\xi = k \sin \theta = k_l(z) \sin \theta_l(z) = k_t(z) \sin \theta_t(z) \quad , \tag{4.3.1}$$

which we shall omit for brevity.

At each boundary separating the solid layers, the four boundary conditions of (1.3.26) should be satisfied. At the boundary of the upper layer continuity of the displacement component u_1 is not required.

A direct method of solution of the problem would be to construct $4n - 1$ algebraic equations with the help of boundary conditions for the amplitudes of $4n-1$ waves, including the reflected wave, and then solving the set of equations by matrix inversion. However, there is another, more reasonable method, which is related to the method applied in Sect. 2.5 and based on the use of recurrence formulas which relate the amplitudes of waves in adjacent layers. This method has been suggested by *Thomson* [4.9] and further developed by *Haskell* [4.10], and is a particular case of the method of the matrix propagator [4.11]. At present the matrix methods are widely used, especially in seismology, in analytical and numerical investigations of the propagation of elastic waves in layered media. Extensive bibliographies of the original works are available in the reviews [4.12, 13] and the monograph [Ref. 4.14, Chaps. 5, 7]. For a detailed discussion and comparison of different variants of the matrix method for layered media we refer the reader to the monographs by *Molotkov* [4.15] and by *Kennett* [4.49]. In the context of numerical modeling using a computer, advantages and disadvantages of the various approaches were analyzed in [4.50].

Let the coordinate of the upper boundary of the layer j, $j = 1, 2, \ldots, n$, be denoted by z_j. In each layer the potentials of elastic waves have the following form

$$\varphi^{(j)} = \varphi_1^{(j)} \exp\left[i\alpha(z - z_{j-1})\right] + \varphi_2^{(j)} \exp\left[-i\alpha(z - z_{j-1})\right] \quad ;$$

$$z_{j-1} \le z \le z_j \quad ;$$

$$\psi^{(j)} = \psi_1^{(j)} \exp\left[i\beta(z - z_{j-1})\right] + \psi_2^{(j)} \exp\left[-i\beta(z - z_{j-1})\right] \quad ;$$

$$\mathrm{Im}\,\{\alpha\} \ge 0 \quad , \quad \mathrm{Im}\,\{\beta\} \ge 0 \quad , \tag{4.3.2}$$

with α, β, $\varphi_{1,2}^{(j)}$ and $\psi_{1,2}^{(j)}$ constant within the layer. Assuming that the field in the jth layer is known, we need to find it in the $(j + 1)$th layer. For this purpose it is convenient to characterize the field by the so-called *displacement-stress vector* f instead of the vector $\varphi = (\varphi_1, \varphi_2, \psi_1, \psi_2)^T$ since components of the latter, as we have seen in the previous section, transform in a complicated manner when crossing a boundary. The vector f is defined by

$$f(z) = (u_1, u_3, \sigma_{33}, \sigma_{31})^T \quad . \tag{4.3.3}$$

Here, superscript T denotes the operation of transposition. (By definition, for the matrix $[B] = \{b_{ik}\}$ composed of elements b_{ik}, $[B]^T = \{b_{ki}\}$). Transposition of the matrix converts a matrix row into a column, i.e., into a vector. On the boundaries the vector $f(z)$ is continuous (contrary to the vector of potentials φ), by virtue of the boundary conditions of (1.3.26). The connection between the vectors $\varphi(z)$ and $f(z)$ is given by (4.0.2, 3). In matrix notation this relationship is

$$f(z) = [B(z, z_{j-1})]\varphi \quad , \tag{4.3.4}$$

where

$$[B(z_j, z_{j-1})] = [L][a, a^{-1}, b, b^{-1}] \quad ,$$

$$[L] = \begin{pmatrix} i\xi & i\xi & -i\beta & i\beta \\ i\alpha & -i\alpha & i\xi & i\xi \\ 2\mu\xi\gamma & 2\mu\xi\gamma & -2\mu\xi\beta & 2\mu\xi\beta \\ -2\mu\xi\alpha & 2\mu\xi\alpha & -2\mu\xi\gamma & -2\mu\xi\gamma \end{pmatrix} \quad . \tag{4.3.5}$$

We denote, for brevity,

$$a = \exp\left[i\alpha(z_j - z_{j-1})\right] \quad , \quad b = \exp\left[i\beta(z_j - z_{j-1})\right] \quad . \tag{4.3.6}$$

The symbol $[a_1, a_2, a_3, a_4]$ is defined as the diagonal matrix with the elements $c_{ij} = a_i \delta_{ij}$.

Using the fact that φ is constant within a layer, from (4.3.4) one can easily find the connection between the displacement-stress vectors on neighboring boundaries

$$f(z_j) = [A^{(j)}] f(z_{j-1}) \quad ,$$

$$
\begin{aligned}
[A^{(j)}] &= [B(z_j, z_{j-1})][B(z_{j-1}, z_{j-1})]^{-1} \\
&= [L][a, a^{-1}, b, b^{-1}][L]^{-1} \quad .
\end{aligned} \tag{4.3.7}
$$

The elements of the matrix are obtained in a usual way from (4.3.5). After simple, though involved, operations we get

$$a_{11} = a_{44} = 2\sin^2\theta_t \cos P + \cos 2\theta_t \cos Q \quad , \qquad P \equiv \alpha(z_j - z_{j-1}) \quad ,$$
$$\qquad\qquad\qquad\qquad\qquad\qquad\qquad\qquad\qquad\qquad Q \equiv \beta(z_j - z_{j-1}) \quad ,$$

$$a_{12} = a_{34} = i(\tan\theta_l \cos 2\theta_t \sin P - \sin 2\theta_t \sin Q) \quad ,$$

$$a_{13} = a_{24} = i\sin\theta_t(\cos Q - \cos P)/\omega\varrho c_t \quad ,$$

$$a_{14} = (\tan\theta_l \sin\theta_t \sin P + \cos\theta_t \sin Q)/\omega\varrho c_t \quad ,$$

$$a_{21} = a_{43} = i(2\cot\theta_l \sin^2\theta_t \sin P - \tan\theta_t \cos 2\theta_t \sin Q) \quad ,$$

$$a_{22} = a_{33} = \cos 2\theta_t \cos P + 2\sin^2\theta_t \cos Q \quad ,$$

$$a_{23} = (\cot\theta_l \sin\theta_t \sin P + \sin\theta_t \tan\theta_t \sin Q)/\omega\varrho c_t \quad ,$$

$$a_{31} = a_{42} = -2i\omega\varrho c_t \sin\theta_t \cos 2\theta_t(\cos Q - \cos P) \quad ,$$

$$a_{32} = -\omega\varrho c_t(\tan\theta_l \cos^2 2\theta_t \sin P + \sin^2 2\theta_t \tan\theta_t \sin Q)/\sin\theta_t \quad ,$$

$$a_{41} = -\omega\varrho c_t[4\cot\theta_l \sin^3\theta_t \sin P + (\cos^2 2\theta_t/\cos\theta_t)\sin Q] \quad . \tag{4.3.8}$$

The values of the parameters ϱ, c_t, θ_l, θ_t, α, and β, corresponding to the jth layer should be substituted into (4.3.8). For $c_l/c > 1$ and $\sin\theta > c/c_l$ we shall have $\sin\theta_l > 1$ according to (4.3.1), i.e., θ_l is complex. In this case it is convenient to set $\theta_l = \frac{\pi}{2} + i\zeta$, $\sin\theta_l = \cosh\zeta$, and $\cos\theta_l = -i\sinh\zeta$ and apply the quantity ζ. For $c_t/c > 1$, this also pertains to the angle θ_t. A more compact representation of the layer matrix $[A^{(j)}]$ as a *matrix exponent* will be derived in Sect. 7.2.2.

Using (4.3.7) successively, we can connect the values of the displacement-stress vector on the boundaries of media 1 and 2 and n and $n + 1$:

$$f(z_n) = [A] f(z_1) \quad , \quad [A] = [A^{(n-1)}] \cdot \ldots \cdot [A^{(3)}] \cdot [A^{(2)}] \quad . \tag{4.3.9a}$$

Within any homogeneous layer one can artificially introduce an additional boundary at an arbitrary level. Then (4.3.9a) becomes valid for *any* z_n and z_1:

$$f(z) = [A(z, \tilde{z})] f(\tilde{z}) \quad . \tag{4.3.9b}$$

The matrix $[A(z, \tilde{z})]$ is called a *matrix propagator*. It has several remarkable properties and allows "propagation" of the field from the horizon $\tilde{z}$ to z. From (4.3.9b) it follows directly that

$$[A(z, \tilde{z})] = [A(\tilde{z}, z)]^{-1} \quad , \tag{4.3.10}$$

$$[A(z, \tilde{z})] = [A(z, z_0)][A(z_0, \tilde{z})] \quad , \quad z \leq z_0 \leq \tilde{z} \quad . \tag{4.3.11}$$

In addition

$$\det [A] = 1 \quad . \tag{4.3.12}$$

Indeed, in accord with (4.3.7) $\det [A^{(j)}] = \det [a, a^{-1}, b, b^{-1}] = 1$, and the identity in (4.3.12) follows from the definition in (4.3.9a).

4.3.2 Reflection Coefficient of the Sound Wave

Now we proceed directly to the problem of the determination of the reflection coefficient. We set the origin of the coordinates (as in Fig. 2.5) on the boundary of media 1 and 2. Then the total field of the incident and reflected waves in the fluid half-space can be written as

$$\varphi^{(n+1)} = \exp [- i\alpha(z - z_n)] + V \exp [i\alpha(z - z_n)] \quad , \quad z \geq z_n \quad , \tag{4.3.13}$$

where $\alpha = \omega c^{-1} \cos \theta$ and V is the reflection coefficient. In the lower elastic half-space there will only be waves departing from the boundary, the potentials of which can be written as

$$\varphi^{(1)} = W_l \exp(-i\alpha_1 z) \quad , \quad \alpha_1 = \omega c_{l1}^{-1} \cos \theta_l \quad ,$$

$$\psi^{(1)} = W_t \exp(-i\beta_1 z) \quad , \quad \beta_1 = \omega c_{t1}^{-1} \cos \theta_t \quad . \tag{4.3.14}$$

The coefficients V, W_l, and W_t can be found from the set of equations

$$[B(z_n, z_n)](V, 1, 0, ,0)^T = [A][B(0, 0)](0, W_l, 0, W_t)^T \quad , \tag{4.3.15}$$

which are obtained by expressing $f(z_n)$ through the potentials with the use of (4.3.4 and 3.9a). Using the explicit form of (4.3.5) of the matrix $[B]$ and the equality $\mu\gamma = -\omega^2 \varrho/2\xi$ which is valid for the fluid, we obtain from (4.3.15)

$$\begin{pmatrix} i\xi(1+V) \\ i\alpha(V-1) \\ -\omega^2 \varrho(1+V) \\ 0 \end{pmatrix} = [A] \cdot \begin{pmatrix} i\xi W_l + i\beta_1 W_t \\ -i\alpha_1 W_l + i\xi W_t \\ 2\mu_1 \xi(\gamma_1 W_l + \beta_1 W_t) \\ 2\mu_1 \xi(\alpha_1 W_l - \gamma_1 W_t) \end{pmatrix} \quad . \tag{4.3.16}$$

The index 1 refers to the quantities describing the elastic half-space.

In deriving formulas for $W_{l,t}$ and V we could not use an equality of the first elements of the vectors in the left-hand and right-hand sides of (4.3.16), as this equality corresponds to an x-component continuity of displacement and may be violated at a fluid-solid interface. The ratio $q \equiv W_t/W_l$ can be found by equating the values of the fourth elements in the vectors of the left-hand and right-hand sides in (4.3.16):

$$q = \frac{A_{41} - \cot \theta_l \cdot A_{42} + i\mu_1 k_{t1} \cos 2\theta_t \sin^{-1} \theta_t \cdot A_{43} - 2i\mu_1 k_{t1} \sin \theta_t \cot \theta_l \cdot A_{44}}{-\cot \theta_t \cdot A_{41} - A_{42} + 2i\mu_1 k_{t1} \cos \theta_t \cdot A_{43} + i\mu_1 k_{t1} \cos 2\theta_t \sin^{-1} \theta_t \cdot A_{44}}$$

$$\tag{4.3.17}$$

Then we find the coefficients W_l and W_t by using the second and the third components of the vector equality given in (4.3.16):

$$
\begin{aligned}
W_l = & -2\varrho\omega^2 \cot\theta\{(\varrho\omega^2 A_{21} + i\alpha A_{31})(1 + q\cot\theta_t) \\
& -(\varrho\omega^2 A_{22} + i\alpha A_{32})(\cot\theta_l - q) + \mu_1 k_{t1}\ \sin^{-1}\theta_t \\
& \times\ \cos 2\theta_t [(i\varrho\omega^2 A_{23} - \alpha A_{33})(1 - q\tan 2\theta_t) \\
& -(i\varrho\omega^2 A_{24} - \alpha A_{34})(q + 2\ \sin^2\ \theta_t\ \cot\theta_l/\cos 2\theta_t)]\}^{-1} \ .
\end{aligned}
\tag{4.3.18}
$$

Now it is easy to find the reflection coefficient. We shall express it in the terms of the input impedance of the set of solid layers Z_{in}:

$$
V = \frac{Z_{in} - Z}{Z_{in} + Z} \ ,
\tag{4.3.19}
$$

where

$$
Z_{in} \equiv \left(\frac{i\sigma_{33}}{\omega u_3}\right)_{z=z_n} \ , \quad Z = \frac{\varrho c}{\cos\theta}
\tag{4.3.20}
$$

is the impedance of the fluid half-space. For the input impedance of the set of layers we find from (4.3.16–18)

$$
Z_{in} = -\frac{E_3}{i\omega E_2} \ ,
$$

$$
\begin{aligned}
E_j \equiv\ & \alpha_1 M_{j2} - i\omega^2\varrho_1(\cos 2\theta_t \cdot M_{j3} + 2\ \sin^2\ \theta_t \cdot \cot\theta_l \cdot M_{j4}) - q[\xi M_{j2} \\
& + i\omega^2\varrho_1(\sin 2\theta_t \cdot M_{j3} - \cos 2\theta_t \cdot M_{j4})] \ , \quad j = 2, 3 \ ,
\end{aligned}
\tag{4.3.21}
$$

where

$$
M_{jk} \equiv A_{jk} - A_{j1}A_{4k}/A_{41} \ , \quad j = 2, 3 \ ; \quad k = 2, 3, 4 \ .
\tag{4.3.22}
$$

Equations (4.3.17–21) completely solve the problem of sound-wave reflection from a set of elastic layers (of arbitrary number).

4.3.3 Scattering Matrix for the Elastic Waves

The problem of reflection of a plane wave incident from an *elastic* half-space on a set of layers can be solved quite similar to that considered in Sect. 4.3.2 (see [4.16]).

In this case it is necessary to determine the scattering matrix which relates the amplitudes of waves incident from infinity upon the boundaries of the set of layers, $z = z_n$ and $z = z_1$, to the amplitudes of waves going away to infinity:

$$
(\varphi_1^{(n+1)}, \psi_1^{(n+1)}, \varphi_2^{(1)}, \psi_2^{(1)})^T = [S](\varphi_2^{(n+1)}, \psi_2^{(n+1)}, \varphi_1^{(1)}, \psi_1^{(1)})^T \ .
\tag{4.3.23}
$$

The matrix relation between these amplitudes can be obtained by expressing the potential vectors in terms of the displacement-stress vectors with the use of (4.3.4):

$$
(\varphi_1^{(n+1)}, \varphi_2^{(n+1)}, \psi_1^{(n+1)}, \psi_2^{(n+1)})^T = [C](\varphi_1^{(1)}, \varphi_2^{(1)}, \psi_1^{(1)}, \psi_2^{(1)})^T \ ,
\tag{4.3.24}
$$

where

$$[C] = [L(z_n)]^{-1}[A(z_n, z_1)][L(z_1)] \tag{4.3.25}$$

is the known matrix. Let us regroup the terms into four linear equations, which follow from (4.3.24), in such a way that the left-hand side would contain the quantities $\varphi_1^{(n+1)}$, $\psi_1^{(n+1)}$, $\varphi_2^{(1)}$, and $\psi_2^{(1)}$ and the right-hand side the quantities $\varphi_2^{(n+1)}$, $\psi_2^{(n+1)}$, $\varphi_1^{(1)}$, and $\psi_1^{(1)}$, cf. (4.2.7),

$$[N_1](\varphi_1^{(n+1)}, \psi_1^{(n+1)}, \varphi_2^{(1)}, \psi_2^{(1)})^T$$
$$= [N_2](\varphi_2^{(n+1)}, \psi_2^{(n+1)}, \varphi_1^{(1)}, \psi_1^{(1)})^T \quad, \tag{4.3.26}$$

where

$$[N_1] = \begin{pmatrix} -1 & 0 & c_{12} & c_{14} \\ 0 & 0 & c_{22} & c_{24} \\ 0 & -1 & c_{32} & c_{34} \\ 0 & 0 & c_{42} & c_{44} \end{pmatrix} \quad,$$

$$[N_2] = \begin{pmatrix} 0 & 0 & c_{11} & c_{13} \\ -1 & 0 & c_{21} & c_{23} \\ 0 & 0 & c_{31} & c_{33} \\ 0 & -1 & c_{41} & c_{43} \end{pmatrix} \quad. \tag{4.3.27}$$

We find the scattering matrix $[S]$ from (4.3.23, 26):

$$[S] = [N_1]^{-1}[N_2] \quad. \tag{4.3.28}$$

4.3.4 Some Special Cases

In the particular case considered in [4.9] where the upper and the lower half-spaces are fluids, we have $q \equiv W_t/W_l = 0$, $\theta_t = 0$, and, using the notation $Z_1 \equiv \omega\varrho_1/\alpha_1 = \varrho_1 c_{l1}/\cos\theta_l$, we get from (4.3.21) the comparatively simple formula:

$$Z_{\text{in}} = \frac{\mathrm{i}(M_{32} - \mathrm{i}\omega Z_1 M_{33})}{\omega(M_{22} - \mathrm{i}\omega Z_1 M_{23})} \quad. \tag{4.3.29}$$

In this case the expressions for the reflection and transmission coefficients will be

$$V = \frac{M_{32} - \mathrm{i}\omega Z_1 M_{33} + (M_{22} - \mathrm{i}\omega Z_1 M_{23})\mathrm{i}\omega Z}{M_{32} - \mathrm{i}\omega Z_1 M_{33} - (M_{22} - \mathrm{i}\omega Z_1 M_{23})\mathrm{i}\omega Z} \quad, \tag{4.3.30}$$

$$W = \frac{-2\mathrm{i}\omega Z_1 \varrho_1^{-1}}{M_{32} - \mathrm{i}\omega Z_1 M_{33} - (M_{22} - \mathrm{i}\omega Z_1 M_{23})\mathrm{i}\omega Z} \quad. \tag{4.3.31}$$

The important case of a plate (that is a homogeneous elastic layer) immersed in a fluid is considered in detail in the monographs ([Ref. 4.3, Chap. 5], [Ref. 4.17, Sects. 9–11], [Ref. 4.18, Chap. 4]).

Based on the formulas obtained above and proceeding to the limit $\mu_j \to 0$ for corresponding j, we can analyze reflection from a discretely layered medium in the case when some layers are fluid. The peculiarity of converting the problem to the case of a fluid layer is that not all the matrix components a_{ij} in (4.3.8) tend to a certain value as $\mu_j \to 0$. The elements $a_{11} = a_{44}$, $a_{13} = a_{24}$, and a_{14} retain the dependence on Q, and the limits of $\sin Q$ and $\cos Q$ as $k_t \to \infty$ do not exist. However, in the final expressions for the reflection and transmission coefficients (4.3.17–21), the terms which contain Q are cancelled and the conversion proceeds without difficulties. Great caution is required, though, in numerical implementations [4.51–54].

When considering the particular case of a liquid layer of thickness d between two liquid half-spaces, Eq. (4.3.29) must coincide with (2.4.7). In fact, taking into account that for a liquid layer $c_t = 0$ and $\theta_t = 0$, in this case we obtain

$$M_{32} = A_{32} = -\omega \varrho_2 c_{l2} \sin P / \cos \theta_l \quad , \quad P = k_{l2} d \cos \theta_l \quad ,$$
$$M_{33} = A_{33} = M_{22} = A_{22} = \cos P \quad ,$$
$$M_{23} = A_{23} = \cos \theta_l \cdot \sin P / \omega \varrho_2 c_{l2} \quad . \tag{4.3.32}$$

In the notation of Chap. 2 we have $\varrho_2 c_{l2} / \cos \theta_l = Z_2$ and $P = \varphi$. Taking this into account, substitution of (4.3.32) into (4.3.29) leads immediately to (2.4.7).

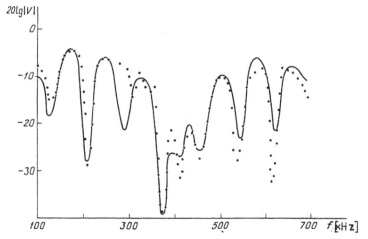

Fig. 4.6. Theoretical (*solid curve*) and experimental (*dotted curve*) results for dependences of the modulus of the reflection coefficient of a sound wave on frequency. The sound wave is normally incident on a three-layered construction, consisting of two plastic plates separated by a water layer. From [4.19]

Now we present some results of calculating $|V|$ in concrete examples and compare with experiments. Figure 4.6, taken from [4.19], shows the frequency dependence of the modulus of the reflection coefficients for a sound wave which is normally incident on a three-layered structure. This structure consists of two plastic layers with parameters $c_l = 2100 \, \text{m/s}$, $\varrho = 1.08 \, \text{g/cm}^3$, $d_2 = d_4 = 0.254 \, \text{cm}$. The space between these layers, with $d_3 = 0.706 \, \text{cm}$, is filled with water ($c = 1500 \, \text{m/s}$, $\varrho = 1 \, \text{g/cm}^3$). In Fig. 4.6 the ordinate is $20 \cdot \log |V|$ and the abscissa is the wave frequency in kHz. The solid line shows the value of the reflection coefficient calculated by a formula analogous to (4.3.30). The points correspond to the experimental curve. The interference of the waves reflected at different boundaries is quite obvious. These effects are responsible for the strongly pronounced minima repeating each $\Delta f = 84 \, \text{kHz}$ above $f = 126 \, \text{kHz}$. In general, the theoretical and experimental

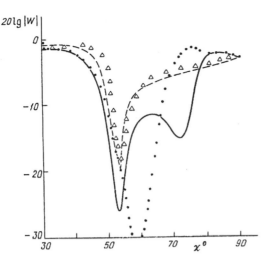

Fig. 4.7. Dependence of the transmission coefficient on the grazing angle of a sound wave for a plexiglas plate placed in water. *Triangles*: experimental data. *Points, dashed curve, solid curve*: theoretical results using different shear wave velocities. From [4.20]

curves are in agreement. The deviations of the experimental data from the theoretical curve may be partly because the boundaries are not completely plane, the incident wave not precisely plane, etc. The main effect, however, is probably errors in the specification of the media parameters. The reflection and transmission coefficient are fairly sensitive to the values of these parameters. This is illustrated in Fig. 4.7, taken from [4.20], where the dependence of the transmission coefficient W on the *grazing angle* $\chi \equiv \frac{\pi}{2} - \theta$ for a wave of frequency $f = 193\,\text{kHz}$ for a plexiglas plate ($c_l = 2650\,\text{m/s}$, $d = 0.325\,\text{cm}$, $\varrho = 1.19\,\text{g/cm}^3$) placed into water is shown. (Here θ is the angle of incidence for a plane wave.) The triangles indicate the experimental data. The other three curves are theortical based on a relation analogous to (4.3.31) for the three values of shear wave velocities: $c_t = 1100\,\text{m/s}$ (points), $c_t = 1250\,\text{m/s}$ (dashed curve), $c_t = 1160\,\text{m/s}$ (solid curve). One can see that a change of c_t by about 7 % significantly changes the behavior of the curve, but that it retains its qualitative form. At some χ, with changing c_t the values of the modulus of transmission coefficient change drastically (by 10 dB and more). This makes it possible, by comparing theoretical and experimental dependences of the reflection and transmission coefficients, to determine the material parameters of plates, among them those which are difficult to measure directly [4.20, 55] (e.g., absorption coefficient for longitudinal and transverse waves, velocity of shear waves).

4.4 Surface and "Leaky" Waves

4.4.1 A Simple Example of Surface Waves

Waves whose amplitude decays rapidly with increasing distance from some surface (a horizontal plane in a layered medium) are called surface waves. Such waves are of great importance in seismology because their amplitudes decrease (due to geometrical spreading) with increasing horizontal separation of a source and a receiver much more slowly than the amplitudes of the usual "volume" wave. Surface waves are also important in acousto-electronics, where the possibility of influencing such a wave in a solid along its entire path of propagation and the slowness of surface waves relative to volume waves are used in new kinds of electronic components. In addition, surface waves are widely used in techniques of undestructive control of surface and surface layers of specimens. Technical applications of surface waves are described in detail in [4.21–25].

In this section we consider surface waves with a planar front. The simplest surface acoustical wave can exist in a homogeneous liquid half-space $z > 0$ if its boundary $z = 0$ has certain properties. The acoustic pressure in this wave is written as

$$p = \exp(-\alpha z + i\xi x - i\omega t) \quad , \quad \alpha > 0 \quad , \quad \xi^2 - \alpha^2 = k^2 \quad . \tag{4.4.1}$$

It decreases exponentially with distance away from the boundary, therefore the surface wave is inhomogeneous one considered in Sect. 2.1. The wave's phase propagates along the boundary at velocity

$$c_{\mathrm{ph}} = \omega/\xi = c(1 + \alpha^2 k^{-2})^{-1/2} \quad , \tag{4.4.2}$$

which is less than the sound velocity c in a free space. This is why surface waves are frequently called retarded.

Some conditions are necessary for a surface wave to exist. For example, it does not exist when the boundary $z = 0$ is a rigid wall. Indeed, the impedance of the wave (4.4.1) at $z = 0$ is

$$Z = -i\omega\varrho\left(\frac{p}{\partial p/\partial z}\right)_{z=0} = \frac{i\omega\varrho}{\alpha} \quad , \tag{4.4.3}$$

whereas the impedance of a rigid wall is infinite.

We denote the impedance of the boundary by Z_1. To satisfy the boundary conditions, according to (4.4.3) we need to have $Z_1 = i\omega\varrho/\alpha$. The boundary with such an impedance can be produced in different ways. In particular, as seen from (2.3.3), the comblike structure considered in Sect. 2.3 has such an impedance when $\tan kh > 0$. Substituting (2.3.3) into (4.4.3), we find the main parameter of the surface wave for a comblike structure

$$\alpha = k \tan kh \quad . \tag{4.4.4}$$

4.4.2 Rayleigh Wave

Various kinds of surface waves can exist near the boundaries of elastic solids. The surface wave near the free boundary of a solid was first described by *Rayleigh* [4.26] and is named after him. Rayleigh waves are often observed in seismology. Let us examine their main properties. Expression (4.0.1) for the potentials of elastic waves shows that the wave field in the half-space $z > 0$ concentrating near the free boundary $z = 0$ arises when the following conditions are fulfilled:

$$\mathrm{Im}\,\{\alpha\} > 0 \quad , \quad \mathrm{Im}\,\{\beta\} > 0 \quad , \quad \varphi_2 = \psi_2 = 0 \quad . \tag{4.4.5}$$

In this case (4.1.1) is transformed into

$$\beta\psi_1 = \gamma\varphi_1 \quad , \quad \alpha\varphi_1 = -\gamma\psi_1 \quad , \tag{4.4.6}$$

which has the nonzero solutions only under the condition that

$$\alpha\beta = -\gamma^2 \quad . \tag{4.4.7}$$

This relation is called the *characteristic equation* or dispersion relation, and enables us to find the horizontal wave number of the Rayleigh wave, ξ_R.

The characteristic equation can be also obtained in quite a different way. It is seen from (4.4.5) that in order for the surface wave to exist, the amplitudes of the waves reflected from the boundary φ_1 and ψ_1 must be finite at zero amplitude of the incident waves. Therefore, the components of the scattering matrix (4.1.2) found in Sect. 4.1 must become infinite at $\xi = \xi_R$. This requirement, with account of (4.1.3, 4), again leads to (4.4.7).

We would like to know the real solutions ξ_R of the characteristic equations. In this case, according to (4.0.1) and (4.4.5) $\alpha = \mathrm{i}(\xi_R^2 - k_t^2)^{1/2}$, $\beta = \mathrm{i}(\xi_R^2 - k_t^2)^{1/2}$. Using the notation

$$q = \frac{c_t^2}{c_l^2} = \frac{\mu}{\lambda + 2\mu} \quad , \quad s = \frac{k_t^2}{\xi_R^2} = \frac{v_R^2}{c_t^2} > 0 \quad , \tag{4.4.8}$$

where v_R is the velocity of the Rayleigh wave, we rewrite (4.4.7) in the form

$$4\sqrt{1-s}\sqrt{1-sq} = (s-2)^2 \quad . \tag{4.4.9}$$

After squaring, the equation becomes algebraic

$$f(s,q) = 0 \quad , \quad f(s,q) \equiv s^3 - 8s^2 + 8(3 - 2q)s - 16(1 - q) \quad . \tag{4.4.10}$$

Note that (4.1.9) also reduces to (4.4.10) after squaring (if we set $s = k_t^2/\xi_0^2$). The positive roots of (4.4.10) that are <1 are the roots of (4.4.7) and determine the values of the velocity of a Rayleigh wave. The roots of (4.4.10) that are >1 are the solutions of (4.1.7) and give (as is seen in Sect. 4.1) the angle of incidence for an elastic wave at which exchange of polarization takes place.

By virtue of inequality (1.3.14), parameter q in (4.4.10) takes the values $0 \le q < \frac{1}{2}$. Therefore, one can easily prove that

$$f(0,q) < 0 \quad ; \quad f(1,q) > 0 \quad ; \quad \frac{\partial f}{\partial s}(s,q) > 0 \quad \text{at} \quad s \le 1 \quad . \tag{4.4.11}$$

Thus, all roots of (4.4.10) are positive, and only one root of the equation is in the interval $0 < s < 1$. Thus, the velocity of the Rayleigh wave is a single-valued function of parameters of an elastic half-space. Since the cubic equation has either one or three real roots depending on the value of q there will be two or no values of ξ_0 at which polarization exchange exists. From (4.4.11) and the inequality $\partial f/\partial q < 0$ for $s < 1$, it also follows that the velocity of the Rayleigh wave is a monotonically increasing function of q.

Consider the particular case $\lambda = \mu$ where according to (4.4.8) $q = \frac{1}{3}$. The cubic equation (4.4.10) has the roots $s = 4$, 3.1547, and 0.8453. The first two roots correspond to exchange of polarization and the third root gives $v_R = 0.9194$ for the velocity of the Rayleigh wave.

The velocity of the Rayleigh wave does not depend on its frequency and is close to but somewhat less than the velocity c_t of shear waves in unbounded media. One can obtain from (4.4.10)

$$\frac{v_R}{c_t} = 1 - \frac{1}{2}\delta - \frac{5}{8}\delta^2 + \frac{27}{16}\delta^3 + O(\delta^4) \quad , \tag{4.4.12}$$

where

$$\delta \equiv \frac{1}{4(3-4q)} = \frac{1-\sigma}{4(1+\sigma)} \leq \frac{1}{4} \quad , \tag{4.4.13}$$

σ is Poisson's ratio. Figure 4.8, according to *Knopoff* [4.27], shows the ratio v_R/c_t as a function of σ. The extreme values of this ratio are

$$\sigma = 0 \quad , \quad q = 0.5 \quad , \quad \frac{v_R}{c_t} = 0.8741 \quad ;$$

$$\sigma = 0.5 \quad , \quad q = 0 \quad , \quad \frac{v_R}{c_t} = 0.9554 \quad . \tag{4.4.14}$$

Note that the approximation in (4.4.12) gives, in the most unfavorable case $\sigma = 0$ ($\delta = \frac{1}{4}$), the value $v_R/c_t = 0.862$, which is close to the exact result of (4.4.14).

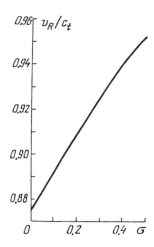

Fig. 4.8. Ratio of the Rayleigh wave velocity v_R to the shear waves velocity c_t as a function of the Poisson ratio σ. After [4.27]

From (4.0.1), taking into account (4.4.5, 8), we find the potentials of the Rayleigh wave

$$\varphi = \varphi_1 \exp\left(-\sqrt{1-qs}\,\xi_R z\right) \quad ,$$

$$\psi = \frac{i}{2}(s-2)(1-s)^{-1/2}\varphi_1 \exp\left(-\sqrt{1-s}\,\xi_R z\right) \quad , \tag{4.4.15}$$

where φ_1 is an arbitrary constant (wave's amplitude). We emphasize that the potential φ falls off with increasing z more rapidly than ψ and at $\xi_R z \gg 1$ the wave becomes purely shear. According to (4.0.2), by separating out the real part we obtain for the horizontal and vertical components of the displacement:

$$u_1 = \varphi_1 \xi_R \left[\left(1 - \frac{s}{2}\right) \exp\left(-\sqrt{1-s}\,\xi_R z\right) \right.$$

$$\left. - \exp\left(-\sqrt{1-qs}\,\xi_R z\right) \right] \sin\left(\xi_R x - \omega t\right) \quad ,$$

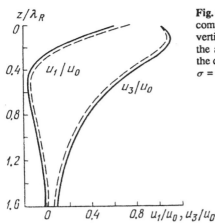

Fig. 4.9. Dependence of the horizontal (u_1) and vertical (u_3) components of the displacement in the Rayleigh wave on the vertical coordinate z. The displacements are referenced to u_0, the amplitude of vertical displacement on the boundary, and the coordinate z to λ_R, the Rayleigh wavelength. *Solid curve:* $\sigma = 0.34$, *dashed curve*: $\sigma = 0.25$. After [4.21]

$$u_3 = \varphi_1 \xi_R \sqrt{1 - qs} \left[\left(1 - \frac{s}{2} \right)^{-1} \exp\left(-\sqrt{1 - s} \xi_R z \right) \right.$$

$$\left. - \exp\left(-\sqrt{1 - qs} \xi_R z \right) \right] \cos\left(\xi_R x - \omega t \right) \quad . \qquad (4.4.16)$$

It is apparent that u_1 and u_3 differ in the phase by a quarter period. Consequently, the trajectories of the particles are ellipsoidal, with the principal axes parallel to the x- and z-axes.

Figure 4.9, according to *Viktorov* [4.21], shows a plot of u_1 and u_3 divided by $\sin(\omega t - \xi_R x)$ and $\cos(\omega t - \xi_R x)$, respectively, referenced to the amplitude u_0 of the vertical displacement at the boundary $z = 0$. The quantity z/λ_R is plotted along the ordinate, where $\lambda_R = 2\pi/\xi_R$ is the Rayleigh wavelength. For the solid curve $\sigma = 0.34$, and for the dashed curve $\sigma = 0.25$. We see that the vertical displacement first increases upon departure from the boundary, reaches a maximum and then falls gradually to zero. The horizontal component of the displacement decreases upon departure from the boundary, goes to zero, changes the sign, reaches a maximum, and then returns to zero asymptotically. The expression for the energy flux in Rayleigh wave and analysis of its dependence on Poisson's ratio can be found in [4.28].

4.4.3 Stoneley and Other Waves at Fluid-Solid and Solid-Solid Interfaces

Now we consider surface waves near the interface of a fluid ($z > 0$) and an elastic ($z < 0$) half-space. The characteristic equation for the horizontal component of the wave vector is obtained from the condition $V = \infty$ [the reflection coefficient is given by (4.2.18)]:

$$4m\alpha\xi^2(\alpha_1\beta_1 + \gamma_1^2) + \alpha_1 k_{t1}^4 = 0 \quad . \qquad (4.4.17)$$

We use the notation analogous to that in (4.4.8):

$$q = \frac{c_{tl}^2}{c_{ll}^2} \quad , \quad r = \frac{c_{tl}^2}{c^2} \quad , \quad s = \frac{k_{tl}^2}{\xi^2} = \frac{v^2}{c_{tl}^2} \quad , \tag{4.4.18}$$

where v is the velocity of the surface wave, c_{ll} and c_{tl} are the velocities of the longitudinal and transverse waves in the elastic half-space, and c is the sound velocity in fluid. Then

$$\alpha = i\omega v^{-1}\sqrt{1-sr} \quad , \quad \alpha_1 = i\omega v^{-1}\sqrt{1-sq} \quad ,$$
$$\beta_1 = i\omega v^{-1}\sqrt{1-s} \tag{4.4.19}$$

and (4.4.17) may be written in a form similar to (4.4.9):

$$4\sqrt{1-s}\sqrt{1-qs} - (s-2)^2 = \frac{s^2}{m}\sqrt{(1-sq)/(1-sr)} \quad . \tag{4.4.20}$$

Reference [4.29] shows that (4.4.20) always has a solution for which $v < c$, $v < c_{tl}$ and that α, α_1, and β_1 are positive imaginary quantities. Then the potentials of the wave fall off while leaving the boundary at $z > 0$ and also at $z < 0$; hence the wave is a surface wave.

We now find the solution of (4.4.20) explicitly for the case where a comparatively rarefied medium, e.g., a gas, is above the elastic half-space, so that one can assume the condition $m \equiv \varrho_1/\varrho \gg 1$, $r = c_{tl}^2/c^2 \gg 1$ to be satisfied. (See Ref. 4.30 on the existence of such waves near the earth's surface and in ices drifting in the ocean). Since we are seeking for the root $v < c$, then $s = v^2/c_{tl}^2 \ll 1$. We divide both parts of (4.4.20) by $(1-sq)^{1/2}$ and expand the left-hand side of the resulting equation in powers of s. Limiting ourselves to the first term and then squaring the entire equation, we obtain $1 - sr = s^2/4m^2(1-q)^2$. By assuming that the right-hand side is initially zero, we obtain $s = 1/r$. In the next approximation we set $s = 1/r$ in the right-hand side and then obtain

$$rs = 1 - [2mr(1-q)]^{-2} \quad . \tag{4.4.21}$$

This gives for the velocity of surface wave

$$v = c_{tl}\sqrt{s} \approx c\left[1 - \frac{1}{8m^2r^2(1-q)^2}\right] \quad , \tag{4.4.22}$$

which is somewhat smaller than the sound velocity in the upper medium.

From (4.4.19) we now find

$$\alpha \approx ik/2m(1-q) \quad , \quad \alpha_1 \approx \beta_1 \approx ik \quad , \quad \text{where} \quad k = \omega/c \quad . \tag{4.4.23}$$

Consequently, the decrease of the amplitude of the potentials in the upper and lower media with increasing distance from the boundary will be described by the exponentials

$$\exp\left[-\frac{kz}{2mr(1-q)}\right] \quad , \quad z > 0 \quad ; \quad \exp(kz) \quad , \quad z < 0 \quad . \tag{4.4.24}$$

Thus, in a gas, since $mr \gg 1$, the amplitude falls off very slowly as one goes away from the boundary, while in the elastic half-space the entire wave field is concentrated in a layer with thickness of the order of a wavelength in the upper medium. The wave is not damped in the horizontal direction if we do not take the absorption of energy into account.

The wave described by the characteristic equation (4.4.20) is sometimes called the *Stoneley wave* [4.31, 32] since it is a particular case of the surface wave at the interface of two elastic media which has been found by *Stoneley* [4.35].

Another type of wave can also exist at the interface of a fluid and an elastic half-space. Its nature is easily understood if we again assume that the upper half-space is filled with a rarefied medium. If the medium were a vacuum, then the wave at the boundary would be a Rayleigh wave. Such a wave will also exist in the case, when the upper half-space is filled with a rarefied fluid. But wave's velocity will be somewhat altered because of the reaction of the upper medium. However, if its velocity is larger than c, i.e., the sound velocity in the upper medium, then the wave will be partially radiated into the upper half-space, and it belongs to the class of *leaky waves* (see [4.33] for a discussion of these waves). The phase velocity of the leaky wave, in contrast to an ordinary surface wave, is not parallel to the boundary, and the amplitude of such a wave will fall off as it travels along the boundary. Figure 4.10 shows the wave fronts in the fluid and the solid (the left side of the figure) and normal to them the real parts of wave vectors (at the right). It is assumed that the damping of the wave in the horizontal direction (from left to right) is small. The leakage of energy from the boundary in the fluid is responsible for the imaginary part of the wave vector, which can be easily found from (4.4.20). The influx of energy to the boundary and energy balance are ensured by the slope of the elastic wave fronts in the lower half-space. By convention, the thickness of the lines that show the wave fronts indicates the amplitude of the wave. It is interesting to note that as one goes away from the boundary into the fluid, along the normal to the boundary one observes an increase in the amplitude of the wave. This is explained by the fact that at points further away from the boundary the wave field is due to radiation of the more left-hand parts of the boundary where the amplitude of the wave is greater than at points lying to the right.

The leaky wave itself cannot exist alone since its field is unbounded at $z \to \infty$. However, such a type of wave can be isolated in the field of a point source in a layered medium. Here as an observation point leaves the boundary the increase in wave's amplitude ceases at a certain distance and it begins to decrease [4.34].

The condition for the existence of a surface wave at the plane interface between two solid homogeneous half-spaces is the equality

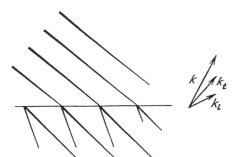

Fig. 4.10. Sketch of a leaky wave at a boundary between a fluid (upper half-space) and a solid (lower half-space). Wave fronts are shown in the left half of the drawing and the real parts of the wave vectors in the right

$$\Delta = 0 \quad , \tag{4.4.25}$$

where Δ is determined by (4.2.15). Equation (4.4.25) can be obtained from the requirement that the components of the scattering matrix of (4.2.2) become infinte, see (4.2.8–12), or from the condition for the existence of a nonzero solution for the set of equations (4.2.3–6) when $\varphi_2 = \psi_2 = \bar{\varphi}_2 = \bar{\psi}_2 = 0$.

If (4.4.25) yields a real solution for ξ in which the conditions $\xi > k_t$ and $\xi > k_{t1}$ are satisfied, then α, β, α_1, and β_1 are purely imaginary quantities, and such a solution will correspond to a surface wave. The main properties of this wave were studied for the first time by *Stoneley* [4.35]. Later, this question and, in particular, the condition for the existence of a Stoneley wave were examined by many authors, see [4.4, 4.36–38, 4.56, 4.57]. Reference [4.39] contains a graph for the determination of the velocity of a Stoneley wave at different parameters of the media in contact. It is easy to show [in a way similar to the conversion of the reflection coefficient given by (4.2.8) to that of (4.2.18)], that when one of the half-spaces is a fluid, (4.4.25) transforms into (4.4.17). As a result, we obtain the surface waves considered above at the boundary of a solid and a fluid.

Different types of leaky waves can propagate along the boundaries of solids and also along the interfaces of a solid and a fluid. Although their amplitude falls of exponentially with distance, taking these waves into account is sometimes important in addressing the problem of a field generated by a point source. Analysis of these waves for a number of cases can be found in [4.40].

In the notation of Sect. 4.3, the equation for the determination of the velocity (and dispersion) of surface and leaky waves when the set of layers is between the fluid and elastic half-spaces, has the form, see (4.3.19),

$$Z_{\text{in}} + Z = 0 \quad . \tag{4.4.26}$$

A detailed analysis of these waves has been given by *Keilis-Borok* [4.41]; see also [4.9] and [Ref. 4.14, Chap. 7, 4.42, 4.43, 4.58].

We assumed above the contact between the bordering solids to be welded. Here the boundary conditions given by (1.3.26) (continuity of the displacements and of the corresponding components of the stress tensor) are satisfied. In some cases other types of contact, in particular, slipping, can occur. It is natural, of course, that changing the boundary conditions strongly affects the scattering matrix or the possibility of the existence of surface waves, and their characteristic equation.

In calculating the reflection coefficient from a set of layers with slip along some solid-solid interfaces, it is often advisable to approach the problem by using formulas obtained above with the assumption that infinitely thin liquid layers exist at these interfaces. The possibility of the existence of a Stoneley wave at the boundary of two unbound elastic half-spaces in contact with slip was studied in [4.44]. References [4.45, 46, 59, 60] consider for several cases how the type of boundary conditions influences the dependence of surface wave velocity on the wave frequency.

5. Reflection of Sound Pulses

In preceeding chapters we studied the reflection of monochromatic waves. Such waves are an idealization. This chapter is aimed to show how the previously obtained results can be applied to waves with *arbitrary* time dependence.

5.1 General Relations. Law of Conservation of Integrated Pulse

5.1.1 Integral Representation of Sound Field

We consider the reflection of a pulse with a plane front from a plane boundary. We approach the problem by expanding the pulse into harmonic waves with the same angle of incidence as the pulse. Let us restrict ourselves to the case of a medium at rest.

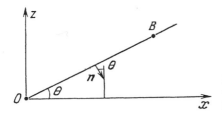

Fig. 5.1. Incidence of a plane pulse upon a plane boundary, which is the plane $z = 0$. OB is the front of the incident pulse, θ is the angle of incidence

We assume that the plane of incidence coincides with the xz-plane (Fig. 5.1). The pulse is incident from a half-space $z > 0$ upon the boundary $z = 0$, and θ is the incidence angle. Then in the general expression for the plane wave (1.1.17) we have $n = (\sin\theta, 0, -\cos\theta)$ and

$$p_i = f(\zeta) \quad , \quad \zeta = \frac{x\sin\theta - z\cos\theta}{c} - t \quad . \tag{5.1.1}$$

The real function f gives the shape of the pulse, i.e., the time dependence of the sound pressure at a fixed point. We expand the pulse into harmonic waves

$$p_i = \int\limits_{-\infty}^{+\infty} \Phi(\omega) \exp(i\omega\zeta) d\omega \quad . \tag{5.1.2}$$

The function $\Phi(\omega)$ is the spectrum density of a pulse. It is related to the function $f(\zeta)$ by the inverse Fourier transformation

113

$$\Phi(\omega) = (2\pi)^{-1} \int_{-\infty}^{+\infty} f(\zeta) \exp(-i\omega\zeta) d\zeta \quad . \tag{5.1.3}$$

Since f is a real function we get from (5.1.3)

$$\Phi^*(\omega) = \Phi(-\omega) \quad . \tag{5.1.4}$$

As usual, the asterisk denotes the complex conjugate.

Because only nonnegative frequencies are physically real, in previous chapters we assumed $\omega \geq 0$. In (5.1.2) we could also limit the integration to positive ω and simultaneously add the complex-conjugated quantity. For convenience, however, we retain the negative frequencies in the complex representation of wave fields.

As before, we denote the reflection coefficient of a plane harmonic wave by V and write the pressure of the reflected pulse in the form

$$p_r = \int_{-\infty}^{+\infty} \Phi(\omega) V(\omega, \theta) \exp(i\omega\zeta_1) d\omega \quad , \quad \zeta_1 = \frac{x \sin\theta + z \cos\theta}{c} - t \quad . \tag{5.1.5}$$

The pressure p_r is real. Hence, we find in the same way as in obtaining (5.1.4)

$$V(-\omega) = V^*(\omega) \quad . \tag{5.1.6}$$

In general, the reflected pulse has a shape different from that of the incident pulse, except when V is independent of ω. (Then from (5.1.6) we infer that V is real.) In this case we can put the reflection coefficient outside the integral in (5.1.5) and we obtain:

$$p_r(x, z, t) = V f(\zeta_1) \quad . \tag{5.1.7}$$

We emphasize that this is not always the case, however. The reflection coefficient from the interface of two homogeneous media is given by (2.2.14) for $\omega > 0$ and is frequency independent. However, at $n < 1$ and $\sin\theta > n$ we have total internal reflection, V is a complex quantity, and $V(-\omega) \neq V(\omega)$. As a result the shape of the pulse changes. The same occurs when a wave is reflected from an inhomogeneous medium.

Assume the half-space $z < 0$ to be stratified, but at rather large $-z$ the medium becomes homogeneous again. Then we can introduce the transmission coefficient W and write the transmitted pulse in the form similar to (5.1.5)

$$p_t = \int_{-\infty}^{+\infty} \Phi(\omega) W(\omega, \theta) \exp(i\omega\zeta_2) d\omega \quad ,$$

$$\zeta_2 = \frac{x \sin\theta_1 - z \cos\theta_1}{c_1} - t \quad , \tag{5.1.8}$$

where θ_1 is the refraction angle which is related to the angle of incidence and the sound velocity c_1 at large $-z$ by $\sin\theta_1 = c_1 c^{-1} \sin\theta$. An expression analogous to (5.1.6) is also valid for the transmission coefficient W.

5.1.2 Conservation of Integrated Pulse

Following [5.1] we shall now prove the theorem of conservation of the total pulse which states that the total (integrated) pulse at any point of the upper medium is equal to the total pulse at any point of the lower medium. This law has also been formulated in [5.2] for several cases. It has recently been proved again in [5.3]. Mathematically the theorem is expressed by the identity

$$\int\limits_{-\infty}^{+\infty} (p_i + p_r)dt = \int\limits_{-\infty}^{+\infty} p_t dt \quad , \tag{5.1.9}$$

which remains valid regardless of the shape of the incident pulse and the point in space at which the integrals are taken on the right and left sides of the equality.

Thus, for example, for the case of total internal reflection of a pulse the maximum value of the pressure in the lower medium decreases with increasing distance from the boundary. However, the pulse will stretch out in time in such a way that the area given by the integral on the right side of (5.1.9) remains the same at any arbitrarily large distance from the boundary.

We begin the proof of identity (5.1.9) by examining the expression for the integrated value of the incident pulse. When the limits of the integration are $\pm\infty$, integral over t is equivalent to the integral over ζ. Therefore,

$$\int\limits_{-\infty}^{+\infty} p_i dt = \int\limits_{-\infty}^{+\infty} p_i d\zeta = \iint\limits_{-\infty}^{+\infty} \Phi(\omega)\exp(i\omega\zeta)d\zeta\, d\omega \quad . \tag{5.1.10}$$

But, as is known, the following is true [Ref. 5.4, Sect. 9]:

$$\int\limits_{-\infty}^{+\infty} \exp(i\omega\zeta)d\zeta = 2\pi\delta(\omega) \quad , \tag{5.1.11}$$

where $\delta(\omega)$ is Dirac's δ-function which is equal to zero everywhere except at $\omega = 0$ where it becomes infinite. The Dirac function possesses the following property:

$$\int\limits_{-\infty}^{+\infty} \varphi(\omega)\delta(\omega)d\omega = \varphi(0) \quad , \tag{5.1.12}$$

where $\varphi(\omega)$ is any function that is continuous at $\omega = 0$. Now we obtain from (5.1.10)

$$\int\limits_{-\infty}^{+\infty} p_i dt = 2\pi\Phi(0) \quad . \tag{5.1.13}$$

This is a quite natural result, since it is known that the area under the curve is given by the constant term (corresponding to $\omega = 0$) of the expansion of this curve in a Fourier series or integral.

As is seen from (5.1.6), the real part of the reflection coefficient is continuous at $\omega = 0$ and the imaginary part is an odd function of ω with a discontinuity at $\omega = 0$. Taking into account (5.1.4–6) we present the integrated pulse of the reflected wave in the form

$$p_r = \int\limits_{-\infty}^{+\infty} \mathrm{Re}\,\{\varPhi V\}\exp\,(i\omega\zeta_1)d\omega - 2\int\limits_{0}^{+\infty} \mathrm{Im}\,\{\varPhi V\}\,\sin\,\omega\zeta_1 d\omega \quad . \qquad (5.1.14)$$

The second term on the right-hand side of (5.1.14) is an odd function of ζ_1, and upon integrating over ζ_1 gives zero. The integrand in the first term is a continuous function at $\omega = 0$, therefore the integral is calculated in the same way as for the incident wave and equals $2\pi\,\mathrm{Re}\,\{\varPhi(0)V(0)\}$. A similar expression is also obtained for the integral of the refracted pulse, by replacement of V by W.

Thus, to prove (5.1.9) it suffices to establish the validity of the equality

$$\varPhi(0) + \mathrm{Re}\,\{V(0)\varPhi(0)\} = \mathrm{Re}\,\{W(0)\varPhi(0)\} \quad .$$

According to (5.1.4) $\varPhi(0)$ is real. Therefore, the latter equality becomes

$$1 + \mathrm{Re}\,\{V(0)\} = \mathrm{Re}\,\{W(0)\} \quad . \qquad (5.1.15)$$

However, at $\omega \to 0$ (where $\omega > 0$ is assumed) the more general equation

$$1 + V = W \qquad (5.1.16)$$

is valid, the real part of which yields (5.1.15). In the case of reflection from the interface of two homogeneous media, the coefficients V and W do not depend on the frequency, and (5.1.16) follows simply from the condition of continuity of the sound pressure. In more complicated cases when the reflection occurs from a layer or a set of layers, as $\omega \to 0$ (i.e., for infinite wavelength) the entire set of layers will have no effect on the reflection process and reflection will take place just as if the media, although formally separated by the layers, were in direct contact with one another. For a discretely layered medium, this can be seen from (2.5.3, 4 and 11), which show that the case where $\omega \to 0$ ($k_j\cos\theta_j \to 0$) is equivalent to the case $d_j \to 0, j = 2,\ldots,n$, i.e., the effect of all intermediate layers is eliminated.

The law of conservation of the integrated pulse is valid under much more general conditions than those assumed in the above proof. Identity (5.1.9) also holds for the region where the medium is inhomogeneous. Generally, for a *finite-time perturbation* in a three-dimensional inhomogeneous fluid the integrated pulse $\int_{-\infty}^{+\infty} p(\boldsymbol{r}, t)dt$ does not depend on $\boldsymbol{r}$. To prove this we shall use (1.1.9) which, after integration over t, yields

$$-\varrho v\bigg|_{t=-\infty}^{t=+\infty} = \nabla\left[\int\limits_{-\infty}^{+\infty} p(\boldsymbol{r}, t)dt\right] \quad , \qquad (5.1.17)$$

where v is the oscillatory velocity of the particles. When it is assumed that the sound pulse is of finite duration, this velocity is zero at infinity. Therefore, the gradient of the integral is also zero, from where the conservation of the integrated pulse follows.

5.1.3 Energy Conservation

The law of conservation of energy for reflection of a plane wave pulse in a layered medium can be written in the form

$$S_z^{(i)} + S_z^{(r)} = S_z^{(t)} \quad , \qquad (5.1.18)$$

where $S_z^{(i)}$, $S_z^{(r)}$, and $S_z^{(t)}$ are the components along the z-axis of the vectors of the power flux density, integrated over time, in the incident, reflected, and refracted waves. The quantities $S_z^{(i)}$ and $S_z^{(r)}$ have opposite signs. In the case of an incident pulse, taking into account (2.1.8) for the vector of the power flux density, we have

$$S_z^{(i)} = \int_{-\infty}^{+\infty} p_i v_{iz}\, dt \quad , \tag{5.1.19}$$

where p_i and v_{iz} are the acoustic pressure and the component of the velocity along the z-axis in the incident wave. $S_z^{(r)}$ and $S_z^{(t)}$ can be written in a similar way.

For the proof of (5.1.18) we note that the vertical components of the energy fluxes in the incident and reflected waves are additive. Actually, it follows from (1.1.9) that in the plane wave of (5.1.1)

$$v_{iz} = -\frac{p_i \cos\theta}{\varrho c} \quad . \tag{5.1.20}$$

Therefore, in the upper medium, the vertical component of the vector of the power flux density given by (2.1.8) is

$$I_z = p v_z = \frac{(p_i + p_r)(p_i - p_r)\cos\theta}{\varrho c} = p_i v_{iz} + p_r v_{rz} \quad . \tag{5.1.21}$$

Consequently, the left-hand side of (5.1.18) is the integrated energy density flux through the boundary from the upper medium. Then in the case of reflection from the interface of two homogeneous media, (5.1.18) follows from the condition of contiuity of the power flux density at the boundary which is an apparent consequence of the continuity conditions for p and v_z. If there is a set of layers between the homogeneous half-spaces $z > 0$ and $z < z_1$, then it should be noted that, by virtue of the horizontal symmetry of the problem, the total energy flux through any part of the plane $x = x_0$, bounded by the horizons $z = 0$ and $z = z_1$, does not depend on x_0. It is also clear that the quantities $S_z^{(i)}$, $S_z^{(r)}$, and $S_z^{(t)}$ do not depend on horizontal coordinates. Then from the law of acoustical energy conservation it follows that the integrated fluxes through the planes $z = z_1$ and $z = 0$ are equal which, as is shown above, is equivalent to (5.1.18).

The validity of (5.1.18) can be also established by representing the pulse as a superposition of plane harmonic waves, and summation of the energy fluxes of these waves.

In the case of total internal reflection, $S_z^{(t)} = 0$, since the refracted pulse propagates along the interface and does not carry away energy in the direction $z \to -\infty$. Then we obtain from (5.1.18) that $S_z^{(i)} + S_z^{(r)} = 0$. For a detailed proof of the latter see [5.3, 5, 6].

5.2 Change of Pulse Shape upon Total Internal Reflection from a Boundary Between Two Homogeneous Media

Following [5.2] by *Fischer*, we consider a pulse whose shape is given by the function (Fig. 5.2)

$$f(\zeta) = \frac{A\tau}{\tau^2 + \zeta^2} \quad , \tag{5.2.1}$$

where A is an amplitude, $\tau > 0$, τ has dimension of time and specifies the width of the pulse. The spectrum density of the pulse is easily determined by (5.1.3) and equals

$$\Phi(\omega) = \tfrac{1}{2} A \exp\left(-|\omega|\tau\right) \quad . \tag{5.2.2}$$

It is seen from (5.1.3, 12) that the spectrum density of the δ-pulse $f(\zeta) = \delta(\zeta)$ is constant and equals $1/2\pi$. Therefore, the pulse given by (5.2.1) transforms into a δ-pulse when $\tau \rightarrow 0$ to an accuracy within a nonessential factor.

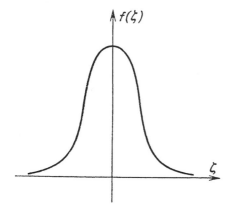

Fig. 5.2. Shape of the incident pulse used in calculation of pulse shape distortion after reflection and refraction at a boundary

If the usual (i.e., partial) reflection rather than total reflection occurs, than, as we have seen, the shape of the reflected and refracted pulses will be the same as the shape of the incident pulse. Taking into account (5.1.1, 5 and 8), we obtain

$$p_i(x, z, t) = \frac{A\tau}{\tau^2 + [(x \sin\theta - z \cos\theta)/c - t]^2} \quad , \tag{5.2.3}$$

$$p_r(x, z, t) = \frac{AV\tau}{\tau^2 + [(x \sin\theta + z \cos\theta)/c - t]^2} \quad , \tag{5.2.4}$$

$$p_t(x, z, t) = \frac{AW\tau}{\tau^2 + [(x \sin\theta_1 - z \cos\theta_1)/c_1 - t]^2} \quad , \tag{5.2.5}$$

where V and W are real coefficients of reflection and transmission.

In the case of total reflection V and W are complex quantities. Using (2.2.9, 14), we separate the real and imaginary parts (for $\omega > 0$):

$$V = B + iC \quad , \quad W = (B+1) + iC \quad , \quad B = \frac{m^2 \cos^2 \theta - s^2}{m^2 \cos^2 \theta + s^2} \quad ,$$

$$C = \frac{-2sm \cos \theta}{m^2 \cos^2 \theta + s^2} \quad , \quad s = \sqrt{\sin^2 \theta - n^2} \geq 0 \quad . \tag{5.2.6}$$

Then using (5.1.14, 2.2) we can write the reflected pulse in the form

$$p_r(\zeta_1) = \frac{1}{2} AB \int_{-\infty}^{+\infty} \exp(-\tau|\omega| + i\omega\zeta_1) d\omega$$

$$- AC \int_{0}^{+\infty} \exp(-\tau\omega) \sin \omega\zeta_1 d\omega \quad . \tag{5.2.7}$$

Both integrals in the last expression can be evaluated without difficulty. By substituting the value of ζ_1 from (5.1.5), we obtain

$$p_r = \frac{AB\tau}{\tau^2 + [(x \sin \theta + z \cos \theta)/c - t]^2} - \frac{AC[(x \sin \theta + z \cos \theta)/c - t]}{\tau^2 + [(x \sin \theta + z \cos \theta)/c - t]^2} \quad . \tag{5.2.8}$$

Thus, the reflected pulse consists of two parts, one of which has the same shape as the incident pulse. In particular, for a δ-pulse as $\tau \to 0$ we obtain

$$p_r = \pi AB\delta[(x \sin \theta + z \cos \theta)/c - t)] - AC[(x \sin \theta + z \cos \theta)/c - t]^{-1} \quad . \tag{5.2.9}$$

Let us now consider the sound wave penetrating into the lower medium. In the expression for the refracted wave given in (5.1.8) we have (for $\omega > 0$)

$$\zeta_2 = \frac{x \sin \theta_1}{c_1} - t - \frac{isz}{c} \quad . \tag{5.2.10}$$

At negative frequencies the complex conjugate expression should be taken for ζ_2. By substituting (5.2.6, 10) into (5.1.8), we obtain

$$p_t = \frac{1}{2} A(1 + B) \int_{-\infty}^{+\infty} \exp\left[-|\omega|\left(\tau - \frac{sz}{c}\right) + i\omega g\right] d\omega$$

$$- AC \int_{0}^{+\infty} \exp\left[-\omega\left(\tau - \frac{sz}{c}\right)\right] \sin \omega g \, d\omega \tag{5.2.11}$$

$$g = t - \frac{x \sin \theta_1}{c_1} \quad . \tag{5.2.12}$$

After simple transformations (5.2.11) yields

$$p_t = \frac{A\{(1 + B)(\tau - sz/c) + C[(x/c_1) \sin \theta_1 - t]\}}{(\tau - sz/c)^2 + [(x/c_1) \sin \theta_1 - t]^2} \quad . \tag{5.2.13}$$

In the case where $\tau = 0$ (a δ-pulse is incident), we have from (5.2.13)

$$p_t = \frac{A\{C[(x/c_1)\sin\theta_1 - t] - (sz/c)(1+B)\}}{(sz/c)^2 + [(x/c_1)\sin\theta_1 - t]^2}. \qquad (5.2.14)$$

We see that the shape of the refracted pulse has nothing in common with the shape of the incident pulse. Furthermore in (5.2.13, 14) the time t enters only in the combination $(x\sin\theta_1)/c_1 - t$. Hence, in the lower medium the pulse propagates along the interface with the velocity $c_1/\sin\theta_1$ which, according to the law of refraction, is equal to $c/\sin\theta$, that is the propagation velocity of the pulse in the upper medium along the interface. It is interesting to note that on the straight line given by

$$\tau - \frac{sz}{c} + \left(t - \frac{x}{c_1}\sin\theta_1\right)\frac{s}{m\cos\theta} = 0 \qquad (5.2.15)$$

the acoustic pressure of the refracted pulse is zero and is of opposite signs on opposite sides of this line.

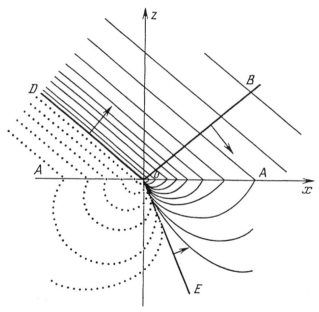

Fig. 5.3. Reflection and refraction of a δ-pulse at a boundary between two media AA. OB: front of incident pulse, OD: front of reflected pulse, OE: front of refracted pulse. *Arrows* indicate direction of propagation. *Solid* $(p > 0)$ and *dotted lines* $(p < 0)$ are the lines of constant sound pressure p. Lines density is proportional to field strength

The incident, reflected, and refracted pulses are shown schematically in Fig. 5.3 at the instant $t = 0$. The incident pulse is assumed to be a δ-function. In this figure AA is the boundary between the media, OB is the front of the incident pulse given by the equation $x\sin\theta - z\cos\theta = 0$, and OD is the front of the part of the reflected pulse which corresponds to the first term in the right-hand side of (5.2.9). The second term in the right-hand side of (5.2.9) is shown by the lines of constant sound pressure, namely, by the solid lines above OD (positive pressure) and the dotted lines (negative pressure) below OD. The decreasing lines density with increasing distance from the line OD shows the decreasing field strength. The field of the refracted pulse OE has a finite value everywhere except at the origin O. It vanishes on the line OE given (5.2.15), and has opposite signs in the opposite sides of OE. The solid ($p_t > 0$) and the dotted ($p_t < 0$) lines in the lower medium are the lines of constant acoustic pressure, the form of which is determined by (5.2.14). The arrows indicate the directions of propagation of the incident, reflected, and refracted pulses.

We note that the dependence of the field of the refracted wave on the z-coordinate at $x = c_1 t / \sin \theta_1$, in accordance with (5.2.13), is $p_t = A(1+B)/(\tau - sz/c)$. The same relation holds at any given x and t as $z \to -\infty$. Thus, the field decreases with increasing distance from the boundary much more slowly than the exponential decrease observed for harmonic plane waves, namely, the amplitude is inversely proportional to the distance. This could be also deduced from the following considerations. The penetration depth of the plane monochromatic wave in the case of total reflection is inversely proportional to the frequency. Thus, the field of the pulse at large distances from the boundary is determined by behavior of its spectrum density $\Phi(\omega)$ at low frequency. Representing the field (5.1.8) in the lower medium in the form

$$p_t = 2 \operatorname{Re} \left\{ W \int_0^\infty \Phi(\omega) \exp \{ \omega s z/c + i\omega[(x \sin \theta_1)/c_1 - t]\} d\omega \right\}$$

and expanding Φ in power series of ω, we can easily establish that the amplitude of the acoustic pressure as $z \to -\infty$ is proportional to $|z|^{-1}$, if $\Phi(0) \neq 0$. If $\Phi(0) = 0$ but $\Phi'(0) \neq 0$, we have $|p_t| \sim |z|^{-2}$, etc. The exponential decay of the amplitude upon increasing $|z|$ occurs for pulses which have lower cutoff frequency $\omega_0 > 0$, such that $\Phi(\omega) = 0$ at $0 \leq \omega \leq \omega_0$.

If the transmitted $\tilde{p}_t$ and reflected $\tilde{p}_r$ pulses are determined for an incident pulse described by a δ-function $[f(\zeta) = A\delta(\zeta)]$, then for an incident pulse of an *arbitrary form* the reflected and transmitted sound fields are expressed by the convolution integrals

$$p_t(x, z, t) = A^{-1} \int_{-\infty}^{+\infty} dt_1 \tilde{p}_t(x, z, t - t_1) p_i(0, 0, t_1) \quad , \tag{5.2.16}$$

$$p_r(x, z, t) = A^{-1} \int_{-\infty}^{+\infty} dt_1 \tilde{p}_r(x, z, t - t_1) p_i(0, 0, t_1) \quad . \tag{5.2.17}$$

These relations follow from the equality

$$p_i(x, z, t) = p_i \left(0, 0, t - \frac{x}{c} \sin \theta + \frac{z}{c} \cos \theta \right)$$

$$= \int_{-\infty}^{+\infty} dt_1 \delta \left(\frac{x}{c} \sin \theta - \frac{z}{c} \cos \theta - t + t_1 \right) p_i(0, 0, t_1) \quad ,$$

taken together with the superposition principle. Using (5.2.9, 14), Eqs. (5.2.16, 17) allow us to determine the reflected and transmitted pulses in the general case by integrating over time. In numerical calculations, in some cases (for short pulses, for example) integration over time is simpler [5.7, 8] than integration over frequency using (5.1.3, 5, 8). As a rule, however, integrating over frequency is preferable, especially when using the algorithm of fast Fourier transformation.

Expressions (5.2.17) and (5.2.9) show that in the case of total reflection the pulse p_r always consists of two parts: one with the same shape as the incident pulse and the second is its *Hilbert transform* [Ref. 5.9, Chap. 8]

$$\pi^{-1} \int_{-\infty}^{+\infty} [(x \sin \theta + z \cos \theta)/c - t + t_1]^{-1} p_i(0, 0, t_1) dt_1 \quad . \tag{5.2.18}$$

Due to a singularity at $t_1 = t - (x \sin \theta + z \cos \theta)/c$ the principal value of the integral should be implied in (5.2.18).

It is important to note that the formalism developed above can be applied to other cases of reflection from boundaries of homogeneous media (two elastic half-spaces, elastic and liquid half-spaces, reflection from a free boundary of a solid) when the reflection and transmission coefficients do not depend on frequency for $\omega > 0$, as for the case of an interface of two liquids.

Several studies of the change in pulse shape upon total reflection are available for a number of cases where the Hilbert transformation (5.2.18) can be easily performed. In [5.6, 10] for example, the case of a stepwise, or "plateau", pulse (sound field has a constant value over some time interval and is equal to zero outside this interval) was considered. *Arons* and *Yennie* [5.5] have examined the reflection of an exponential pulse described at $x = z = 0$ by the equation

$$p(0, 0, t) = 0 \quad , \quad t < 0 \quad ;$$
$$p(0, 0, t) = A \exp(-\lambda t) \quad , \quad t \geq 0 \quad , \quad \lambda > 0 \quad . \tag{5.2.19}$$

This equation gives a good description of the shape of the head of a pulse in the case of an underwater explosion. *Arons* and *Yennie* compared the theory with experimental results obtained by recording explosive pulses in a layer of water bounded by a free surface above and a solid bottom below. Reference [5.11] considers the reflection of a pulse in which during some time a linear increase precedes an exponential decrease, as in (5.2.19). *Cron* and *Nuttal* [5.12] studied the transformation of the shape of a quasimonochromatic pulse with stepwise or Gaussian envelopes. *Tjøtta* and *Tjøtta* [5.8] gave a detailed analysis of the penetration of a transient plane wave into the lower medium under conditions of total internal reflection. This work contains many illustrations of the sound field for plateau and Gaussian pulses as well as for quasimonochromatic pulses of different shapes: a sinusoidal pulse of finite duration or with a Gaussian envelope and a pulse with $p(0, 0, t) = 0$ at $t < 0$ and $p(0, 0, t) = A \sin \omega t$ for $t \geq 0$.

Several papers consider reflection from interfaces of the other kind. Thus, reference [5.13] is devoted to the analysis of distortion of a pulse in an inhomogeneous elastic medium. Reflection and transmission of an exponential pulse through a plate at a normal incidence is considered in [5.14]. A more complicated case of reflection of a sound pulse from an absorbing layer dividing two homogeneous half-spaces was analyzed with numerous examples in [5.15]. In [5.7] the reflected and transmitted sound signals for a plateau pulse incident on a set of elastic layers (with absorption) are found by the method of calculation of coefficients of reflection and transmission for a monochromatic plane wave, similar to that presented in Sect. 4.3, and by (5.2.16, 17).

The problem of distortion of the shape of a pulse upon reflection is closely related to that of deformation of a signal propagating in a medium with dispersion. A review of this problem can be found in [Ref. 5.16; Chaps. 21, 22, 24] and in [5.17].

Equation (5.2.9) shows that under the conditions of total reflection the sound pressure in the upper medium is different from zero at any t even before the arrival of the incident pulse. In other words, a "precursor" wave propagates in the upper half-space. However, this does not contradict the causality principle. The plane incident pulse has contact with the boundary at all times. At the place of contact, a *lateral wave* is excited [Ref. 5.18, Chap. 4; 5.29, Chap. 3] which propagates along the interface with a velocity higher than that of the contact point of the incident pulse with the boundary. The lateral wave is responsible for the precursor. This behavior becomes more understandable, if we consider the reflection of a pulse with a nonplanar front [5.19, 26]. Suppose the pulse has no contact with the boundary at the initial moment and reaches it at some moment t_1. The reflected wave is nonzero for $t \geq t_1$ only. For the special case where the pulse coincides with the plane wave (5.1.1, 2.1) at $z > l > 0, t = 0$ and is zero at $z < l, t = 0$, the reflected and transmitted waves can be expressed in elementary functions. This has been shown in [5.19] where it is demonstrated that at some time after the pulse reaches the boundary, the forward front of the lateral wave is far ahead of the front of the incident perturbation, and the back front of the lateral wave forms the precursor in the upper half-space.

5.3 Total Reflection of a Pulse in Continuously Layered Media

The above analysis of reflection of plane pulses from an interface of homogeneous media enables us to make a number of significant conclusions about the reflection of pulses from a continuously layered medium when the sound velocity and medium density change little over a distance of the order of the wavelength for all the significant monochromatic components of the incident pulse. In this case the approximation of geometrical acoustics can be applied (which is considered in detail in Chaps. 8, 9). The reflection coefficient for a plane monochromatic wave from the half-space $z < 0$, where the only turning point is at $z = z_m$, is equal to $V = \exp(i\varphi)$, where the phase is, see (9.2.10),

$$\varphi = 2k_0 \int_{z_m}^{0} n(z) \cos \theta(z) dz - \pi/2 \quad , \quad k_0 = \omega/c_0 \quad . \tag{5.3.1}$$

Here $\omega > 0$, c_0 is the sound velocity in the half-space $z > 0$, $n(z) \equiv c_0/c(z)$ is the refraction index, and $\theta(z)$ is the angle which the wave vector makes with z-axis. The first term in the right-hand side of (5.3.1) gives the phase change in the geometrical approximation while the wave propagates from the boundary $z = 0$ to the turning plane $z = z_m$ (where $\cos \theta(z_m) = 0$), and back. Equation (5.3.1) shows that except for small vicinity of the plane $z = z_m$ the wave propagates without reflection with an ordinary geometric advancement of phase, but in the turning plane there is a phase loss of $\frac{\pi}{2}$ regardless of frequency.

We now consider a pulse propagating from $z = 0$ to the turning point. We again have an integral representation as in (5.1.2), where

$$\zeta = \frac{x}{c(z)} \sin \theta(z) - \frac{1}{c_0} \int_0^z n(z_1) \cos \theta(z_1) dz_1 - t \quad . \tag{5.3.2}$$

According to Snell's law (2.2.6) the quantity $c^{-1} \sin \theta$ does not depend on z and equals $c_0^{-1} \sin \theta_0$, where θ_0 is the incidence angle of the pulse at the boundary $z = 0$.

We note that for an arbitrary pulse represented as the superposition of harmonic waves given by (5.1.2), multiplication of the spectrum density $\Phi(\omega)$ by the exponent $\exp(i\omega\tau)$, where τ is independent of the frequency, does not change the shape of the pulse and only shifts it in time by an amount τ. Indeed, the introduction of such a factor into the integral in (5.1.2) transforms $f(\zeta)$ into some $f_1(\zeta)$, where

$$f_1(\zeta) = \int_{-\infty}^{+\infty} \Phi(\omega) \exp[i\omega(\tau + \zeta)] d\omega = f(\zeta + \tau) \quad .$$

Consequently, while propagating from $z = 0$ to $z = z_m$ the pulse does not change shape. Similarly on the return path from $z = z_m$ to $z = 0$ the pulse will also keep the shape which it acquired during the passage through the turning point $z = z_m$. Thus, it remains for us to analyze only the change of the pulse shape at the turning point due to the loss of $\frac{\pi}{2}$ in the phase in each of the elementary harmonic waves.

The corresponding reflection coefficient $\tilde{V} = \exp(-i\frac{\pi}{2})$ is independent of frequency (for $\omega > 0$). This case has already been discussed in the previous section. We set $B = 0$, $C = -1$ in (5.2.6). Then we find, similarly to (5.2.9) for the case of the δ-shaped pulse,

$$p_i = \frac{A}{\zeta_1} \quad ,$$

$$\zeta_1 = \frac{x}{c} \sin \theta - \frac{1}{c_0} \left[\int_{z_m}^0 n(z_1) \cos \theta(z_1) dz_1 + \int_{z_m}^z n(z_1) \cos \theta(z_1) dz_1 \right] - t \quad . \tag{5.3.3}$$

Note that the result obtained is valid for any $n(z)$ if only total reflection takes place and the angle of incidence is not very close to $\frac{\pi}{2}$. It is valid for a plane incident wave case but caution is necessary in applying it to the case of waves produced by a bounded beam or point source.

In [5.20] *Tolstoy* attempted to explain on the basis of (5.3.3) the change of shape of a δ-pulse which is excited by a point source and is propagating in a surface waveguide. The sound velocity in the waveguide increases with increasing distance from the surface and a certain class of rays leaving the source turns in the medium and returns again to the boundary. Figure 5.4 shows one of these rays turning at the level $z = z_m$.

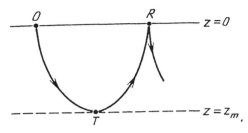

Fig. 5.4. Ray showing the propagation of a δ-pulse generated by a point source in a surface waveguide. *Solid line $z = 0$*: waveguide's boundary, *dashed line $z = z_m$*: turning plane

Tolstoy assumed that the pulse does not change its shape from the source to the turning point, and takes the form given in (5.3.3) in the vicinity of the turning point due to a $-\pi\omega/2|\omega|$ phase shift in all harmonic components. Note that according to (5.3.3) the acoustic pressure differs from zero at any t. If the point source initially turns on at $t = 0$ then the causality principle is not satisfied. In [5.20], to avoid this contradiction it is assumed that (5.3.3) is valid only for $\zeta_1 < 0$, i.e., after the incident pulse reaches the level $z = z_m$, whereas for $\zeta_1 > 0$ one has $p_r = 0$. This assumption, however, appears to be ineffective. In particular the law of conservation of integrated pulse is violated: Before the point T in Fig. 5.4 one has $\int_{-\infty}^{+\infty} p\, dt = A$ whereas along the arc TR this integral is infinite. The cause of these contradictions is the misapplication of results obtained for plane pulses to a point source. Moreover, the phase jump of $-\frac{\pi}{2}$ at the ray takes place not in the vicinity of the turning point, but rather at the point where the ray is tangent to a caustic (Refs. 5.18, Chap. 6; 5.29, Chap. 6].

A consistent analysis of the sound field produced by a point impulsive source in a layered medium, with a discussion of various caustics as well as under conditions of waveguide propagation, was presented in [5.21–25, 30–32]. Spherical pulse reflection at an interface of two (fluid or solid) half-spaces was considered, in particular, in ([5.26], [Ref. 5.27; Chap. 6], [5.28]).

6. Universal Properties of the Plane-Wave Reflection and Transmission Coefficients

The reflection and transmission coefficients of plane monochromatic waves have several universal properties which do not depend on the type of medium stratification. We begin this chapter with a discussion of the symmetry properties of the transmission coefficient which are retained upon reversion of the direction of wave propagation [6.1].

6.1 Symmetry with Respect to Reversion of the Wave Path

6.1.1 Sound Waves in Fluids

Let the inhomogeneous layer $z_2 < z < z_1$ with the sound velocity $c(z)$, density $\varrho(z)$, and flow velocity $v_0(z)$ be placed between the homogeneous half-spaces with parameters c_1, ϱ_1, v_{01} $(z > z_1)$ and c_2, ϱ_2, v_{02} $(z < z_2)$. When dissipation occurs, the wave number and density may be complex values. All the parameters are assumed to be piece-wise continuous with respect to z. According to (1.2.10, 12) the wave that is incident from the upper half-space is given by

$$p_i = A \exp\left[i\boldsymbol{\xi} \cdot \boldsymbol{r} - i\nu_1(z - z_1)\right] \quad,$$
$$\nu_1 = \sqrt{k_1^2 \beta_1^2 - \xi^2} \quad, \quad \mathrm{Im}\{\nu_1\} \geq 0 \quad, \tag{6.1.1}$$

where $\beta = 1 - \boldsymbol{\xi} \cdot \boldsymbol{v}_0/\omega$, as above. The condition on the sign of the square root in (6.1.1) ensures that the amplitude of the wave does not increase in the propagation direction. In the lower half-space the plane wave that is generated has the same value of the horizontal wave vector $\boldsymbol{\xi}$:

$$p_t = A W_{12}(\boldsymbol{\xi}) \exp\left[i\boldsymbol{\xi} \cdot \boldsymbol{r} - i\nu_2(z - z_2)\right] \quad,$$
$$\nu_2 = \sqrt{k_2^2 \beta_2^2 - \xi^2} \quad, \quad \mathrm{Im}\{\nu_2\} \geq 0 \quad. \tag{6.1.2}$$

The quantity W_{12} is the transmission coefficient of the layer (with respect to acoustic pressure) when the wave is incident from the upper half-space.

In the case under consideration it is convenient to take the wave equation in the form of (1.2.25). Denoting some pair of continuously differentiable linearly independent solutions of this equation by $f_{1,2}(\zeta)$, we obtain for the acoustic pressure in an inhomogeneous medium

$$p = A \exp\left(i\boldsymbol{\xi} \cdot \boldsymbol{r}\right)[B_1 f_1(\zeta) + B_2 f_2(\zeta)] \quad, \quad \zeta(z_2) \equiv \zeta_2 \leq \zeta \leq \zeta_1 \equiv \zeta(z_1) \quad, \tag{6.1.3}$$

where $B_{1,2}$ are unknown constants which can be determined from the boundary conditions at $\zeta = \zeta_{1,2}$. From (6.1.1–3) and (1.2.26) we have

$$B_1 f_1(\zeta_1) + B_2 f_2(\zeta_1) = 1 + V_1 \quad ; \quad B_1 f_1(\zeta_2) + B_2 f_2(\zeta_2) = W_{12} \quad ;$$
$$B_1 f_1'(\zeta_1) + B_2 f_2'(\zeta_1) = -i(1 - V_1)\nu_1 \varrho_0/\varrho_1 \beta_1^2 \quad ;$$
$$B_1 f_1'(\zeta_2) + B_2 f_2'(\zeta_2) = -i W_{12}\nu_2 \varrho_0/\varrho_2 \beta_2^2 \quad . \tag{6.1.4}$$

Eliminating $B_{1,2}$ and the reflection coefficient V_1 we find

$$W_{12}(\xi) = -2iw\varrho_2\beta_2^2(\varrho_0\nu_2)^{-1}\{[f_1(\zeta_1)f_2(\zeta_2) - f_1(\zeta_2)f_2(\zeta_1)] + [f_1'(\zeta_1)f_2'(\zeta_2)$$
$$- f_1'(\zeta_2)f_2'(\zeta_1)]\varrho_1\varrho_2\beta_1^2\beta_2^2/\varrho_0^2\nu_1\nu_2] + i[(f_2(\zeta_2)f_1'(\zeta_1)$$
$$- f_1(\zeta_2)f_2'(\zeta_1))\varrho_1\beta_1^2/\varrho_0\nu_1 + (f_1'(\zeta_2)f_2(\zeta_1)$$
$$- f_1(\zeta_1)f_2'(\zeta_2))\varrho_2\beta_2^2/\varrho_0\nu_2]\}^{-1} \quad , \tag{6.1.5}$$

where $w = f_1(\zeta_2)f_2'(\zeta_2) - f_1'(\zeta_2)f_2(\zeta_2)$ is the Wronskian of the solutions f_1 and f_2 which does not depend on the point at which the values of functions $f_{1,2}$ and their derivatives are evaluated.

It is clear from the symmetry of the problem (and can be confirmed by direct calculation) that the expression for the transmission coefficient of the layer for the plane wave incident from the lower half-space, W_{21}, can be obtained from (6.1.5) by subscript interchange $1 \leftrightarrow 2$ and multiplication of $\nu_{1,2}$ by (-1). This does not change the quantities in square brackets in (6.1.5), and w transfers into $(-w)$. Thus we obtain the identity

$$\frac{W_{12}(\xi)\nu_2}{\varrho_2\beta_2^2} = \frac{W_{21}(\xi)\nu_1}{\varrho_1\beta_1^2} \quad , \tag{6.1.6a}$$

which expresses the symmetry of the transmission coefficients for an arbitrary inhomogeneous layer. If one introduces the impedances of the half-spaces $Z_j = \omega\varrho_j\beta_j^2/\nu_j$, $j = 1, 2$, as in Sect. 2.6, then (6.1.6a) can be written more compactly:

$$W_{12}(\xi)Z_1(\xi) = W_{21}(\xi)Z_2(\xi) \quad . \tag{6.1.6b}$$

For an unmoving layered medium, for (6.1.6) to be valid it is sufficient to require that the moduli of the horizontal wave vectors of waves incident from both upper and lower half-spaces be equal, since the transmission coefficient is independent of the vector orientation in the horizontal plane. The direction of the horizontal wave vector becomes significant when a flow is present, however. Figure 6.1a shows for the case of a moving medium the total wave vectors of plane waves incident from the upper and lower half-spaces for which the transmission coefficients are related by (6.1.6), where the solid curve shows the flow velocity distribution in the layer. An analogous expression is true for $W_{12}(\xi)$ and $W_{21}(-\xi)$ (Fig. 6.1b), if for the wave incident from the lower half-space the distribution of the current velocity $v_0(z)$ is replaced by $-v_0(z)$ (in other words, if the flow direction is reversed).

Now consider the energy transmission coefficients of the layer for a wave incident from above and from below R_{12} and R_{21}, respectively. (The definition of the energy transmission coefficient is given in Sect. 2.2). The z-component I_z of the acoustic power flux density vector in the moving medium, averaged over one

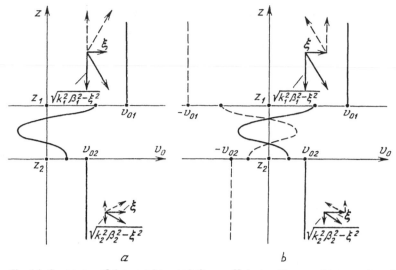

Fig. 6.1. Symmetry of the sound transmission coefficients with respect to reversion of the wave path in a moving medium (**a**) without and (**b**) with flow reversion. Wave vectors of plane waves for which the transmission coefficients are related through (6.1.6) are shown by the *solid* (*dashed*) *arrows* for the wave incident from above (from below). In **a** the flow velocity profile $v_0(z)$ is the same for both waves; in **b** the wave incident from above propagates in the medium with the flow velocity profile shown by the *solid curve* and the wave incident from below with the flow velocity profile shown by the *dashed curve*

period, equals ([6.2], [Ref. 6.3; Sect. 1.7]), cf. (2.1.11),

$$I_z = 0.5 \, \text{Re} \, \{p^* w / \beta\} = (2\omega \varrho_0)^{-1} \, \text{Im} \, \{p^* p_\zeta\} \quad , \tag{6.1.7}$$

where w is the vertical component of the sound particle velocity. Calculating the densities of the power flow for incident and transmitted waves at the points (x, y, z_1) and (x, y, z_2) with the help of (6.1.1, 2) we find

$$
\begin{aligned}
R_{12}(\xi) &= \frac{I_z^{(t)}}{I_z^{(i)}} = \frac{|W_{12}(\xi)|^2 \, \text{Re} \, \{v_2 / \varrho_2 \beta_2^2\}}{\text{Re} \, \{v_1 / \varrho_1 \beta_1^2\}} \\
&= \frac{|W_{12}(\xi)|^2 \, \text{Re} \, \{1/Z_2(\xi)\}}{\text{Re} \, \{1/Z_1(\xi)\}} \quad .
\end{aligned} \tag{6.1.8a}
$$

Similarly

$$R_{21}(\xi) = \frac{|W_{21}(\xi)|^2 \, \text{Re} \, \{v_1 / \varrho_1 \beta_1^2\}}{\text{Re} \, \{v_2 / \varrho_2 \beta_2^2\}} = \frac{|W_{21}(\xi)|^2 \, \text{Re} \, \{Z_1^{-1}(\xi)\}}{\text{Re} \, \{Z_2^{-1}(\xi)\}} \quad . \tag{6.1.8b}$$

When the quantities ϱ_j, β_j, and v_j for $j = 1, 2$ are real, then from (6.1.6) we obtain the identity

$$R_{12}(\xi) = R_{21}(\xi) \quad . \tag{6.1.9}$$

Consequently, if the incident and transmitted waves in the input and output media are homogeneous plane waves, then the value of the energy transmission coefficient is not changed when the wave path is reversed. Equations (6.1.6, 9) were obtained in Sect. 2.2 for the interface of homogeneous media.

Let us make an additional assumption: that there is no absorption in the entire medium. Then because of energy conservation the sum of the vertical components of the power flux density vectors in the reflected and transmitted waves is equal to the vertical component of the power flux density vector in the incident wave, and from (6.1.9) it follows that $|V_1| = |V_2|$. Thus, for $\varrho(z)$, $c(z)$, and $\nu_{1,2}$ real, the modulus of the reflection coefficient of a plane wave reflected from arbitrary inhomogeneous layer is not changed when the wave path is reversed.

For total internal reflection, one of the quantities ν_j is real and other imaginary and (6.1.9) becomes invalid. When the half-spaces are absorbing, (6.1.9) is satisfied only under the condition

$$|Z_1| \, \mathrm{Re} \, \{Z_1^{-1}\} = \pm |Z_2| \, \mathrm{Re} \, \{Z_2^{-1}\} \quad . \tag{6.1.10}$$

Thus, the visual result (6.1.9) cannot be as widely applied as the expressions (6.1.6) which relate the transmission coefficients of a layer with respect to acoustic pressure.

Relation (6.1.9) was obtained in [6.4] for an unmoving discretely layered fluid. Related problems were also considered in [6.5; 6.22, Chaps. 2 and 12; 6.23].

6.1.2 Elastic Waves in Solids

For SH waves in a solid the analysis is similar to that above. The symmetry relation for the transmission coefficients with respect to particle displacement is written in the form

$$W_{12}(\xi)\mu_2 \sqrt{\varrho_2\omega^2/\mu_2 - \xi^2} = W_{21}(\xi)\mu_1 \sqrt{\varrho_1\omega^2/\mu_1 - \xi^2} \quad , \tag{6.1.11}$$

where μ is the shear modulus. When using the notation $Z_j = 1/(\varrho_j\mu_j - \mu_j^2\xi^2/\omega^2)^{1/2}$, $j = 1,2$, then the condition required for (6.1.9) to be valid will still be (6.1.10).

The components of the scattering matrix of a P-SV type plane wave in an arbitrary layered inhomogeneous solid also possess some symmetry properties. These properties were examined in [6.6, 7, 24] for a single interface and in [6.4, 8, 9, 25] for a multilayered medium.

Let us consider the set of $(n - 1)$ homogeneous elastic layers enclosed between two elastic half-spaces, as shown in Fig. 6.2. For the displacement potentials we keep the notation used in (4.3.2). Reflection of plane waves is completely described by the scattering matrix $[S]$ which, in the present case, relates the potentials of elastic waves incident upon the boundaries of the half-spaces with those going away to infinity (Fig. 6.2):

$$(\varphi_1^{(n+1)}, \psi_1^{(n+1)}, \varphi_2^{(1)}, \psi_2^{(1)})^T = [S](\varphi_2^{(n+1)}, \psi_2^{(n+1)}, \varphi_1^{(1)}, \psi_1^{(1)})^T \quad . \tag{6.1.12}$$

The $[S]$-matrix components have the same physical meaning of the reflection, transmission, and transformation coefficients as in (4.2.2).

For fields that are harmonically dependent on horizontal coordinates and time, the values of the displacement-stress vectors at two boundaries of a homogeneous layer are related by (4.3.7): $f(z_j) = [A^{(j)}]f(z_{j-1})$. Due to the property of the prop-

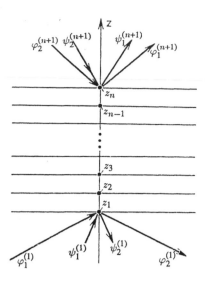

agator matrix given by (4.3.10), elements of $[A^{(j)}]^{-1}$ matrix, which is the inverse of $[A^{(j)}]$, can be obtained from (4.3.8) by changing the signs of P and Q. Noting this and the presence of identical elements in $[A^{(j)}]$, it is not difficult to check that the matrix equality

$$[A^{(j)}]^T[M] = [M][A^{(j)}]^{-1} \tag{6.1.13}$$

holds, where

$$[M] = \begin{pmatrix} 0 & 0 & 0 & 1 \\ 0 & 0 & -1 & 0 \\ 0 & 1 & 0 & 0 \\ -1 & 0 & 0 & 0 \end{pmatrix} . \tag{6.1.14}$$

Let $f_1(z)$ and $f_2(z)$ be some diplacement-stress vectors. Then the scalar quantity

$$F(z) = [f_1(z)]^T[M]f_2(z) \tag{6.1.15}$$

remains constant inside a layer. Indeed we have, according to (6.1.13):

$$\begin{aligned} F(z_j) &= \{[A^{(j)}]f_1(z_{j-1})\}^T[M][A^{(j)}]f_2(z_{j-1}) \\ &= [f_1(z_{j-1})]^T[A^{(j)}]^T[M][A^{(j)}]f_2(z_{j-1}) = F(z_{j-1}) \quad . \end{aligned} \tag{6.1.16}$$

Since the quantity F as well as the vectors $f_{1,2}$ are continuous at the layer boundaries, it follows from (6.1.16) that F is constant throughout the medium. In terms of the propagator matrix this property is written as, see (4.3.9b),

$$\begin{aligned} [f_1(z)]^T[M]f_2(z) &= [f_1(\tilde{z})]^T[A(z,\tilde{z})]^T[M][A(z,\tilde{z})]f_2(\tilde{z}) \\ &= [f_1(\tilde{z})]^T[M]f_2(\tilde{z}) \quad . \end{aligned} \tag{6.1.17}$$

The vectors $f_{1,2}(\tilde{z})$ may be chosen arbitrarily. Consequently, in any discretely layered medium the identity

$$[A]^T[M][A] = [M] \quad , \tag{6.1.18}$$

is true, which generalizes the equality in (6.1.13).

It is convenient to express the invariant F in terms of potentials of elastic waves. It follows from (4.3.4) and (6.1.15) that

$$F = [\varphi_1(z)]^T[N]\varphi_2(z) \quad , \tag{6.1.19a}$$

where

$$[N(z)] = [L(z)]^T[M][L(z)] \quad . \tag{6.1.19b}$$

The matrix $[L]$ is defined by (4.3.5). A direct calculation gives

$$[N] = 2i\varrho\omega^2 \begin{pmatrix} 0 & \alpha & 0 & 0 \\ -\alpha & 0 & 0 & 0 \\ 0 & 0 & 0 & -\beta \\ 0 & 0 & \beta & 0 \end{pmatrix} \quad . \tag{6.1.20}$$

A dilatational plane wave incident from the upper half-space with an amplitude equal to unity corresponds to a vector of potentials (Fig. 6.2)

$$\varphi_1 = (V_{ll}, 1, V_{lt}, 0)^T \quad , \quad z \geq z_n \quad ; \quad \varphi_1 = (0, W_{ll}, 0, W_{lt})^T \quad , \quad z \leq z_1 \quad . \tag{6.1.21a}$$

Similarly, for vectors of potentials of a shear wave incident from the upper medium φ_2 and of dilatational and shear waves incident from the lower medium φ_3 and φ_4 we have

$$\begin{aligned}
\varphi_2 &= (V_{tl}, 0, V_{tt}, 1)^T \quad , \\
\varphi_3 &= (\tilde{W}_{ll}, 0, \tilde{W}_{lt}, 0)^T \quad , \quad z \geq z_n \quad ; \\
\varphi_4 &= (\tilde{W}_{tl}, 0, \tilde{W}_{tt}, 0)^T \quad , \\
\varphi_2 &= (0, W_{tl}, 0, W_{tt})^T \quad , \\
\varphi_3 &= (1, \tilde{V}_{ll}, 0, \tilde{V}_{lt})^T \quad , \quad z \leq z_1 \quad , \\
\varphi_4 &= (0, \tilde{V}_{tl}, 1, \tilde{V}_{tt})^T \quad .
\end{aligned} \tag{6.1.21b}$$

By calculating the invariant F for different pairs of the vectors φ_j, $j = 1, 2, 3, 4$ by (6.1.19, 20) and setting the F values equal at points $z = z_n$ and $z = z_1$, we find the symmetry relations for the plane wave reflection, transmission, and transformation coefficients:

$$\alpha_n V_{tl} + \beta_n V_{lt} = 0 \quad , \quad \alpha_1 \tilde{V}_{tl} + \beta_1 \tilde{V}_{lt} = 0 \quad ; \tag{6.1.22}$$

$$\alpha_1 \varrho_1 W_{ll} = \alpha_n \varrho_n \tilde{W}_{ll} \quad ; \tag{6.1.23}$$

$$\beta_1 \varrho_1 W_{tt} = \beta_n \varrho_n \tilde{W}_{tt} \quad ; \tag{6.1.24}$$

$$\beta_1 \varrho_1 W_{lt} = -\alpha_n \varrho_n \tilde{W}_{tl} \quad , \quad \alpha_1 \varrho_1 W_{tl} = -\beta_n \varrho_n \tilde{W}_{lt} \quad . \tag{6.1.25}$$

The symmetry relations of (6.1.18) and (6.1.22–25) were proved above for a discretely layered medium. However, they are also valid when the z-dependence of the medium parameters in the layer $z_1 \leq z \leq z_n$ is arbitrary. This is because one

131

can consider continuously varying parameters to be the limit of discrete variations in multiple homogeneous layers as the thickness of the layers tends to zero. In deriving (6.1.18) and (6.1.22–25) it was never assumed that ϱ and the wavenumbers of dilatational and shear waves were real. Hence, the symmetry relations are also valid in an absorbing medium.

In the particular case of reflection at an interface of two homogeneous half-spaces the scattering matrix was found explicitly in Sect. 4.2. Equation (6.1.22) shows that the relation between the transformation coefficients V_{tl} and V_{lt} given by (4.2.10) is universal. Using (4.2.11, 12) one can check, after cumbersome transformations, that (6.1.23–25) are satisfied in the given case. Equation (6.1.23) is analogous to (6.1.6) obtained for sound waves in a fluid. In the case of a fluid at rest it is easy to express both (6.1.6 and 23) in the same form if one keeps in mind that these equations are written in terms of transmission coefficients with respect to pressure p and potential φ, respectively, which are related by (4.2.17).

6.2 Analytic Properties

One widely used method to study the field of a point source in a layered medium is by decomposing it into plane waves. That is why it is important to know the properties of solutions of one-dimensional wave equations as well as the reflection and transmission coefficients as functions of the angle of incidence or the horizontal component of a wave vector. These properties are considered in detail in mathematical literature for a one-dimensional reduced wave equation [6.5, 10, 11]. In the more general case of sound wave propagation in a motionless fluid with stratification of sound speed c and density ϱ, a number of universal properties of the reflection and transmission coefficients were proved by *Brekhovskikh* [6.12], see also [6.26].

Let $c(z)$ and $\varrho(z)$ be smooth functions tending to c_1, ϱ_1 and c_2, ϱ_2 as $z \to +\infty$ and $z \to -\infty$, respectively. We assume also that for any z $\varrho \neq 0$. It was shown in [6.12] that under these conditions the sound pressure $p(\xi, z)$ caused by plane incident wave in a layered medium is an analytic function of ξ. (For a discussion of properties of analytic functions see [6.13, 14] or any other handbook on the theory of functions of complex variables.) The reflection and transmission coefficients $V(\xi)$ and $W(\xi)$ of a plane wave incident from a homogeneous medium onto a layered half-space are also analytical functions of ξ. They have no essential singular points in a finite part of the complex plane. By approximating the discontinuities of c and ϱ to be limits of rapid changes of smooth functions one can use the above derived results for media with piece-wise smooth density and sound speed. In this case, at a number of points p has no even first derivatives with respect to z but does remain an analytic function of ξ. An illustration of these properties of the sound field is given by the results for discretely layered media in Chap. 2. For example, in the case of reflection of a plane wave at an interface of two homogeneous fluids the sound field given by (2.2.2, 3) has a discontinuity in the derivative $\partial p / \partial z$ at the interface, but p, as well as the reflection and transmission coefficients given by (2.2.14, 18) are analytic functions of the variable $\xi = k_1 \sin \theta_1 = k \sin \theta$ for all z. A proof of analyticity will be given in Chap. 10 for the general case.

6.2.1 Poles of the Reflection and Transmission Coefficients

In the ξ complex plane the functions $p(\xi, z)$, $V(\xi)$, and $W(\xi)$ may have isolated singular points, namely, poles and branch points. It was shown in Chap. 4 that reflection-coefficient poles are related to surface and "leaky" waves. The acoustic energy conservation law confines the poles to be in some regions of the ξ complex plane. The vertical components of wave vectors of reflected and transmitted waves must have, at the poles $\xi = \xi_p$, positive and negative imaginary parts, respectively, and hence, these waves decay to zero as $|z| \to \infty$. Otherwise the reflected and transmitted waves would carry an infinite energy flux away from the boundary while the incoming energy flux of an incident wave would be finite.

We shall now show that in a nonabsorbing medium with $k = \omega/c(z)$ and $\varrho(z)$ real all the poles of V and W are on the real ξ axis. Let the medium be homogeneous for $z > 0$. A plane wave $\exp[i(\xi x - \nu_1 z)]$ of amplitude 1 is incident on a boundary at $z = 0$ of a layered half-space. Let $f(\xi, \zeta)$ be the solution to the wave equation

$$\frac{d^2 p}{d\zeta^2} + \varrho_1^2 \varrho^{-2}(k^2 - \xi^2)p = 0 \quad , \quad \zeta = \varrho_1^{-1} \int_0^z \varrho(\tilde{z})d\tilde{z} \quad , \tag{6.2.1}$$

which reduces to a plane wave of amplitude 1 propagating to infinity as $z \to -\infty$, i.e.,

$$\lim_{z \to -\infty} [f(\xi, \zeta) \exp(i\nu_2 z)] = 1 \quad . \tag{6.2.2}$$

Then in the lower half-space $p = W f(\xi, \zeta) \exp(i\xi x)$. When $z > 0$, we have $\zeta = z$ and

$$p = [\exp(-i\nu_1\zeta) + V \exp(i\nu_1\zeta)] \exp(i\xi x) \quad . \tag{6.2.3}$$

For brevity the following designations are used

$$\nu_j(\xi) = (k_j^2 - \xi^2)^{1/2} \quad , \quad \text{Im}\{\nu_j\} \geq 0, \quad j = 1, 2 \quad . \tag{6.2.4}$$

To determine the reflection and transmission coefficients V and W we write the boundary conditions at $\zeta = 0$:

$$W f(\xi, 0) = 1 + V \quad , \quad W f'(\xi, 0) = -i\nu_1(1 - V) \quad , \tag{6.2.5}$$

where the prime indicates differentiation with respect to ζ. From (6.2.5) it follows that

$$W(\xi) = \frac{2}{f(\xi, 0) + i f'(\xi, 0)/\nu_1} \quad , \quad V(\xi) = f(\xi, 0)W(\xi) - 1 \quad . \tag{6.2.6}$$

Hence, $\xi = \xi_p$ is the pole of the transmission coefficient if

$$Q(\xi_p) \equiv f(\xi_p, 0) + i f'(\xi_p, 0)/\nu_1 = 0 \quad . \tag{6.2.7}$$

As long as $f(\xi_p, 0) \neq 0$ ξ_p is simultaneously the pole of the reflection coefficient. [From the assumption that $f(\xi_p, 0) = 0$ and (6.2.7), it follows that $f'(\xi_p, 0) = 0$. Then $f(\xi_p, \zeta) \equiv 0$ which contradicts (6.2.2)].

The function f^*, the complex conjugate of $f(\xi_p, \zeta)$, is a solution of

$$\frac{d^2 f^*}{d\zeta^2} + \varrho_1^2 \varrho^{-2}[k^2 - (\xi_p^2)^*]f^* = 0 \quad .$$

Multiplying this equation by $(-f)$ and then adding it to (6.2.1) multiplied by f^* [where p is expressed through $f(\xi_p, \zeta)$], we obtain, after integration over ζ, the identity

$$[f^* f' - (f^*)' f]\Big|_{-\infty}^{+\infty} = [\xi_p^2 - (\xi_p^*)^2] \int_{-\infty}^{+\infty} |f|^2 \varrho_1^2 \varrho^{-2} d\zeta \quad . \tag{6.2.8}$$

From (6.2.3 and 6) it follows that $V(\xi_p) = \infty$ and

$$f(\xi_p, \zeta) = f(\xi_p, 0) \exp(i\nu_1 \zeta) \quad , \quad \zeta > 0 \quad . \tag{6.2.9}$$

It was shown above that $\mathrm{Im}\{\nu_{1,2}(\xi_p)\} > 0$, as ξ_p is a pole. Then, according to (6.2.2, 9), $\lim\limits_{|\zeta| \to \infty} (f^* f' - (f^*)' f) = 0$. Hence, (6.2.8) gives $\xi_p^2 = (\xi_p^*)^2$. We also have $\xi_p^2 > k_{1,2}^2$ because $\mathrm{Im}\{\nu_{1,2}\}$ are positive. It's clear that if $\xi = \xi_p$ is a pole, then $\xi = -\xi_p$ is a pole, too. Thus the poles of the reflection and transmission coefficients are on a real ξ-axis and are distributed in pairs symmetric about the point $\xi = 0$. In addition $\nu_j(\xi_p) = i|\nu_j(\xi_p)|$, $j = 1, 2$.

Proceeding with this discussion of the poles, we would like to show that all the poles of the reflection and transmission coefficients are simple, i.e., the quantities $(\xi - \xi_p)V$ and $(\xi - \xi_p)W$ tend to finite limits as $\xi \to \xi_p$. Obviously, a derivative $dQ/d\xi$ is significant. It follows from (6.2.7) that

$$\frac{dQ(\xi_p)}{d\xi} = \varphi(0) + i\nu_1^{-1}\varphi'(0) + i\xi_p \nu_1^{-3} f'(\xi_p, 0) \quad . \tag{6.2.10}$$

Here the function $\varphi(\zeta) \equiv \partial f(\xi_p, \zeta)/\partial\xi$ obeys the equation

$$\frac{d^2\varphi}{d\zeta^2} + \varrho_1^2 \varrho^{-2}(k^2 - \xi_p^2)\varphi = 2\xi_p \varrho_1^2 \varrho^{-2} f(\xi_p, \zeta) \quad . \tag{6.2.11}$$

Let f_1 be the solution of (6.2.1) and have the following asymptotic behavior:

$$\lim_{z \to -\infty} [f_1(\xi, \zeta) \exp(-i\nu_2 z)] = 1 \quad . \tag{6.2.12}$$

The functions f and f_1 are linear independent. We shall use them to cast a solution to (6.2.11) in the form

$$\varphi = A(\zeta) f(\xi_p, \zeta) + A_1(\zeta) f_1(\xi_p, \zeta) \quad . \tag{6.2.13}$$

Let

$$A' f + A_1' f_1 = 0 \quad . \tag{6.2.14}$$

Then (6.2.11) reduces to

$$A' f' + A_1' f_1' = 2\xi_p \varrho_1^2 \varrho^{-2} f(\xi_p, \zeta) \quad . \tag{6.2.15}$$

Solving (6.2.14, 15) with respect to A' and A_1' we obtain the general solution of (6.2.11)

$$\varphi = Bf + B_1 f_1 + 2\xi_p \varrho_1^2 w^{-1} \left(f \int_0^\zeta \varrho^{-2} f f_1 d\zeta - f_1 \int_{-\infty}^\zeta \varrho^{-2} f^2 d\zeta \right) \quad , \tag{6.2.16}$$

where B and B_1 are arbitrary constants and

$$w = f'f_1 - ff_1' = \text{const} \tag{6.2.17}$$

is a Wronskian of the solutions f and f_1.

For $\zeta \to -\infty$ φ should tend to zero as f does. Hence, $B_1 = 0$. Substituting (6.2.16) into (6.2.10) and using (6.2.7, 17), we obtain

$$\frac{dQ(\xi_p)}{d\xi} = -2\xi_p[f'(\xi_p,0)]^{-1}\left\{ \int_{-\infty}^{0} \left(\frac{\varrho_1 f}{\varrho}\right)^2 d\zeta + \frac{i}{2\nu_1}f^2(\xi_p,0) \right\} \ . \tag{6.2.18}$$

It was shown above that $i/\nu_1(\xi_p)$ is a positive number. The function $f(\xi_p,\zeta)$ is a real-valued function of ζ since the coefficients in both (6.2.1) and (6.2.2) are real for $\xi = \xi_p$. Hence, the quantity in braces in (6.2.18) is positive and $dQ(\xi_p)/d\xi = 0$ only if $\xi_p = 0$. But this point is not a pole because Re $\{\nu_1(0)\} = k_1 \neq 0$. Thus, it is proven that all the zeros of the denominator in the expressions for W and V are simple, i.e., all the poles of the reflection and transmission coefficients are simple.

One can also show that these properties of the poles, namely, that $|\xi_p| > k_1$, that all the poles are simple and that they fall on the real ξ-axis, are also valid when the lower medium is restricted by a surface $z = z_1 < 0$ with a real impedance Z independent of ξ. In particular, this surface may be pressure-release or absolutely rigid boundary.

As an example, consider the poles of the transmission coefficient of an Epstein layer. In this case, according to (3.4.12), $W = \Gamma(1 + \alpha - \gamma)\Gamma(1 - \beta)/\Gamma(\alpha + 1 - \beta)\Gamma(1 - \gamma)$. The quantities α, β, γ are related to the layer and to the incident wave parameters by (3.4.7). The gamma function $\Gamma(y)$ never equals zero and has only simple poles at the points $y = -m$, $m = 0, 1, 2\ldots$ [Ref. 6.15; Chap. 6]. To determine the poles of W one should solve the equations $1 - \beta = -m$, $1 + \alpha - \gamma = -m$ which, due to (3.4.7), become

$$1 - \sqrt{1 - 16Mk_0^2b^{-2}} - 2ib^{-1}\left[\sqrt{k_0^2 - \xi^2} + \sqrt{k_0^2 - \xi^2 - k_0^2N}\right] = -2m \quad , \tag{6.2.19a}$$

$$1 + \sqrt{1 - 16Mk_0^2b^{-2}} - 2ib^{-1}\left[\sqrt{k_0^2 - \xi^2} + \sqrt{k_0^2 - \xi^2 - k_0^2N}\right] = -2m \quad . \tag{6.2.19b}$$

As usual it is supposed that the imaginary parts of all the square roots are nonnegative. The quantity $Mk_0^2b^{-2}$ is real. That is why the real part of the left-hand side of (6.2.19b) is not less than 1, and this equation has no solutions. Inspection of the real and imaginary parts of the left-hand side of (6.2.19a) shows that solutions exist only when the quantities

$$\sqrt{1 - 16Mk_0^2b^{-2}} \equiv 2d_2 \ , \quad \sqrt{\xi^2 - k_0^2} \ , \quad \text{and} \quad \sqrt{\xi^2 + k_0^2N - k_0^2}$$

are real. Hence, in accordance with the general theory, all the poles of $W(\xi)$ are simple and lie on the real ξ-axis, satisfying the inequalities $\xi^2 \geq k_0^2$, $\xi^2 \geq k_0^2(1 - N)$. It will be shown below that the points $\xi^2 = k_0^2$ and $\xi^2 = k_0^2(1 - N)$ are also not poles.

Solving an irrational algebraic equation (6.2.19a) gives the positions of the poles:

$$\xi_p^2 = k_0^2 + \tfrac{1}{4}(B - k_0^2 N/B)^2 \quad , \qquad B = b(d_2 - m - 0.5) \quad . \tag{6.2.20}$$

The solutions exist for m values satisfying the inequality

$$2m \le 2d_2 - 2k_0\sqrt{|N|}/b - 1 \quad . \tag{6.2.21}$$

If the right-hand side is negative (for example, if $M = 0$, $N \ne 0$), then W has no poles at all. For some combinations of the layer's parameters, ξ_p^2 in (6.2.20) may be equal to k_0^2 or $k_0^2(1 - N)$. In the first case $\Gamma(1 - \gamma) = \Gamma[2b^{-1}(\xi^2 - k_0^2)^{1/2}]$ in (3.4.12) and is infinite simultaneously with $\Gamma(1 - \beta)$, while W remains finite. In the second case, note that for $\xi^2 \ne k_0^2$ the denominator in (3.4.12) is bounded. In (6.2.20) $\xi_p^2 = k_0^2(1 - N)$ and $\xi_p^2 \ne k_0^2$ when $N = -b^2 k_0^2(m - 0.5 - d_2)^2 < 0$. Because the poles of a Γ-function are simple, W tends to infinity at a rate prescribed by $W \sim [\xi^2 - k_0^2(1 - N)]^{-1/2}$. Hence, these singularities are not poles. Thus we encounter another type of singularity in the reflection and transmission coefficients, namely, branch points. These are closely related to the lateral waves [Refs. 6.16, Chaps. 4, 5; 6.27, Chap. 3] which are a sort of diffraction component of a field of a localized sound source.

6.2.2 Branch Points

For an unbounded layered fluid there are four branch points: $\xi = \pm k(\pm\infty)$. We shall first prove this statement for the reflection coefficient from a discretely layered medium. According to (2.5.7), V depends on ξ through the vertical components of the wave vectors $\nu_j = (k_j^2 - \xi^2)^{1/2}$ in all the layers ($j = 2, \ldots, n$) and two half-spaces ($j = 1, n + 1$). The wave vector ν_j is included in the impedance $Z_j = \omega\varrho_j/\nu_j$ and in the quantity $s_j = \tan(\nu_j d_j)$. [For details see (2.5.1–4).] The dependence of V on ν_j is analytic, hence V has no other branch points except for branch points $\xi = \pm k_j$ of the ν_j themselves. The quantities ν_j of the layers enter in an even manner into (2.5.4) (which relates the input impedances at the upper and lower boundaries of a layer), and hence into the input impedance $Z_{in}^{(n)}$ of the entire system. It means, that a change of the sign of ν_j causes $Z_j \to -Z_j$, $s_j \to -s_j$, but does not affect $Z_{in}^{(n)}$. That is why in the vicinity of the point $\xi = k_j$ one can represent the reflection coefficient as a series in ν_j^2, i.e., in powers of $(k_j^2 - \xi^2)$, and branching doesn't occur. On the contrary, the vertical components of the wave vectors in the half-spaces ν_1 and ν_{n+1} enter (2.5.4) only through Z_1 and Z_{n+1}. Therefore, in the vicinities of the points $\xi = \pm k_1$, a series expansion contains even as well as odd powers of $(k_1^2 - \xi^2)^{1/2}$. Hence, $\xi = \pm k_1$ are the branch points of the function $V(\xi)$, and by analogy $\xi = \pm k_{n+1}$ are also branch points. Consider, for example, the Fresnel reflection coefficient given by (2.2.14). In terms of the designations used in the present section

$$V = \frac{\varrho_1 \sqrt{k_2^2 - \xi^2} - \varrho_2 \sqrt{k_1^2 - \xi^2}}{\varrho_1 \sqrt{k_2^2 - \xi^2} + \varrho_2 \sqrt{k_1^2 - \xi^2}} \quad .$$

The location of branch points in this case is clear and illustrates the general statements proved above. Note that the branch points correspond to incidence angles of wave which are equal to $\frac{\pi}{2}$ and the critical angle for total reflection.

If a medium is restricted from below by a plane $z = z_1$ possessing an impedance $Z(\xi)$ which has no branch points (for example, the $z = z_1$ surface may be a pressure-release or an absolutely rigid boundary), then the impedance Z_1 does not enter into the expression for $V(\xi)$, and only the branch points $\xi = \pm k_{n+1}$ remain. In a layered medium which is restricted by two impedance boundaries, the quantity $p(\xi, z)/p(\xi, z_0)$ where $z_0 = \text{const}$, has no branch points as a function of ξ. By proceeding to the limit one can prove that the results obtained above are valid for media with arbitrary dependencies of $c(z)$ and $\varrho(z)$ in a layer restricted by two homogeneous half-spaces or impedance boundaries.

A similar discussion of the existence of branch points is presented in [Ref. 6.17; Chap. 4] and [Ref. 6.18; Chap. 5]. There is, however, another approach which is also of interest, in which both continuous and discrete variation in the fluid parameters are allowed from the beginning. We shall begin with the expression for the transmission coefficient given in (6.1.5) and retain the designations used there. (The proof is analogous for the reflection coefficient.) As long as the medium is at rest, $\beta(z) \equiv 1$. Let the linearly independent solutions of the wave equation $f_{1,2}(\zeta)$ obey the initial conditions

$$f_1(\zeta_1) = 0 \quad , \quad f_1'(\zeta_1) = k_0 \quad ;$$
$$f_2(\zeta_1) = 1 \quad , \quad f_2'(\zeta_1) = 0 \quad , \quad k_0 = \text{const} \quad . \tag{6.2.22}$$

In the wave equation of (1.2.25), the coefficient before p is an analytic function of ξ. Therefore the $f_{1,2}(\zeta)$, which are solutions of initial value problems (1.2.25) and (6.2.22), are regular functions of the parameters [Ref. 6.19, Part I, Sect. 5]. Equations (6.1.5) and (6.2.22) give

$$W_{12}(\xi) = \frac{2ik_0\varrho_2}{\varrho_0\nu_2} \left\{ \frac{k_0\varrho_1\varrho_2}{\varrho_0^2\nu_1\nu_2} f_2'(\zeta_2) \right.$$
$$\left. - f_1(\zeta_2) + \frac{i}{\varrho_0} \left[\frac{k_0\varrho_1}{\nu_1} f_2(\zeta_2) + \frac{\varrho_2}{\nu_2} f_1'(\zeta_2) \right] \right\}^{-1} \quad . \tag{6.2.23}$$

From (6.2.23) it's obvious that $W_{12}(\xi)$ has exactly four branch points.

In a discretely layered *moving* fluid the impedance is $Z_j = \omega\varrho_j\beta_j^2/\nu_j$, where $\nu_j = (k_j^2\beta_j^2 - \xi^2)^{1/2}$, and $s_j = \tan(\nu_j d_j)$ (Sect. 2.6). Quantity β_j analytically depends on the x- and y-components of the vector $\xi = (\xi_1, \xi_2, 0)$. Therefore, the branch points can only be exclusively at the zeros of ν_j. The same reasoning as was applied to the fluid at rest leads to the conclusion that the branching of $V(\xi)$ and $W(\xi)$ occurs when one of the following conditions is satisfied:

$$k_j^2\beta_j^2 = \xi^2 \quad , \quad j = 1, n+1 \quad . \tag{6.2.24}$$

Let $\xi_2 = 0$ and v_0 be parallel to the x-axis in the lower half-space. Then branching occurs when $\xi_1 = k_1/(k_1 v_{01}/\omega \pm 1)$. In general, the dependence of the modulus of the vector ξ, obeying (6.2.24), on an angle φ_j between ξ and v_{0j}, is given by

$$\xi_j = \frac{k_j}{k_j v_{0j}\omega^{-1}\cos\varphi_j \pm 1} \quad . \tag{6.2.25}$$

For SH waves in a layered solid the results are analogous to those obtained for a fluid at rest. Branch points are at $\xi = \pm k_{t1}, \pm k_{tn+1}$, where k_t is the wave number for the shear waves.

For elastic P-SV waves the scattering matrix $[S]$ is given by (4.3.25, 27, 28), with the matrix propagator $[A(z_{n+1}, z_1)]$ equal to the product of the matrices of the individual layers given by (4.3.8). Branching in the $[S]$-matrix components may occur only due to their dependence on the quantities α_j, β_j, which are the vertical components of the wave vectors for dilatational and shear waves in the layers and in the two half-spaces. Using the relations $\cos \theta_l = \alpha/k_l$, $\cos \theta_t = \beta/k_t$, and (4.3.1) to express the components of the matrix of the layer in terms of ξ, α, and β, one can check that all the components of the matrix are even functions of α and β. Matrix $[L]$ in (4.3.5) contains α and β terms to the first power. The $[L]$ matrices corresponding to both half-spaces enter into the expression for the $[S]$-matrix in (4.3.25). Hence, the components of the scattering matrix for P-SV waves in a layered solid have four pairs of branch points: $\xi = \pm k_{l1}, \pm k_{t1}, \pm k_{ln+1}$, and $\pm k_{tn+1}$. Equations (4.2.8–15) for scattering at an interface of two homogeneous elastic media can be used for illustration of the obtained result.

As was shown above, in a layered medium bounded by homogeneous half-spaces all the branch points are of second order. (Recall that a singular point ξ_0 of a function $(\xi - \xi_0)^{1/n}$ is called a branch point of nth order). If the half-spaces are inhomogeneous but their parameters approach their limiting values sufficiently rapidly as $|z| \to \infty$, then the same type of branching occurs. For example, for the Epstein profile in which $\varrho = $ const and $c(z)$ proceeds exponentially to its limiting value as $|z| \to \infty$, the branch points of the reflection and transmission coefficients are of second order. In cases where the elastic parameters approach their limiting values more slowly other orders of branching are possible.

To demonstrate this possibility, consider sound-wave reflection by a layered half-space $(z < 0)$ in which

$$ k^2(z) = k_2^2 + \alpha_2(1 - z/z_1)^{-2} \quad , \quad z_1 > 0 \quad ; \quad \varrho(z) = \varrho_2 = \text{const} \quad . \qquad (6.2.26) $$

Let k_1 and ϱ_1 be parameters of the homogeneous half-space $z > 0$. The input impedance of the lower medium was found in Sect. 3.2, and according to (3.2.15, 17, 22) it equals

$$ Z = i\omega\varrho_2 \left\{ \frac{1}{2z_1} + \frac{d}{dz_1} \ln H_m^{(1)}[\sqrt{k_2^2 - \xi^2}\, z_1] \right\}^{-1} \quad , \qquad (6.2.27) $$

where

$$ m = \sqrt{\tfrac{1}{4} - \alpha_2 z_1^2} \quad . \qquad (6.2.28) $$

For simplicity, only the case where $0 < m < 1$ will be considered.

To figure out the type of $Z(\xi)$ branching at the points $\xi = \pm k_2$ it is convenient to use the known representation of the Hankel function [Ref. 6.15, Chap. 9]:

$$ H_m^{(1)}(z) = \frac{1}{\sin m\pi} [\exp(-i\pi m) J_m(z) - J_{-m}(z)] \quad , $$

$$ J_m(z) = \left(\frac{z}{2}\right)^m \sum_{n=0}^{\infty} \frac{(-z^2/4)^n}{n!\,\Gamma(m + n + 1)} \quad . \qquad (6.2.29) $$

Substitution of (6.2.29) into (6.2.27) leads to the expression

$$Z = i\omega\varrho_2 \left\{ \frac{1 - 2m}{2z_1} - m\frac{\Gamma(1 - m)}{\Gamma(1 + m)}\exp\left(-i\pi m\right)\left(\frac{z_1}{2}\right)^{2m-1} \right.$$

$$\left. \times (k_2^2 - \xi^2)^m + O[(k_2^2 - \xi^2)^{2m}] \right\}^{-1} . \qquad (6.2.30)$$

In particular, for $\alpha_2 = 0$, when the lower half-space is homogeneous we have $m = \frac{1}{2}$ and from (6.2.30) it follows that

$$Z = \omega\varrho_2(k_2^2 - \xi^2)^{-1/2}\left[1 + O(\sqrt{k_2^2 - \xi^2})\right] ,$$

which is in agreement with the exact value

$$Z = \frac{\varrho_2 c_2}{\cos\theta_2} = \omega\varrho_2(k_2^2 - \xi^2)^{-1/2} .$$

Expression (6.2.30) shows that the input impedance and the reflection coefficient can have branch points of any order, depending on the parameter m. Through the parameter $\alpha_2 \sim \omega^2$, the type of branching depends on the variation of the sound speed in the half-space and on the frequency of the wave.

6.3 Nonreflecting Layers

6.3.1 For Any Fixed Frequency and Angle of Incidence a Reflectionless Layer Exists

By studying sound wave reflection from a homogeneous layer confined between two homogeneous half-spaces, in Sect. 2.4 we have seen that certain combinations of media impedances can produce a nonreflecting system, as in the cases of the half-wave and the quater-wave layers. When the frequency and angle of incidence of the wave are fixed, one can find an entire class of layered media for which $V = 0$ [6.20].

Let us use results obtained in Sect. 3.1. The density of the medium is assumed constant. As a model equation (3.1.7), we shall take the wave equation for a homogeneous medium, that is,

$$g(\eta) = k_0^2 = \text{const} . \qquad (6.3.1)$$

We make the change of variables:

$$\eta(z) = \int_{z_0}^{z} \sqrt{n^2(z) - \sin^2\theta_0}\, dz , \qquad (6.3.2)$$

where $n(z)$ is a smooth function. It is assumed that $\min\,[n^2(z)] > \sin^2\theta_0$. According to (3.1.8–10),

$$\Phi(z) = A_1\Phi_1 + A_2\Phi_2 ,$$

$$\Phi_{1,2} = (n^2 - \sin^2\theta_0)^{-1/4} \cdot \exp\left[\pm ik_0\eta(z)\right] , \quad A_{1,2} = \text{const} \qquad (6.3.3)$$

is a general solution to the equation

$$\frac{d^2\Phi}{dz^2} + k_0^2[\mu^2(z) - \sin^2\theta_0]\Phi = 0 \quad , \quad \mu^2 = n^2 + P(z) \quad , \tag{6.3.4}$$

where

$$P(z) = \frac{4[\ln(n^2 - \sin^2\theta_0)]'' - [(\ln(n^2 - \sin^2\theta_0))']^2}{16k_0^2} \quad . \tag{6.3.5}$$

The function $\Phi(z)$ represents the vertical dependence of the sound field in a medium with refraction index $\mu(z)$ for $\xi = k_0 \sin\theta_0$, and θ_0 is the angle of incidence at that horizon, where $\mu = 1$.

Let us suppose that $n(z)$ tends to constants as $z \to \pm\infty$ and that its derivatives tend to zero sufficiently rapidly. Then for large $|z|$, the solutions Φ_1 and Φ_2 are plane waves; $P(z) = 0$. Obviously, the layer with refraction index $\mu(z)$ is nonreflecting for the given ω and θ_0 since the waves Φ_1 and Φ_2 propagate in it independently in opposite directions. The function $P(z)$ depends on the frequency and incidence angle of the wave, that is, a layer which is nonreflecting for some θ_0 and ω generates reflected waves for other values of θ_0 or ω. The difference between $\mu(z)$ and $n(z)$ tends to zero with increasing frequency.

As an example, consider the case where $\theta_0 = 0$, and $n^2(z)$ is given by the Epstein law, that is $n^2 = k^2(z)/k_0^2$, where $k^2(z)$ is taken from (3.4.1). Then (6.3.5) gives

$$P(z) = -a(z)S^{-2}[2 - N - 4M + \exp(bz) + (1 - N)\exp(-bz)]^{-2} \quad , \tag{6.3.6}$$

where

$$a(z) = \frac{N^2}{4} + 8M^2 + 8M(N - 2) + (N + 4M)\exp(bz)$$

$$+ (N - 1)(N - 4M)\exp(-bz) - 6M\tanh\frac{bz}{2}\left(N + \frac{2}{3}\tanh\frac{bz}{2}\right) \quad . \tag{6.3.7}$$

The quantity S, which is proportional to layer thicnkess, is defined in (3.4.5).

Figure 6.3 shows the additional term $P(z)$ in (6.3.4), which reduces reflection to zero for transition ($M = 0$) and symmetric ($N = 0$) Epstein layers considered in

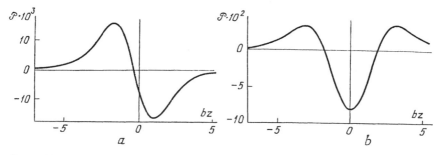

Fig. 6.3. The function $P(z)$ corresponding to transition (**a**) and symmetric (**b**) Epstein layers

Sect. 3.4, and when $S = 2$ that is, when the effective thickness of the layer is close to half the wavelength. For the transition layer (Fig. 6.3a) $N = \frac{1}{2}$, for the symmetric (Fig. 6.3b) $M = -2$. $P(z)$ can be obtained from the figure for any other S values by multiplication of the ordinates by $(2/S)^2$. Note that additions to $n^2(z)$ aren't large for both the symmetric and transition layers for $S = 2$ and are even smaller at larger S.

An example of a nonreflecting (for fixed values of frequency and incidence angle) *moving* medium was given in Sect. 3.7.2.

6.3.2 An Example of Reflectionless Layer for Arbitrary Angle of Incidence

Layered media which are nonreflecting in the entire interval of incidence angles for a given frequency are also available. Such a medium may be derived from the results of Chap. 3. The square of the modulus of the reflection coefficient for a symmetric Epstein layer is given by (3.4.24) if $M \le 0$. In this case the sound speed is an even function of z and has a minimum value at $z = 0$. For the wave frequency

$$\omega = 0.5bc_0\sqrt{|L(L+1)/M|} \quad , \quad L = 0, 1, 2, \ldots \quad , \tag{6.3.8}$$

we have $\cos \pi d_2 = 0$ in (3.4.24) and $V = 0$ for all real values of the incidence angle θ_0. Here c_0 is the sound speed for $|z| \rightarrow \infty$.

It follows from (6.2.20, 21) that in the case under consideration, the transmission coefficient W has poles at the points

$$\sin \theta_0 = \xi/k_0 = \pm\sqrt{1 + \frac{(L-m)^2|M|}{L(L+1)}} \quad , \quad m = 0, 1, \ldots, L-1 \quad . \tag{6.3.9}$$

These points are also poles of the function $V(\xi)$, see (3.4.21). Note that in this case the reflection coefficient could not be an analytic function of ξ. Indeed, it follows from the fact that $V = 0$ at every ξ in the interval $(-k_0, k_0)$ that in the case of analyticity $V \equiv 0$, and there are no poles. Nonanalyticity of $V(\xi)$ does not contradict the general properties of the reflection coefficient stated in Sect. 6.2, where we considered reflection of plane waves incident upon half-space from a *homogeneous* medium. From (3.3.17) one can show that according to general theory, V is an analytic function of ξ when the sound speed satisfies the Epstein law in the half-space $-\infty < z < z_0$ and $c = \text{const}$ for $z > z_0$.

Further examples of nonreflecting layered media for a given frequency can be found in [6.28].

6.3.3 An Example of Reflectionless Layer for Arbitrary Frequency

Inhomogeneous media which are nonreflecting for waves of arbitrary frequency for a given fixed angle of incidence are also of interest. In [6.21] it is asserted that at normal incidence $V = 0$ for any ω in the case of a medium with constant density and sound speed profile, which is given by parametric relation

$$c(z) = c_0 \coth^2 \eta(z) \quad , \quad z/\alpha c_0 = \coth \eta - \eta \quad , \quad -\infty < \eta < 0 \quad , \tag{6.3.10}$$

where α and c_0 are positive constants. When η changes from $-\infty$ to 0, z decreases monotonically and takes all real values. Note that $c \to c_0$, as $z \to +\infty$ and $c \to +\infty$ as $z \to -\infty$.

The substitution $p = u \coth \eta$ transmutes the equation $d^2p/dz^2 + \omega^2 c^{-2}(z)p = 0$ for the sound pressure into the form

$$\frac{d^2u}{d\eta^2} + (\omega^2 \alpha^2 + 2\cosh^{-2} \eta)u = 0 \quad . \tag{6.3.11}$$

Equation (6.3.11) is a particular case of an equation for the sound field in the nonreflecting Epstein layer considered above. In the case of nonreflecting layer, hypergeometric functions used to solve the wave equation reduce to elementary functions. By direct substitution it is easy to check that

$$u = A \exp(i\omega\alpha\eta)(i\omega\alpha - \tanh \eta) + B \exp(-i\omega\alpha\eta)(i\omega\alpha + \tanh \eta) \tag{6.3.12}$$

is a general solution of (6.3.11). Here A and B are arbitrary constants. Assuming $B = 0$, we obtain

$$p = A(i\omega\alpha \coth \eta - 1) \exp(i\omega\alpha\eta) \quad . \tag{6.3.13}$$

As $z \to +\infty$ we have $\eta \approx 1 - z/\alpha c_0$ and the asymptotic form of (6.3.13) will be

$$p \approx A_1 \exp(-i\omega z/c_0) \quad , \quad A_1 = A(i\omega\alpha - 1)\exp(i\omega\alpha) \quad .$$

This corresponds to the plane wave incident from $z = +\infty$ in the absence of a reflected wave. In [6.21] this served as a reason to conclude that there will be no reflection from a medium with sound speed stratification (6.3.10). But (6.3.13) leads to unlimited sound pressure values as $z \to -\infty$: $|p| \sim |\omega\alpha A/\eta| \to \infty$. The particle velocity $v = (-i/\omega\varrho)\partial p/\partial z$ also tends to infinity as $z \to -\infty$. The physically meaningful solution corresponds to the requirement that $u = 0$ when $\eta = 0$. It is

$$p = 2A[\cos(\omega\alpha\eta) - \omega\alpha \coth \eta \cdot \sin(\omega\alpha\eta)] \quad . \tag{6.3.14}$$

From an asymptotic representation of (6.3.14) valid for $z \to +\infty$, it is clear that *total* reflection take place for all ω values. Physically, this result is obvious since the medium becomes incompressible as $z \to -\infty$.

An example of a layered medium which is nonreflective for sound of arbitrary frequency under some angle of incidence $\theta = \theta_1$, was constructed in [6.1]. Let a half-space $z > 0$, from which a plane wave is incident, be homogeneous and have sound speed $c \equiv c_1$, and $\varrho \equiv \varrho_1$. Let the density stratification in a lower medium $z < 0$ be described by some piece-wise smooth function $\varrho(z)$. We define the coordinate ζ as in (6.2.1). The function $c(z)$ is chosen so that in the wave equation

$$\frac{\partial^2 p}{\partial \zeta^2} + \left(\frac{\omega\varrho_1}{c_1\varrho}\right)^2 [c_1^2 c^{-2}(z) - \sin^2 \theta_1]p = 0 \tag{6.3.15}$$

the coefficient before p is constant for $-\infty < z < +\infty$, namely,

$$c(z) = c_1[\sin^2 \theta_1 + \varrho^2 \varrho_1^{-2} \cos^2 \theta_1]^{-1/2} \quad . \tag{6.3.16}$$

Then the solution of (6.3.15) which obeys the radiation condition as $z \to -\infty$ is

$$p = A \exp[-i\omega c_1^{-1} \cos \theta_1 \cdot \zeta(z)] \quad . \tag{6.3.17}$$

Because $\zeta = z$ for $z > 0$, in the upper medium (6.3.17) represents a plane wave incident from $z = +\infty$; $V = 0$.

A reflected field will also be absent in the case of a plane sound *pulse* with arbitrary time dependence and with an incidence angle θ_1. In the particular case where $\theta_1 = 0$ it follows from (6.3.16) that $\varrho(z)c(z) \equiv \varrho_1 c_1$, i.e., at normal incidence an arbitrary layered medium with uniform wave impedance is reflectionless. This fact is well known.

The result obtained above allows obvious interpretation in the case of discretely layered media. Consider an interface of some homogeneous layers, with the parameters of the upper (lower) layer labeled by subscript 2 (3). Let θ_2 be the angle of incidence in layer 2. Writing (6.3.16) for each layer and taking into account that, according to Snell's law, $\sin \theta_2 = (c_2/c_1) \sin \theta_1$, after simple transformation we find [cf. (2.2.20)]

$$\tan^2 \theta_2 = \frac{\varrho_3^2 \varrho_2^{-2} - c_2^2 c_3^{-2}}{c_2^2 c_3^{-2} - 1} \quad .$$

Hence, at each interface, the angle of incidence is equal to the angle of total transparency defined in Sect. 2.2. There is no reflection at any boundary. Thus, absence of the reflected field is not due to interference and therefore occurs at any frequency.

7. Acoustic Waves in Absorbing Anisotropic Media

Up to now we have neglected dissipation of the wave energy. In reality there are always some irreversible processes which cause *absorption of the wave energy* and its transformation into internal energy of the medium. In this chapter we consider effects related to this absorption (Sect. 7.1). It causes not only decay of the amplitude of the acoustic signal but also its change in shape. The absorption can also considerably influence the reflection and transmission coefficients.

In considering the propagation of waves in an elastic body we have assumed that this body is locally isotropic. *Anisotropy*, i.e., a difference of the properties of the medium in different directions, must be taken into account in acoustics of crystals as well as in seismology, for example. In the latter the anisotropy of rocks is caused mainly by the gravitational force. It is not our purpose to elaborate on the acoustics of anisotropic solids but rather to describe basic effects. In the particular case of piezoelectrics we will discuss several effects caused by the crystal anisotropy. Wave propagation in transversally isotropic media will also be considered in some detail (Sect. 7.2). This problem is important in the elastic wave theory of so-called *finely layered media* (Sect. 7.3), that is, media consisting of a large number of comparatively thin layers. Such media are very often found in seismology. Many sound absorbing systems and composite materials can also be described as finely layered media.

7.1 Absorption of Sound

7.1.1 Waves in Dissipative Fluids

In Chap. 1, in deriving the wave equation we assumed that the sound propagation was adiabatic. Adiabaticity is violated due to viscosity and thermal conductivity, however. As a result the sound energy partly irreversibly converts into internal energy of the medium. In mixtures and solutions diffusion also contributes to this process. Its role in sound absorption is small, however, and will be neglected. The medium will be assumed to be at rest in the absence of sound. When viscosity is taken into account, the Euler equation (1.1.9) becomes [Ref. 7.1, Sect. 15]

$$\varrho \frac{\partial v_i}{\partial t} = \frac{\partial \sigma_{ik}}{\partial x_k} \quad ,$$

$$\sigma_{ik} = -p\delta_{ik} + \eta \left(\frac{\partial v_i}{\partial x_k} + \frac{\partial v_k}{\partial x_i} \right) + \left(\zeta - \frac{2}{3}\eta \right) \frac{\partial v_l}{\partial x_l} \delta_{ik} \quad . \tag{7.1.1}$$

Indexes i, k, l assume values 1, 2, 3. Parameters η and ζ are called the viscosity coefficients or the first (shear) and the second (volume) viscosities, respectively. Both are positive and may depend on the sound frequency. If they do, (7.1.1) is valid only for monochromatic waves. In the linear approximation (with respect to the wave amplitude) adopted by us the entropy density $\tilde{S} = S + S_0$ satisfies the equation [Ref. 7.1, Sect. 49]:

$$\varrho T_0 \left(\frac{\partial S}{\partial t} + v \nabla S_0 \right) = \mathrm{div}\,(\kappa \nabla T) \quad . \tag{7.1.2}$$

Here $\tilde{T} = T_0 + T$ is temperature, S_0 and T_0 are entropy density and temperature in the absence of sound. $\kappa > 0$ is the thermal conductivity coefficient. We choose the entropy and the density to be independent thermodynamic variables as in Sect. 1.1. Then, according to the state equation we have

$$p = \left(\frac{\partial \tilde{p}}{\partial \tilde{\varrho}} \right)_{\tilde{S}} \varrho' + \left(\frac{\partial \tilde{p}}{\partial \tilde{S}} \right)_{\tilde{\varrho}} S \quad , \tag{7.1.3}$$

$$T = \left(\frac{\partial \tilde{T}}{\partial \tilde{\varrho}} \right)_{\tilde{S}} \varrho' + \left(\frac{\partial \tilde{T}}{\partial \tilde{S}} \right)_{\tilde{\varrho}} S \quad . \tag{7.1.4}$$

If we assume $\kappa = 0$ in (7.1.2) then (7.1.3) reduces to the state equation (1.1.8) used in Chap. 1, where now $\nabla p_0 = v_0 = 0$. Equations (7.1.1–4) together with the continuity equation of (1.1.7) form a closed linear system from which the seven unknown p, ϱ', v_i, T, and S can be determined.

In a homogeneous medium this system can be satisfied by the plane wave solution. Let the time and spatial dependence of the plane wave be $\exp{(\mathrm{i}k_j x_j - \mathrm{i}\omega t)}$. Then we find, by assuming $\varrho' \neq 0$ and eliminating v_i from (7.1.1) that

$$\frac{p}{\varrho'} = \omega^2 k^{-2} \left[1 + \frac{\mathrm{i}k^2(\zeta + 4\eta/3)}{\omega \varrho} \right] \quad . \tag{7.1.5}$$

On the other hand one can also relate p to ϱ' by using (7.1.2–4) and obtain after simple calculations

$$\frac{p}{\varrho'} = c^2 - \frac{k^2(c^2 - c_T^2)}{k^2 - \mathrm{i}\omega c_T^2/\chi c^2} \quad . \tag{7.1.6}$$

In these calculations we have used the identity

$$\left(\frac{\partial \tilde{T}}{\partial \tilde{\varrho}} \right)_{\tilde{S}} \left(\frac{\partial \tilde{S}}{\partial \tilde{T}} \right)_{\tilde{\varrho}} = - \left(\frac{\partial \tilde{S}}{\partial \tilde{\varrho}} \right)_{\tilde{T}} = 1 - \frac{c_T^2}{c^2}$$

and the notation

$$\chi = \frac{\kappa}{\varrho C_p} \quad , \quad C_p = T_0 \left(\frac{\partial \tilde{S}}{\partial \tilde{T}} \right)_{\tilde{p}} \quad ,$$

$$c = \sqrt{(\partial \tilde{p}/\partial \tilde{\varrho})_{\tilde{S}}} \quad , \quad c_T = \sqrt{(\partial \tilde{p}/\partial \tilde{\varrho})_{\tilde{T}}} \quad . \tag{7.1.7}$$

Here χ is the temperature conductivity coefficient, and C_p is heat capacity at constant pressure. Later we shall see that c and c_T are the sound velocities in the extreme cases when propagation is adiabatic or isothermal, respectively. Both c and c_T are of the same order of magnitude and $c > c_T$.

By equating the right-hand sides of (7.1.5,6), one obtains the dispersion equation for acoustic waves in a homogeneous, viscous medium with nonzero thermal conductivity:

$$k^4[1 - i\omega(\zeta + 4\eta/3)/\varrho c_T^2] - k^2[\omega^2 c_T^{-2} + \omega^2(\zeta + 4\eta/3)/\varrho\chi c^2 + i\omega/\chi]$$
$$+i\omega^3/\chi c^2 = 0 \quad . \tag{7.1.8}$$

The solutions of this biquadratic equation $\pm k_I$, $\pm k_{II}$ correspond to two types of waves. The real part Re $\{k\}$ relates to the phase velocity $c_{ph} = \omega/\text{Re}\{k\}$, whereas the imaginary part determines the rate of absorption of the wave. Namely, the wave amplitude decreases by factor e over the distance $1/\text{Im}\{k\}$ in the direction of the largest attenuation.

Let us consider some specific cases. In the absence of thermal conductivity $(\chi \to 0)$, (7.1.8) has only one finite solution

$$k_I = \omega c^{-1}[1 - i\omega(\zeta + 4\eta/3)/\varrho c^2]^{-1/2} \quad . \tag{7.1.9}$$

In the most common and important examples, both viscosity and thermal conductivity are present but small $(\chi \ll c^2/\omega, \zeta + 4\eta/3 \ll \varrho c^2/\omega)$ and we have

$$k_I = \frac{\omega}{c}\left[1 + \frac{i\omega(\zeta + 4\eta/3)}{2\varrho c^2} + \frac{i\omega\chi}{2}(c_T^{-2} - c^{-2})\right] \quad ,$$

$$k_{II} = \sqrt{\frac{i\omega^3}{\chi c^2}\frac{1}{k_I}} \approx \sqrt{\frac{i\omega}{\chi}} \quad . \tag{7.1.10}$$

If the thermal conductivity is high but the viscosity is small $(\chi \gg c^2/\omega, \zeta + 4\eta/3 \ll \varrho c^2/\omega)$ then

$$k_I = \frac{\omega}{c_T}\left[1 + \frac{i\omega(\zeta + 4\eta/3)}{2\varrho c_T^2} + \frac{ic_T^2}{2\omega\chi}\left(1 - \frac{c_T^2}{c^2}\right)\right] \quad ,$$

$$k_{II} \approx \sqrt{\frac{i\omega}{\chi}\frac{c_T}{c}} \quad . \tag{7.1.11}$$

In all these cases, there is a small positive imaginary part in the wave number k_I which gives rise to sound absorption.

Yet another effect can be caused by irreversible processes, namely, dispersion, which is the dependence of the wave phase velocity on the frequency. If χ, ζ, and η are independent of the frequency, according to (7.1.10, 11) the sound velocity diminishes from c to c_T while ω increases. (The result is the same if thermal conductivity increases while ω remains constant.) In the case of an acoustic pulse, its shape changes in the process of propagation since each monochromatic component propagates at its own velocity and has its own rate of decay.

In addition to sound waves, "thermal" waves with wave number k_{II} and "viscous" waves can exist. The latter are not accompanied by compression of the medium $(\varrho' \equiv 0)$. According to the continuity equation of (1.1.7) $k_l v_l = 0$ for viscous waves, i.e., these waves are transverse. They may have two independent polarizations. It follows from $\varrho' \equiv 0$ and (7.1.3–5) that $p = S = T = 0$. From (7.1.1) one obtains the dispersion relation for viscous waves:

$$k^2 = \frac{i\omega\varrho}{\eta} \quad . \tag{7.1.12}$$

Because $\mathrm{Re}\{k\} \approx \mathrm{Im}\{k\}$ in viscous and thermal waves, these waves are strongly inhomogeneous.

7.1.2 Absorption in Solids

In solids other irreversible processes related to deformation can occur: plasticity, vacancy drift in crystals, interaction with thermal phonons, etc. A general theory of sound absorption in all the diverse solids, from rocks to metals and plastics, does not exist. Usually dissipative processes are described phenomenologically by replacing the elastic constants in Hooke's law by time-dependent operators. The most general relation between small strains and the stress tensor in an isotropic viscoelastic body can be written in the form [7.2]

$$\hat{B}_1(t)\sigma_{ij} = \hat{B}_2(t)\delta_{ij}\frac{\partial u_k}{\partial x_k} + \hat{B}_3(t)\left(\frac{\partial u_i}{\partial x_j} + \frac{\partial u_j}{\partial x_i}\right) \quad . \tag{7.1.13}$$

The integro-differential operators $\hat{B}_j(t)$ become *functions* of $-i\omega$ in the case of monochromatic waves. In the general case, these functions are complex. By designating $\lambda = B_2(-i\omega)/B_1(-i\omega)$, $\mu = B_3(-i\omega)/B_1(-i\omega)$ we can reduce (7.1.13) to the usual form of Hooke's law given in (1.3.1) with complex frequency-dependent λ and μ.

It is not an unusual assumption that as in fluids, viscous forces in solids are proportional to the time derivative of deformation [Ref. 7.3, Sect. 34]. This corresponds to the following choice of the operators $\hat{B}_j$:

$$\hat{B}_1 \equiv 1 \quad , \quad \hat{B}_2 = \lambda + \left(\zeta - \frac{2\eta}{3}\right)\frac{\partial}{\partial t} \quad , \quad \hat{B}_3 = \eta\frac{\partial}{\partial t} \quad .$$

In this case the dissipative parts of the stress tensors in (7.1.1, 13) coincide completely and viscosity is taken into account by replacement of the Lamé constants λ and μ with $\lambda + i\omega(2\eta/3 - \zeta)$ and $\mu - i\omega\eta$, respectively. When dissipation is small ($\eta \ll \mu$, $\zeta \ll \lambda$) we obtain for the wave numbers of longitudinal and transvese waves, as in Sect. 1.3,

$$k_l = \omega\sqrt{\frac{\varrho}{\lambda + 2\mu}}\left(1 + \frac{i\omega(\zeta + 4\eta/3)}{2(\lambda + 2\mu)}\right) \quad ,$$

$$k_t = \omega\sqrt{\frac{\varrho}{\mu}}\left(1 + \frac{i\omega\eta}{2\mu}\right) \quad . \tag{7.1.14}$$

The presence of dissipation does not affect the boundary conditions at interfaces of elastic media, which were discussed in Chap. 1. The equation of motion given in (1.3.2) also holds, of course. Hence the results obtained in Chaps. 1, 4, and 6 remain valid for viscoelastic media, only λ and μ are now complex quantities. In particular, the expressions for the scattering matrix (4.2.8–12) still hold at the interface of two viscoelastic half-spaces. The applicability of the results obtained in Chap. 4 to viscoelastic media has been experimentally proven many times (see, for example, [7.4, 41]). The analytical expressions for the reflection, transmission, and transformation coefficients of plane waves remain unchanged, but their behavior as a function of incidence angle, for example, does change considerably because λ and μ are now complex. A thorough analysis of the dependence of these coefficients on the parameters of viscoelastic media and on incidence angle can be found in [Ref. 7.5, Chap. 1] with many numerical examples.

Analysis of the results of Chap. 4 shows that the modulus of the reflection coefficient may be greater than unity in the case of contact between dissipative media. Some researchers erroneously conclude that this contradicts the energy conservation law (for a debate on this issue see, for example, [7.6], where it is shown that $|V| > 1$ is possible when a wave is incident from a dissipative medium onto an ideal elastic half-space). In fact, this is due to the nonadditivity of energy fluxes in the incident and reflected waves. Indeed, it is easy to verify by using (2.1.11) that only at real k^2, i.e., in nondissipative fluid, is the vertical component of the acoustic power flux density vector equal to the difference of those in the incident and reflected waves (that is proportional to $1 - |V|^2$ for waves described by $\exp[i(\xi x - \omega t)]$ at real ω and ξ. With viscoelastic solids, one has to additionally account for the nonadditivity of power fluxes in P and SV plane waves with the same dependence on time and the horizontal coordinates [7.42].

Consider, after [7.7], reflection of an *SV* plane wave at a plane, free boundary of a viscoelastic body. We assume that λ is real, but $\mu = \mu_0(1 - i\varepsilon)$, where μ_0 and ε are real and $0 < \varepsilon \ll 1$. The reflection coefficient V_{tt} is given by (4.1.3, 4). If absorption is absent ($\varepsilon = 0$) and $k_l < \xi < k_t$, the reflected longitudinal wave is inhomogeneous; $|V_{tt}| = 1$. The correction to V_{tt} due to dissipation is (only the term with the first power of ε is retained)

$$\delta V_{tt} = -i\varepsilon\mu_0 \frac{\partial V_{tt}}{\partial \mu} = \frac{4\varepsilon V_{tt}(1-2a)a[(1-a)(a-b)]^{-1/2}}{16a^2(1-a)(a-b) + (1-2a)^4}$$
$$\times \left[(a-b)(2a-3) - 2b^2(1-2a)(1-a)\right] \quad, \qquad (7.1.15)$$

where $a = \xi^2/k_t^2$, $b = k_l^2/k_t^2$. The values of all quantities in (7.1.15), except ε must be taken at $\varepsilon = 0$. Since $c_l^2 > 2c_t^2$ $b < 0.5$. In the given interval of incidence angles $b < a < 1$. A simple analysis shows that at $0.5 < a < 1$ $\delta V_{tt}/V_{tt} > 0$ and hence, the modulus of the reflection coefficient is greater than unity. For example, at $a = 0.9$, $b = 0.17$, and $\varepsilon = 0.05$, (7.1.15) gives $|V_{tt} + \delta V_{tt}| = 1.34$. Nevertheless, the energy flux is directed to the boundary. Indeed, the vertical component of the power flux density vector in harmonic waves is [Ref. 7.8, Sect. 3.3] $I_z = -0.5 \,\mathrm{Re}\,\{\sigma_{3j}^* \partial u_j/\partial t\}$. At a free boundary $z = 0$ we have $I_z = 0$ due to the boundary conditions $\sigma_{3j} = 0$. Using (4.0.1–3), or (1.3.1, 2), to express $\partial\sigma_{3j}/\partial z$ and $\partial u_j/\partial z$ at an arbitrary horizon in terms of σ_{3j} and u_j at the same horizon, after some rearrangements we obtain for $z \geq 0$

$$\frac{\partial}{\partial z} I_z = -\omega \in \mu_0 \left(\xi^2 |u_1|^2 + \left| \frac{\sigma_{33} - i\xi\lambda u_1}{\lambda + 2\mu} \right|^2 + \frac{1}{2} \left| \frac{\sigma_{31}}{\mu} \right|^2 \right) \quad. \qquad (7.1.16)$$

We see that $\partial I_z / \partial z < 0$. Hence, $I_z < 0$ everywhere in the upper half-space which proves our statement. An analogous analysis for the more involved case of the reflection of SV waves at a boundary between a fluid and a viscoelastic body is given in [7.7]. For the general case of plane-wave reflection at an interface of two viscoelastic solids, related topics have been considered in [7.42, 43].

The propagation of monochromatic sound waves in *absorbing fluid* is often described by using the wave equation (1.2.2) and supposing that k^2 is a complex quantity. This is a correct approach in the case of homogeneous media but not in the general case. Indeed, (1.2.2) is of second order and at interfaces its solutions may satisfy only two boundary conditions. However, in the case of a viscous, thermal conducting liquid there must be eight independent boundary conditions. As for solids, there must be continuous three components of the stress tensor, particle velocity, and also temperature, and the component of heat flux density, which is normal to the boundary, $\kappa \partial T / \partial n$. (Otherwise, according to (7.1.2,3) at the boundary the entropy and pressure would be infinite.) When thermal conductivity can be neglected ($\kappa \to 0$), by using (7.1.1–3 and 1.1.7) we obtain for the stress tensor in a viscous liquid

$$\sigma_{ik} = \left(\varrho c^2 - i\omega\zeta + \frac{2i\omega\eta}{3} \right) \delta_{ik} \frac{\partial u_l}{\partial x_l} - i\omega\eta \left(\frac{\partial u_i}{\partial x_k} + \frac{\partial u_k}{\partial x_i} \right) \quad . \tag{7.1.17}$$

This expression coincides with Hooke's law (1.3.1) with $\lambda = \varrho c^2 + i\omega(2\eta/3 - \zeta)$, $\mu = -i\omega\eta$. Hence a viscous liquid can be treated as an elastic medium with complex λ and imaginary μ. Longitudinal and transverse waves in solids are analogous to sound and viscous waves in liquids.

When a sound wave in a viscous liquid is reflected at a boundary its energy partly transforms into that of viscous waves. This process can be described with the help of the scattering matrices considered in Chap. 4. Hence, sound absorption in layered media occurs through two processes. First, as in a homogeneous medium, it is the usual volume absorption, due to which the wave amplitude decays with the distance L as $\exp(-L \, \mathrm{Im} \, \{k\})$. The quantity in the exponent contains η and ζ in linear form (in the case of small viscosity). This kind of absorption is the most important in continuously layered media.

The other mechanism of absorption is present near boundaries and becomes important when the sound propagates in narrow tubes and along thin layers. The scattering matrix contains the first power of the vertical component of the wave vector. Hence this kind of absorption is proportional, generally speaking, to $|k_t^2 - \xi^2|^{-1/2} \approx |k_t|^{-1} \sim \eta^{1/2}$. Since, as a rule, the viscosity is small, the second mechanism of dissipation may be more important than the first even when the layer thickness is large compared to the wavelength. This mechanism can not be taken into account just by using the wave equation with complex k^2.

An analogous process occurs with heat conductivity. Upon reflecting at the boundary, the sound wave generates strongly decaying thermal waves which consume part of the sound energy. This type of absorption process is more fully discussed in [7.44], in the review [7.9] and references therein. For further references and a more detailed discussion of the acoustic effects of dissipative processes in fluids, see also [Ref. 7.45, Chap. 10; 7.46].

7.2 Anisotropic Elastic Media. Gulyaev-Bluestein Waves

7.2.1 The Christoffel Equation

When deformations are small, we have a linear relation between the stress and strain tensors:

$$\sigma_{ij} = C_{ijkl}u_{kl} \quad , \quad u_{kl} \equiv 0.5\left(\frac{\partial u_k}{\partial x_l} + \frac{\partial u_l}{\partial x_k}\right) \quad . \tag{7.2.1}$$

The fourth order tensor C_{ijkl} is characteristic of a body and is called the elastic modulus tensor. Equation (7.2.1) is the generalization of Hooke's law (1.3.1) for an anisotropic medium. Since the tensors σ_{ij} and u_{kl} are symmetric, the tensor C_{ijkl} can be considered to be invariant with respect to the index interchanges in the first and second pairs: $C_{ijkl} = C_{jikl} = C_{ijlk} = C_{jilk}$. The tensor is also symmetric with respect to interchange of the pairs themselves [Ref. 7.8, Sect. 3.3]: $C_{ijkl} = C_{klij}$. As a result, not more than 21 (from 3^4) components of the elastic modulus tensor appear to be independent in an anisotropic medium. The higher the symmetry of the medium, the lesser the number of independent components.

The equation of motion of an elastic medium is given by (1.3.2). For a plane harmonic wave

$$\boldsymbol{u} = \boldsymbol{v}\exp(ikn_jx_j - i\omega t) \quad , \quad n_jn_j = 1 \tag{7.2.2}$$

we obtain from (7.2.1) and (1.3.2), in the case of a homogeneous anisotropic medium (taking into account the symmetry of the tensor C_{ijkl}),

$$(\Gamma_{jl} - \varrho\omega^2 k^{-2}\delta_{jl})v_l = 0 \quad , \quad \Gamma_{jl} = C_{ijlk}n_in_k \quad . \tag{7.2.3}$$

One can see that the quantity $\varrho\omega^2 k^{-2}$ is the eigenvalue and the corresponding vector $\boldsymbol{v}$ is the eigenvector of the matrix $[\Gamma]$ which depends on the direction of propagation. The condition for the existence of nonzero solutions of the system (7.2.3) is

$$\det(\Gamma_{jl} - \varrho\omega^2 k^{-2}\delta_{jl}) = 0 \quad . \tag{7.2.4}$$

This is the so-called *Christoffel equation*. It determines three permissible values of the wave number k and hence the phase and group velocities of the corresponding waves. These velocities generally have different directions and depend on the orientation of the wave vector $k\boldsymbol{n}$. Some universal properties of the velocities were established in [7.47].

When the wave number is known, the vector $\boldsymbol{v}$ can be found from (7.2.3), aside from a normalization factor, that is, the polarization of the wave propagating in the given direction can be found. It follows from the symmetry and reality of the matrix $[\Gamma]$ that the three plane waves with the same $\boldsymbol{n}$ are always polarized in mutually perpendicular directions.

To describe waves in a viscoelastic medium it is convenient to represent them as a superposition of monochromatic waves. Then, as in the previous section where isotropic media were considered, the influence of dissipation will be to simply give the components of the tensor C_{ijkl} complex, frequency-dependent values.

The general elastic theory of anisotropic media is considered in a number of texts (see, for example, [7.3, 10]). Wave propagation in such media, important in crystal acoustics and seismology, is treated in [7.11–13, 27, 48, 49]. Rayleigh waves in crystals of different symmetry are considered in [7.14, 15, 50]. Differences in directions of phase and group velocities and their consequences were discussed in [7.16]. Geometrical acoustics of anisotropic media is studied particularly in [7.12, 17, 55]. In [7.18, 19, 56, 57] the phase velocity as a function of propagation direction is considered in a homogeneous weakly anisotropic medium; the case of arbitrary anisotropy was analyzed in [7.58]. Propagation of ultrasonic beams in crystals is discussed in [7.20]. Surface waves in discretely layered anisotropic elastic half-spaces with free boundary are investigated in [7.21, 22].

7.2.2 Elastic Waves in Transversally Isotropic Solids. Matrix Exponent

Here we discuss more thoroughly the special case of *transversally isotropic media*. In such media, the properties are isotropic in a plane, for our case let it be the xy-plane, with the z-axis the symmetry axis. The theory of transversally isotropic media is of considerable interest in seismology.

If we change the x-axis (or y-axis) direction for the opposite one then the u_{1j}, $j \neq 1$ (or, respectively, u_{2i}, $i \neq 2$) component of the strain tensor changes sign. The real state of the body remains unchanged. The same is valid for the tensor C_{ijkl} due to the isotropy in xy plane. That is why in the expression for the elastic energy $W = 0.5 C_{jikl} u_{ji} u_{kl}$ [Ref. 7.8, Sect. 3.3] terms with C_{ijkl} containing the indexes 1 and 2 an odd number of times must be absent. In addition, the interchange of indexes 1 and 2 does not change the elasticity modulus. Hence the tensor C_{ijkl} may contain only six different nonzero components: C_{3333}, $C_{1313} = C_{2323}$, $C_{1133} = C_{2233}$, $C_{1111} = C_{2222}$, C_{1212}, C_{1122}. Not all of them are independent, however. The components of the strain tensor u_{ij}, $i, j = 1, 2$ may be present in the energy expression only in the combinations $u_{11} + u_{22}$, $u_{12}^2 - u_{11} u_{22}$ which are invariant with respect to rotation of the coordinate system about the z-axis. As a result, for a transversally isotropic medium we have the equality $C_{1122} = C_{1111} - 2C_{1212}$ and

$$2W = C_{1111}(u_{11} + u_{22})^2 + 4C_{1212}(u_{12}^2 - u_{11}u_{22}) + C_{3333}u_{33}^2$$
$$+ 2C_{1133}(u_{11} + u_{22})u_{33} + 4C_{1313}(u_{13}^2 + u_{23}^2) \quad . \tag{7.2.5}$$

It is easy to show [Ref. 7.3; Sect. 10] that the tensor C_{ijkl} in crystals of hexagonal symmetry where the symmetry axis is of the sixth order (instead of infinite order as above), also has the same properties.

Taking into account the properties of the elastic modulus tensor found above we can now write the generalized Hooke's law (7.2.1) for the case of a transversally isotropic medium as follows:

$$\sigma_{11} = C_{1111}u_{11} + (C_{1111} - 2C_{1212})u_{22} + C_{1133}u_{33} \quad , \tag{7.2.6}$$

$$\sigma_{12} = 2C_{1212}u_{12} \quad , \tag{7.2.7}$$

$$\sigma_{13} = 2C_{1313}u_{13} \quad , \tag{7.2.8}$$

$$\sigma_{33} = C_{3333}u_{33} + C_{1133}(u_{11} + u_{22}) \quad . \tag{7.2.9}$$

Other components σ_{ij} can be obtained by taking into account the symmetry of the medium and (7.2.6–9). It is convenient to write the five elastic moduli of the transversely isotropic medium in the form:

$$C_{1111} = \lambda + 2\mu \quad , \quad C_{3333} = \lambda' + 2\mu' \quad ,$$
$$C_{1133} = \lambda' \quad , \quad C_{1212} = \mu \quad , \quad C_{1313} = \mu'' \quad .$$

The completely isotropic case ($C_{1111} = C_{3333}$, $C_{1313} = C_{1212}$, $C_{1133} = C_{1122}$) then corresponds to the equalities $\lambda = \lambda'$, $\mu = \mu' = \mu''$. It is then easy to see that (7.2.6–9) reduce to Hooke's law (1.3.1) for an isotropic medium.

The propagation of elastic waves in a transversely isotropic medium has much in common with that in an isotropic body. In both cases, horizontally polarized waves (SH waves) which propagate independently from P-SV waves can be considered. This can be proved by using (7.2.3) where at $n = (\sin\theta, 0, \cos\theta)$ we have

$$[\Gamma] = \begin{pmatrix} (\lambda+2\mu)\sin^2\theta + \mu''\cos^2\theta & 0 & (\lambda'+\mu'')\sin\theta\cos\theta \\ 0 & \mu\sin^2\theta + \mu''\cos^2\theta & 0 \\ (\lambda'+\mu'')\sin\theta\cos\theta & 0 & \mu''\sin^2\theta + (\lambda'+2\mu')\cos^2\theta \end{pmatrix} .$$
$$(7.2.10)$$

Here θ is the angle of incidence and xz is the plane of incidence. The solution of (7.2.3) for SH waves is

$$k^2 = \varrho\omega^2(\mu\sin^2\theta + \mu''\cos^2\theta)^{-1} \quad , \quad v = (0,1,0) \quad . \tag{7.2.11}$$

When θ increases from $\theta = 0$ up to $\theta = \frac{\pi}{2}$ the SH waves phase velocity varies from $(\mu''/\varrho)^{1/2}$ to $(\mu/\varrho)^{1/2}$.

Let us denote $k\sin\theta = \xi$, $k\cos\theta = \alpha$. Then we shall have a quadratic equation for α^2 to determine permissible values for the vertical component of the wave vector in P-SV waves. Its solution is

$$\alpha_{1,2}^2 = \frac{-b \mp \sqrt{b^2 - 4a\mu''(\lambda'+2\mu')}}{2(\lambda'+2\mu')\mu''} \quad , \tag{7.2.12}$$

where

$$a = [\xi^2(\lambda+2\mu) - \varrho\omega^2](\mu''\xi^2 - \varrho\omega^2) \quad ,$$
$$b = \xi^2[(\lambda+2\mu)(\lambda'+2\mu') + (\mu'')^2 - (\lambda'+\mu'')^2]$$
$$- \varrho\omega^2(\lambda'+2\mu'+\mu'') \quad . \tag{7.2.13}$$

In the isotropic case $\alpha_1^2 = \varrho\omega^2(\lambda+2\mu)^{-1} - \xi^2$, $\alpha_2^2 = \varrho\omega^2/\mu - \xi^2$, that is, α_1 corresponds to longitudinal, whereas α_2 to transverse waves. We assume in what follows that Im $\{\alpha_{1,2}\} \geq 0$.

The displacement field for P-SV waves in the case of a transversely isotropic medium is

$$u = \exp(i\xi x - i\omega t)[v^{(1)}\varphi_1\exp(i\alpha_1 z) + v^{(2)}\varphi_2\exp(-i\alpha_1 z)$$
$$+ v^{(3)}\varphi_3\exp(i\alpha_2 z) + v^{(4)}\varphi_4\exp(-i\alpha_2 z)] \quad . \tag{7.2.14}$$

Here φ_j are constants whereas the vectors $v^{(j)}$ are determined with the help of (7.2.3) (within an accuracy to a normalizing factor):

$$\boldsymbol{v}^{(1)} = \mathrm{i}\left(\xi, 0, \frac{\varrho\omega^2 - (\lambda + 2\mu)\xi^2 - \mu''\alpha_1^2}{\alpha_1(\lambda' + \mu'')}\right)^T \quad , \qquad \boldsymbol{v}^{(2)} = (v_1^{(1)}, 0, -v_3^{(1)})^T \quad ,$$

$$\boldsymbol{v}^{(3)} = -\mathrm{i}\left(\alpha_2, 0, \frac{\varrho\omega^2 - (\lambda + 2\mu)\xi^2 - \mu''\alpha_2^2}{\xi(\lambda' + \mu'')}\right)^T \quad , \qquad \boldsymbol{v}^{(4)} = (-v_1^{(3)}, 0, v_3^{(3)})^T \quad .$$

$$(7.2.15)$$

In each of the waves, the displacement vectors have components that are parallel as well as perpendicular to the wave vector and therefore, if a displacement is parallel to the wave vector at some z, at some other horizon it will also have a component perpendicular to the wave vector. Hence, even in a homogeneous but anisotropic medium, during propagation longitudinal waves transform into transverse ones, and vice versa.

Let us now consider P-SV waves in a *discretely layered* transversely isotropic medium. As above we assume that the wave field dependence on time and on the horizontal coordinates x, y is harmonic. We also again assume that the medium is isotropic in the xy-plane. The reflection of a plane wave from an arbitrary set of layers can be easy described with the help of matrix method [7.21, 23]. We introduce as in Chap. 4 the displacement-stress vector $\boldsymbol{f} = (u_1, u_3, \sigma_{33}, \sigma_{31})^T$. In obtaining the boundary conditions in Chap. 1 we did not use Hooke's law (7.2.1). Hence, in an anisotropic medium the vector $\boldsymbol{f}$ is again continuous at a boundary with welded contact. Let the horizon z and z_0 be within the same layer. Then we obtain from the (7.2.14, 15, 8, 9) after simple operations

$$\boldsymbol{f}(z) = [L][\exp(\mathrm{i}\alpha_1 z), \exp(-\mathrm{i}\alpha_1 z), \exp(\mathrm{i}\alpha_2 z), \exp(-\mathrm{i}\alpha_2 z)]$$
$$\times (\varphi_1, \varphi_2, \varphi_3, \varphi_4)^T = [L][l][L]^{-1}\boldsymbol{f}(z_0) \quad , \qquad (7.2.16)$$

where symbol $[\dots, \dots, \dots, \dots]$ indicates (as in Chap. 4) a diagonal matrix:

$$[l] = \left[\exp\left[\mathrm{i}\alpha_1(z - z_0)\right] \quad , \quad \exp\left[-\mathrm{i}\alpha_1(z - z_0)\right] \quad , \right.$$
$$\left. \exp\left[\mathrm{i}\alpha_2(z - z_0)\right] \quad , \quad \exp\left[-\mathrm{i}\alpha_2(z - z_0)\right]\right] \quad \text{and} \qquad (7.2.17)$$

$$[L] = \begin{pmatrix} \mathrm{i}\xi & \mathrm{i}\xi \\ v_3^{(1)} & -v_3^{(1)} \\ \mathrm{i}(\lambda' + 2\mu')\alpha_1 v_3^{(1)} - \xi^2\lambda' & \mathrm{i}(\lambda' + 2\mu')\alpha_1 v_3^{(1)} - \xi^2\lambda' \\ \mu''(\mathrm{i}\xi v_3^{(1)} - \alpha_1\xi) & -\mu''(\mathrm{i}\xi v_3^{(1)} - \alpha_1\xi) \end{pmatrix}$$

$$\begin{pmatrix} -\mathrm{i}\alpha_2 & \mathrm{i}\alpha_2 \\ v_3^{(3)} & v_3^{(3)} \\ \mathrm{i}(\lambda' + 2\mu')\alpha_2 v_3^{(3)} - \xi^2\lambda' & -\mathrm{i}(\lambda' + 2\mu')\alpha_2 v_3^{(3)} + \xi^2\lambda' \\ \mu''(\mathrm{i}\xi v_3^{(3)} + \alpha_2^2) & \mu''(\mathrm{i}\xi v_3^{(3)} + \alpha_2^2) \end{pmatrix} \qquad (7.2.18)$$

Equation (7.2.16) is analogous to (4.3.4, 5) obtained for isotropic media. Instead of a vector of potentials, in an anisotropic medium in a given homogeneous layer we have a vector of constants $(\varphi_1, \varphi_2, \varphi_3, \varphi_4)^T$.

The peculiarity of anisotropic media is contained in matrices $[L]$ and $[l]$, which do, however, have features common with the case considered in Chap. 4. The normalization of the vectors $\boldsymbol{v}^{(j)}$ (7.2.15) was chosen in such a way that in an isotropic medium (7.2.17 and 18) coincide with (4.3.5). This can be proved by comparing the matrices element by element.

153

Relations (7.2.16–18) allow us to calculate the components of a displacement-stress vector at a given boundary of a layer if they are known at the other one. The construction of the matrix propagator based on the matrices of the separate layers $[A] = [L][l][L]^{-1}$ and all the following procedures are quite analogous to those for the isotropic case considered in Sect. 4.3 and will not be repeated here.

Consider the matrix formalism for SH waves in a transversely isotropic medium. According to (7.2.11), for the displacements in a homogeneous medium for these waves we have $u_1 = u_3 = 0$,

$$u_2 = [\psi_1 \exp(i\beta z) + \psi_2 \exp(-i\beta z)] \exp(i\xi x - i\omega t) \quad,$$

$$\psi_{1,2} = \text{const} \quad, \quad \beta = \sqrt{(\varrho\omega^2 - \mu\xi^2)/\mu''} \quad. \tag{7.2.19}$$

We have used the relations $\xi = k \sin\theta$, $\beta = k \cos\theta$ to eliminate the incidence angle from the dispersion equation in (7.2.11). From (7.2.8,9) it follows that $\sigma_{33} \equiv 0$, $\sigma_{23} = 2C_{1313}u_{23} = \mu'' \partial u_2/\partial x_3$. Hence the boundary conditions in the case of welded contact of two solids are the continuity of the displacement component u_2 and of the stress tensor component σ_{23}. Taking this into account we choose displacement-stress vector as $f = (u_2, \sigma_{23})^T$. Then we obtain from (7.2.19):

$$f(z) = [L][\exp(i\beta z), \exp(-i\beta z)](\psi_1, \psi_2)^T = [L][l][L]^{-1} f(z_0) \quad, \tag{7.2.20}$$

where $[L]$ and $[l]$ are 2×2 matrices:

$$L_{11} = L_{12} = 1 \quad, \quad L_{21} = -L_{22} = i\mu''\beta \quad,$$

$$[l] = [\exp(i\beta(z - z_0)), \exp(-i\beta(z - z_0))] \quad. \tag{7.2.21}$$

The matrix for a separate layer is $[A] = [L][l][L]^{-1}$ as for P-SV waves but the expression for $[L]$ as well as for $[l]$, including even dimensions, are different.

In some cases it is more convenient to represent the layer matrix $[A]$ as a matrix exponent. Such a representation will also be used when we discuss the properties of finely layered media. Let some function $F(x)$ be expanded in a Taylor's series $\sum_{n=0}^{+\infty} a_n x^n/n!$. Then the function of a matrix is defined as a sum

$$F([B]) = \sum_{n=0}^{+\infty} \frac{a_n[B]^n}{n!} \quad. \tag{7.2.22}$$

For example, the matrix $[l]$ from (7.2.21) can be written as

$$[l] = \exp\{(z - z_0)[m]\} \quad, \quad \text{where} \quad [m] = [i\beta, -i\beta] \quad. \tag{7.2.23}$$

Indeed, we have according to the definition in (7.2.22)

$$\exp\{(z - z_0)[m]\} = \sum_{n=0}^{+\infty} \frac{\{(z - z_0)[m]\}^n}{n!}$$

$$= \sum_{n=0}^{+\infty} \left[\frac{i^n(z - z_0)^n \beta^n}{n!}, \frac{(-i(z - z_0)\beta)^n}{n!} \right]$$

$$= [\exp[i\beta(z - z_0)], \exp[-i\beta(z - z_0)]] \quad.$$

Using the representation in (7.2.23) and the equality (which can be easily proven)

$\{[L][m][L]^{-1}\}^n = [L][m]^n[L]^{-1}$, we obtain for the layer matrix $[A] = [L][l][L]^{-1}$

$$[A] = \exp\{(z - z_0)[\kappa]\} \quad , \quad \text{where} \quad [\kappa] \equiv [L][m][L]^{-1} \quad . \tag{7.2.24}$$

The matrix $[\kappa]$ can be calculated by (7.2.21, 23). Hence, we find for SH waves

$$\kappa_{11} = \kappa_{22} = 0 \quad , \quad \kappa_{12} = 1/\mu'' \quad , \quad \kappa_{21} = \mu\xi^2 - \varrho\omega^2 \quad . \tag{7.2.25}$$

Equation (7.2.24) also holds for P-SV waves with $[m] = [\mathrm{i}\alpha_1, -\mathrm{i}\alpha_1, \mathrm{i}\alpha_2, -\mathrm{i}\alpha_2]$ and $[L]$ defined by (7.2.18). Calculating with the use of (7.2.24) gives the rather compact result, [cf. (7.2.18)],

$$\kappa_{11} = \kappa_{13} = \kappa_{22} = \kappa_{24} = \kappa_{31} = \kappa_{33} = \kappa_{42} = \kappa_{44} = 0 \quad , \quad \kappa_{12} = \kappa_{34} = -\mathrm{i}\xi \quad ,$$

$$\kappa_{14} = \frac{1}{\mu''} \quad , \quad \kappa_{21} = \kappa_{43} = -\frac{\mathrm{i}\xi\lambda'}{\lambda' + 2\mu'} \quad , \quad \kappa_{23} = \frac{1}{\lambda' + 2\mu'} \quad ,$$

$$\kappa_{32} = -\varrho\omega^2 \quad , \quad \kappa_{41} = -\varrho\omega^2 + \xi^2\left(\lambda + 2\mu - \frac{(\lambda')^2}{\lambda' + 2\mu'}\right) \quad . \tag{7.2.26}$$

7.2.3 Harmonic Waves in Piezocrystals

Interesting phenomena occur in anisotropic media when elastic waves interact with other physical fields. Among these phenomena, the most important might be *piezoelectricity* which is employed for the transformation of electromagnetic energy into acoustical and vice versa (sound generation and reception). An electric field E applied to a piezoelectric crystal gives rise to mechanical stress (linearly dependent on E). In addition, the inverse effect also occurs that is, mechanical stresses give rise to an electric field E. Mathematically this can be described by the relations

$$\sigma_{ij} = C_{ijkl}u_{kl} - \beta_{kij}E_k \quad , \tag{7.2.27}$$

$$D_i = D_{0i} + \varepsilon_{ij}E_j + \beta_{ijk}u_{jk} \quad . \tag{7.2.28}$$

Here E and D are the electric field and electric induction, ε_{ij} is dielectric permittivity tensor, $D_0 = $ const is pyroelectric induction vector. Tensor β_{kij} is called piezotensor. Relation (7.2.27) is Hooke's law for piezoelectric crystals. The invariance of the piezotensor with respect to exchange of the second and the third indexes $\beta_{kij} = \beta_{kji}$ follows from the symmetry of the stress tensor. By the same type of reasoning as was used to consider the symmetry properties of the tensor C_{ijkl} for a transversally isotropic medium it is easy to show that any tensor of the third order is zero in the case of a crystal with a center of symmetry. In particular, $\beta_{kij} = 0$ (no piezoelectric effect) for an isotropic medium.

The electromagnetic field induced by elastic waves in a piezoelectric can be described by the equations of electrostatics, curl $E = 0$, div $D = 0$, since the propagation velocity of these waves is much less than the light velocity c. Introducing an electrostatic potential φ so that $E = -\nabla\varphi$, we obtain from (1.3.2), (7.2.27, 28) and div $D = 0$, for a homogeneous medium

$$\varrho\frac{\partial^2 u_i}{\partial t^2} = C_{ijkl}\frac{\partial u_{kl}}{\partial x_j} + \beta_{kij}\frac{\partial^2 \varphi}{\partial x_k \partial x_j} \quad , \quad \varepsilon_{ij}\frac{\partial^2 \varphi}{\partial x_i \partial x_j} = \beta_{ijk}\frac{\partial u_{jk}}{\partial x_i} \quad . \tag{7.2.29}$$

Note that the electrostatic potential has the same periodicity in time and space as a displacement field so that for the plane harmonic wave we have

$$\boldsymbol{u} = \boldsymbol{v} \exp\left(ikn_j x_j - i\omega t\right) \quad , \quad \varphi = \Psi \exp\left(ikn_j x_j - i\omega t\right) \quad , \quad n_j n_j = 1 \quad .$$

$$(7.2.30)$$

From the system of equations, (7.2.29), we obtain $\Psi = \beta_{ijk} v_j n_i n_k / \varepsilon_{lm} n_l n_m$,

$$(\tilde{\Gamma}_{ik} - \varrho \omega^2 k^{-2} \delta_{ik}) v_k = 0 \quad , \quad \tilde{\Gamma}_{ik} = \Gamma_{ik} + \frac{(\beta_{jis} n_j n_s)(\beta_{prk} n_p n_r)}{\varepsilon_{lm} n_l n_m} \quad .$$

$$(7.2.31)$$

The parameter q, defined by $\Gamma_{ik}(1+q^2) = \tilde{\Gamma}_{ik}$ is called the electromechanical coupling coefficient. As a rule $q \ll 1$. If the piezoeffect is absent, $q = 0$ and (7.2.31) becomes identical to (7.2.3). The phase and group velocities of elastic waves in a piezoelectric are determined with the use of the Christoffel equation (7.2.4), where Γ_{ik} must be replaced by $\tilde{\Gamma}_{ik}$.

Thus, the piezoeffect influences the velocities of elastic waves in crystals. Besides that, however, it gives rise to new kinds of surface waves. Consider, for example, a transverse surface wave in a piezocrystal of C_{6v} crystal class. Such crystals are of the hexagonal symmetry type. They have six planes of symmetry intersecting at symmetry axes of the sixth order. (Let it be Oy coordinate axis.) That is why their dielectric tensor is diagonal and $\varepsilon_{11} = \varepsilon_{33} \equiv \varepsilon$. Nonzero components of the peizotensor are β_{222}, $\beta_{112} = \beta_{232} \equiv \beta$, $\beta_{211} = \beta_{233}$. For the elastic modulus tensor we have $C_{1212} = C_{2323} \equiv C$ and any components where some index is contained an odd number of times equals zero. (In terms we have used for transversally isotropic media $C = \mu''$). The vector $\boldsymbol{D}_0$ is directed along the symmetry axes: $D_{01} = D_{03} = 0$.

Let the plane $z = 0$ be the boundary of the crystal, and the wave propagate in the x-direction. We assume that the surface wave is a plane one. In this case the potential φ and sole displacement component $u_2 \equiv u$ do not depend on y. For the strain and stress tensors, the nonzero components are

$$u_{12} = 0.5 \frac{\partial u}{\partial x} \quad , \quad u_{23} = 0.5 \frac{\partial u}{\partial z} \quad ,$$

$$\sigma_{21} = C \frac{\partial u}{\partial x} + \beta \frac{\partial \varphi}{\partial x} \quad , \quad \sigma_{23} = C \frac{\partial u}{\partial z} + \beta \frac{\partial \varphi}{\partial z} \quad .$$

At $z < 0$, i.e., inside the crystal, we obtain from (7.2.29)

$$\varrho \frac{\partial^2 u}{\partial t^2} = \left(C + \frac{\beta^2}{\varepsilon}\right) \Delta u \quad , \quad \Delta \varphi = \beta \varepsilon^{-1} \Delta u \quad .$$

$$(7.2.32)$$

Above the crystal (in the vacuum, $z > 0$) we have $u \equiv 0$ and the Laplace equation for the potential, that is $\Delta \varphi = 0$. Mechanical and electrical boundary conditions must be satisfied at the plane $z = 0$. Stress tensor components σ_{3j} must be zero at this plane. Since in the case under consideration $\sigma_{13} \equiv \sigma_{33} \equiv 0$, only one condition remains

$$\left(C \frac{\partial u}{\partial z} + \beta \frac{\partial \varphi}{\partial z}\right)_{z=0} = 0 \quad .$$

$$(7.2.33)$$

If the piezoelectric surface is free, the tangential component of E and the normal component of D must be continuous, that is,

$$\left[\frac{\partial \varphi}{\partial x}\right]_{z=0} = 0 \quad , \quad \left(\varepsilon\frac{\partial \varphi}{\partial z} - \beta\frac{\partial u}{\partial z}\right)_{z=-0} = \left(\frac{\partial \varphi}{\partial z}\right)_{z=+0} \quad . \tag{7.2.34a}$$

Other interesting case is when the crystal surface is electrically shorted, that is, covered by a thin metal foil. The foil has a negligible influence on the mechanical boundary conditions, and the only electrical condition is

$$\left(\frac{\partial \varphi}{\partial x}\right)_{z=0} = 0 \quad . \tag{7.2.34b}$$

Let us look for surface wave in the form

$$u = v(z)\exp{(i\xi x - i\omega t)} \quad , \quad \varphi = \Psi(z)\exp{(i\xi x - i\omega t)} \quad . \tag{7.2.35}$$

Then the first of the expressions in (7.2.32) gives at $z \leq 0$

$$v = A_1 \exp\left(\sqrt{\xi^2 - k^2}z\right) + A_2 \exp\left(-\sqrt{\xi^2 - k^2}z\right) \quad ,$$

$$k^2 \equiv \varrho\omega^2\left(C + \frac{\beta^2}{\varepsilon}\right)^{-1} \quad . \tag{7.2.36}$$

The amplitude of the surface wave must tend to zero as the distance from the boundary increases. Therefore we require $\xi > k$ and $A_2 = 0$. In the same manner we find the electric potential:

$$\Psi = B_1 \exp{(\xi z)} + \beta\varepsilon^{-1}A_1 \exp\left(\sqrt{\xi^2 - k^2}z\right) \quad , \quad z \leq 0 \quad ;$$

$$\Psi = B_2 \exp{(-\xi z)} \quad , \quad z \geq 0 \quad . \tag{7.2.37}$$

Substitution of (7.2.35–37) into the boundary conditions for electrically free and shorted surfaces yields, respectively,

$$B_1 = -\frac{A_1\beta}{\varepsilon(\varepsilon + 1)} \quad , \quad B_2 = \frac{A_1\beta}{\varepsilon + 1} \quad , \quad \xi = \frac{\omega}{c_1} \quad ,$$

$$c_1^2 = \frac{C[1 - a^2(\varepsilon + 1)^{-2}]}{\varrho(1 - a)} \quad ; \tag{7.2.38a}$$

$$B_1 = -\beta\varepsilon^{-1}A_1 \quad , \quad B_2 = 0 \quad , \quad \xi = \frac{\omega}{c_2} \quad , \quad c_2^2 = \frac{C(1 + a)}{\varrho} \quad ,$$

$$a = \frac{\beta^2}{\beta^2 + \varepsilon C} \quad . \tag{7.2.38b}$$

Here c_1 and c_2 are the velocities of the surface wave. These waves, called *Gulyaev-Bluestein waves*, propagate without dispersion. In general, the velocities of other surface waves in a piezoelectric cannot be found analytically and numerical methods must be used [7.24, 51].

The velocities of elastic waves in an unbounded piezoelectric can be found from the Christoffel equation. By assuming $n_2 = 0$ and taking into account the symmetry of the crystal we easily find from (7.2.3 and 31)

$$\tilde{\Gamma}_{ik} = \Gamma_{ik} \quad , \quad \text{if} \quad i \neq 2 \quad \text{and} \quad k \neq 2 \quad ;$$
$$\tilde{\Gamma}_{12} = \tilde{\Gamma}_{23} = 0 \quad , \quad \Gamma_{22} = C \quad , \quad \tilde{\Gamma}_{22} = C + \beta^2/\varepsilon \quad . \tag{7.2.39}$$

We see that the piezoeffect does not affect the waves polarized in the xz-plane if their wave vector lies in the same plane. These elastic waves do not generate an electric field. The wave where $u_1 = u_3 = 0$ propagates in the crystal with the velocity

$$c_3 = \varrho^{-1/2}\sqrt{C + \beta^2/\varepsilon} = \sqrt{C/\varrho(1-a)} \quad .$$

If $\beta \to 0$ this wave becomes an ordinary shear wave with velocity $c_4 = (C/\varrho)^{1/2}$.

Since $a > 0$, the velocities of bulk and surface waves obey the inequality $c_4 < c_3 < c_2 < c_1$. Usually $a \ll 1$ and all four velocities are nearly equal. The differences between c_1, c_2, and c_3 is of the order of $O(a^2)$ and the difference between c_4 and c_1, c_2, or c_3 is of the order of $O(a)$. The depth of penetration of the surface waves into the piezoelectric given by $l_s \simeq \omega^{-1}(c_3^{-2} - c_s^{-2})^{-1/2}$, $s = 1, 2$, is large compared to the wavelength. The wave along an electrically shorted surface penetrates deeper. When the surface is free, an electrical field exist in the upper medium, too, but only at distances from the surface of the order of the wavelength. The existence of Gulyaev-Bluestein wave is exclusively due to the piezoeffect. When $\beta \to 0$ then $l_s \to \infty$ and these waves become ordinary bulk shear waves. More information about elastic wave propagation in piezoelectrics and their practical applications can be found in [7.25–28].

7.3 Elastic Properties of Finely Layered Media

Consider the propagation of an elastic wave in a medium where the density and Lamé parameters are periodic functions of z, with the period h small compared to the wavelength. This problem has been considered by many researchers (see, for instance, [7.29–32]). A rather elaborate analysis of the case where the medium consists of two kinds of alternating homogeneous layers was made by *Rytov* [7.33]. From our point of view the most general and adequate approach to this problem, and the one which we shall pursue, is the matrix method (see *Molotkov* [7.34]).

7.3.1 Matrix Propagator for Inhomogeneous Solids

We shall need an expression for the matrix for an *inhomogeneous* solid layer. To obtain it, we divide the inhomogeneous layer of thickness h into $n \gg 1$ layers, each of which can be assumed to be homogeneous. The matrix for each homogeneous layer given by (7.2.24) we represent as a series in powers of $[\kappa]$. Then, the matrix of the inhomogeneous layer will be

$$[A] = \prod_{j=n}^{1} \{[E] + h_j[\kappa_j] + h_j^2[\kappa_j]^2/2! + \dots\}$$

$$= [E] + \sum_{j=1}^{n} [\kappa_j]h_j + \frac{1}{2}\sum_{j,l=1}^{n} [\kappa_j][\kappa_l]h_j h_l + \dots \quad , \tag{7.3.1}$$

where h_j is the thickness of the layer j, $j > l$. In the limit $n \to \infty$ we obtain

$$[A] = [E] + \int_0^h [\kappa(z)]dz + \int_0^h [\kappa(z_1)]dz_1 \int_0^{z_1} [\kappa(z_2)]dz_2$$

$$+ \int_0^h [\kappa(z_1)]dz_1 \int_0^{z_1} [\kappa(z_2)]dz_2 \int_0^{z_2} [\kappa(z_3)]dz_3 + \dots \quad . \tag{7.3.2}$$

Since the matrix elements $[\kappa(z)]$ are bounded, the series (7.3.2) converges. Indeed, let $|\kappa_{ik}| < \delta$ at any z, i, and k. Then any element of the product of N matrices $[\kappa]$ is not greater than $m^{N-1}\delta^N$, where m is the dimension of matrix $[\kappa]$ (2 or 4). Hence the series (7.3.2) can be majorized by a power expansion of $\exp(m\delta h)$ for any element of the matrix $[A]$.

The system consisting of n identical layers (periods) with resulting thickness $H = nh$ has the matrix $[A_0] = [A]^n$. We represent the latter as

$$[A_0] \equiv [A]^n \equiv \{[E] + n^{-1}H[B]\}^n \quad,$$

$$[B] = [B_0] + n^{-1}H[B_1] + n^{-2}H^2[B_2] + \dots \quad,$$

$$[B_0] = \int_0^1 [\kappa(hu)]du \quad,$$

$$[B_1] = \int_0^1 [\kappa(hu_1)]du_1 \int_0^{u_1} [\kappa(hu_2)]du_2 \quad,$$

$$[B_2] = \int_0^1 [\kappa(hu_1)]du_1 \int_0^{u_1} [\kappa(hu_2)]du_2 \int_0^{u_2} [\kappa(hu_3)]du_3 \quad, \dots \quad . \tag{7.3.3}$$

We assume that the matrix $[\kappa(z)]$ is a function of the dimensionless ratio z/h. (This is the case, for example, when the period of thickness h consists of a set of homogeneous layers whose thickness changes proportionally to h whereas other parameters do not depend on h.) In this case the matrices $[B_l]$, $l = 0, 1, 2, \dots$ do not depend on h.

If we have a scalar case, where $[A_0]$ and $[B]$ are not matrices but just numbers, we have in (7.3.3) that $[A_0] \to \exp\{H[B_0]\}$ at $n \to \infty$. Let us show that analogous matrix relation also holds. We have according to (7.3.3)

$$[A_0] = [Q]^n + \frac{H}{n} \sum_{j=0}^{n-1} [Q]^j \{[B] - [B_0]\}[Q]^{n-j-1}$$

$$+ \frac{H^2}{n^2} \sum_{j=0}^{n-2} \sum_{l=0}^{n-2-j} [Q]^j \{[B] - [B_0]\}[Q]^l \{[B]$$

$$- [B_0]\}[Q]^{n-j-l-2} + \dots \quad, \tag{7.3.4}$$

where $[Q] = [E] + n^{-1}H[B_0]$. We expand $[Q]^n$ in powers of $[B_0]$ by using the binomial formula and the easily verified property of binomial coefficients:

$$[Q]^n = \sum_{j=0}^{n} C_n^j \left\{ \frac{H}{n}[B_0] \right\}^j \quad ,$$

$$C_n^j = \frac{n(n-1)\cdot \ldots \cdot (n-j+1)}{j!}$$

$$= \frac{n^j}{j!} - \frac{n^{j-1}}{2(j-2)!} + \frac{(3j-1)n^{j-2}}{24(j-3)!} \left[1 + O\left(\frac{1}{n}\right) \right] \quad . \tag{7.3.5}$$

When n is sufficiently large

$$[Q]^n = \sum_{j=0}^{n} \frac{\{H[B_0]\}^j}{j!} - \frac{\{H[B_0]\}^2}{2n} \sum_{j=0}^{n-2} \frac{\{H[B_0]\}^j}{j!} + O\left(\frac{1}{n^2}\right)$$

$$= \exp\{H[B_0]\} \left\{ [E] - \frac{H^2[B_0]^2}{2n} \right\} + O(n^{-2}) \quad . \tag{7.3.6}$$

Since $[B] - [B_0] \sim 1/n$, according to (7.3.4) the difference $[A_0] - [Q]^n$ equals $O(n^{-1})$. This can be estimated as shown above and gives [Ref. 7.34, Sect. 5.1]:

$$[A_0] = \exp\{H[B_0]\} + \frac{H^2}{n} \sum_{j,l=0}^{\infty} \frac{H^{j+l}}{(j+l+1)!}[B_0]^j \left\{ [B_1] - \frac{1}{2}[B_0]^2 \right\}[B_0]^l$$

$$+ O(n^{-2}) \quad . \tag{7.3.7}$$

The series on the right-hand side of (7.3.7) converges. In the particular case when the period is symmetrical, i.e., $[\kappa(z)] = [\kappa(h - z)]$, we have

$$[B_1] = \int_0^1 [\kappa(hu_1)]du_1 \int_{1-u_1}^1 [\kappa(hu_2)]du_2 = \frac{1}{2} \iint_0^1 du_1 du_2 [\kappa(hu_1)][\kappa(hu_2)]$$

$$= 0.5[B_0]^2 \quad . \tag{7.3.8}$$

In this case the correction to the limiting value of $[A_0]$ in (7.3.7) becomes zero. Calculation of the next term of the correction gives [Ref. 7.34, Sect. 5.1]:

$$[A_0] = \exp\{H[B_0]\} + \frac{H^3}{n^2} \sum_{j,l=0}^{\infty} \frac{H^{j+l}}{(j+l+1)!}[B_0]^j \left\{ [B_2] - \frac{1}{6}[B_0]^3 \right\}[B_0]^l$$

$$+ O(n^{-3}) \quad . \tag{7.3.9}$$

If $[B_1] = [B_0]^2/2$, $[B_2] = [B_0]^3/6$, the correction is of the higher order with respect to small value n^{-1} and so on. When the period is homogeneous, all the corrections become zero.

7.3.2 An Effective Medium

By comparing (7.2.24) and (7.3.7, 9) we see that at $n \to \infty$ the system of n identical layers of thickness H is described by the matrix coinciding with that of a layer filled by some homogeneous *effective medium*. Let the original medium be transversally isotropic and isotropy plane horizontal. We shall obtain the parameters of the effective medium for SH waves by equating the elements of the matrix $[\kappa]$ (7.2.25) for the homogeneous layer to the elements of the matrix $[B_0]$. The matrices are equal at any ξ and ω if $1/\tilde{\mu}'' = \langle 1/\mu'' \rangle$, $\tilde{\mu} = \langle \mu \rangle$, $\tilde{\varrho} = \langle \varrho \rangle$, where the tilde denotes a property of the effective medium and $\langle f \rangle$ is averaging according to

$$\langle f \rangle = h^{-1} \int\limits_0^h f(z)dz \quad . \tag{7.3.10}$$

For P-SV waves, we similarly equate the matrix $[\kappa]$ (7.2.26) to the corresponding matrix $[B_0]$ (7.3.3) and obtain three more algebraic equations relating average and effective parameters. Solving these equations with respect to elastic constants we find

$$\tilde{\lambda} = \langle \lambda - (\lambda')^2 (\lambda' + 2\mu')^{-1} \rangle + \frac{\langle \lambda'(\lambda' + 2\mu')^{-1} \rangle^2}{\langle (\lambda' + 2\mu')^{-1} \rangle} \quad ,$$

$$\tilde{\varrho} = \langle \varrho \rangle \quad , \quad \tilde{\mu} = \langle \mu \rangle \quad , \quad \tilde{\mu}' = \frac{\langle \mu'(\lambda' + 2\mu')^{-1} \rangle}{\langle (\lambda' + 2\mu')^{-1} \rangle} \quad ,$$

$$\tilde{\mu}'' = \langle 1/\mu'' \rangle^{-1} \quad , \quad \tilde{\lambda}' = \frac{\langle \lambda'(\lambda' + 2\mu')^{-1} \rangle}{\langle (\lambda' + 2\mu')^{-1} \rangle} \quad . \tag{7.3.11}$$

Up to now, we have considered only harmonic plane waves. However, these results hold for other types of waves since in a transversally isotropic medium any kind of wave can be represented as a superposition of SH and P-SV harmonic waves. The parameters of the effective medium can be calculated from (7.3.11). These parameters do not depend on h. We can determine the conditions under which it is possible to use the approximation of an effective medium by examining the expressions for the correction terms in (7.3.7, 9). If $[B_1] \neq [B_0]^2/2$, the difference $[B_1] - [B_0]^2/2$ in (7.3.7) can be replaced by $-[B_0]^2/2$ without changing the order of magnitude. In this case the correction term reduces to the simple expression given in (7.3.6). As a rule the correction $H^2[B_0]^2/n \simeq H^2[\kappa]^2/n$ is negligible compared to $[E]$ if $h/\lambda \ll n^{-1/2}$, where λ is wavelength, see (7.2.25, 26) for $[\kappa]$. If the difference $[B_1] - [B_0]^2/2$ is small, the correction term must be estimated with (7.3.9). By replacing $[B_0]^3/6 - [B_2]$ by $[B_0]^3$ we obtain $H^3[B_0]^3/n^2$ for the order of magnitude of the correction term. This gives a weaker than usual requirement for the appropriateness of the effective homogeneous medium approximation to the multilayered system $h/\lambda \ll n^{-1/3}$.

It is important that the requirement of strict periodicity in the original, finely layered medium is not a necessary precondition for replacing it by an effective medium. The matrices $[B_l]$, $l = 1, 2, \ldots$ of the separate layers are present only in the correction terms. If the matrices $[B_0]$ given by (7.3.3) are similar for all layers, the system will again be equivalent to an effective medium (7.3.11). In the case where the differences between $[B_0]$ in different layers are on the order of $1/n$, the correction terms will be of the same order as in the strictly periodic system. For example, if each "period" is a set of thin homogeneous or inhomogeneous layers, the parameters of the effective medium do not depend on the arrangement of the individual layers within the boundaries of each period.

7.3.3 The Most Important Special Cases

Let us see what kind of effective medium we shall have when the initial periodic system is an elastic locally isotropic solid or a liquid. In the first case, the displacement-stress vectors for P-SV and SH waves will be the same, as in the more general case of a transversally isotropic medium. The effective values of the elastic constants are again given by (7.3.11), where $\lambda' = \lambda$, $\mu' = \mu'' = \mu$ must be placed inside the averaging sign. Then

$$
\tilde{\lambda} = 2\langle \lambda\mu(\lambda + 2\mu)^{-1}\rangle + \frac{\langle \lambda(\lambda + 2\mu)^{-1}\rangle^2}{\langle (\lambda + 2\mu)^{-1}\rangle} \quad , \quad \tilde{\varrho} = \langle \varrho \rangle \quad ,
$$

$$
\tilde{\mu} = \langle \mu \rangle \quad , \quad \tilde{\mu}' = \frac{\langle \mu(\lambda + 2\mu)^{-1}\rangle}{\langle (\lambda + 2\mu)^{-1}\rangle} \quad , \quad \tilde{\mu}'' = \langle 1/\mu \rangle^{-1} \quad ,
$$

$$
\tilde{\lambda}' = \frac{\langle \lambda(\lambda + 2\mu)^{-1}\rangle}{\langle (\lambda + 2\mu)^{-1}\rangle} \quad .
\tag{7.3.12}
$$

Thus, a finely layered isotropic medium, when averaged over z, can be considered to be homogeneous anisotropic.

It is important to note that the effective medium may have properties quite different from that of the constituent layers. Consider, for example, the case when the period consists of two kinds of homogeneous isotropic layers of equal thickness with parameters $\lambda_1 \ll \lambda_2$, $\mu_1 \ll \mu_2$ and $\lambda_1/\mu_1 = \lambda_2/\mu_2$. Using (7.3.12) we obtain approximately $\tilde{\mu} \approx \mu_2/2$, $\tilde{\mu}' = \tilde{\mu}'' \approx 2\mu_1$. The effective medium appears to be strongly anisotropic since $\tilde{\mu} \gg \tilde{\mu}'$.

The results obtained can also be applied to a liquid periodic system. For this purpose we shall use the analogy between the propagation of sound waves in a layered liquid medium and propagation of SH waves in an isotropic layered elastic medium (Sect. 1.3). Both processes are described by the same equation and boundary conditions if we allow $p \to u_2$, $\varrho \to 1/\mu$, $c^2 \to \mu/\varrho$. The stress tensor component $\sigma_{23} = \mu \partial u_2/\partial x_3$ must be then replaced by $\varrho^{-1}\partial p/\partial x_3$, i.e., $i\omega v_3$ according to the Euler equation, where v_3 is the vertical component of particle velocity.

The sound equivalent to the displacement-stress vector will then be $f = (p, i\omega v_3)^T$ and the matrix of the layer is $[A] = \exp\{(z - z_0)[\kappa]\}$, where

$$
\kappa_{11} = \kappa_{22} = 0 \quad , \quad \kappa_{12} = \varrho \quad , \quad \kappa_{21} = (\xi^2 - k^2)/\varrho \quad ,
\tag{7.3.13}
$$

by analogy with (7.2.25). Following the scheme used in the previous section, (7.3.13) could also be obtained directly from the acoustical equations (1.1.12, 13) with the boundary condition (1.1.22).

The effective homogeneous medium is described by the matrix propagator $[A(z, z_0)] = \exp\{(z - z_0)[B_0]\}$, where

$$(B_0)_{11} = (B_0)_{22} = 0 \quad , \quad (B_0)_{12} = \langle \varrho \rangle \quad ,$$
$$(B_0)_{21} = \xi^2 \langle \varrho^{-1} \rangle - \omega^2 \langle 1/\varrho c^2 \rangle \quad . \tag{7.3.14}$$

Since, in general $\langle \varrho \rangle \neq 1/\langle \varrho \rangle^{-1}$, the effective medium is not a liquid. We mentioned a similar case where the effective elastic medium was not isotropic although the original medium consisted of isotropic layers. The effective medium of (7.3.14) can be treated as liquid, however, but in another coordinate system $(x, y, \tilde{z})$, where

$$\tilde{z} = qz \quad , \quad q = \sqrt{\langle \varrho \rangle \langle \varrho^{-1} \rangle} \quad . \tag{7.3.15}$$

Due to Cauchy-Schwartz inequality [Ref. 7.35, Sect. 4.6] we have $q \geq 1$, so that the new coordinate system is obtained from the original one by stretching along the vertical. By differentiation of $f(z) = [A(z, z_0)]f(z_0)$ with respect to z we obtain

$$\frac{df(z)}{dz} = [B_0][A(z, z_0)]f(z_0) = [B_0]f(z) \quad . \tag{7.3.16}$$

In the new coordinate system the displacement-stress vector is $\tilde{f} = (p, i\omega q v_3)^T$, hence (7.3.16) is now $d\tilde{f}/d\tilde{z} = [\tilde{\kappa}]\tilde{f}$, or $\tilde{f}(\tilde{z}) = \exp\{(\tilde{z} - \tilde{z}_0)[\tilde{\kappa}]\}\tilde{f}(\tilde{z}_0)$, where

$$\tilde{\kappa}_{11} = \tilde{\kappa}_{22} = 0 \quad , \quad \tilde{\kappa}_{12} = \langle \varrho^{-1} \rangle^{-1} \quad ,$$
$$\tilde{\kappa}_{21} = \xi^2 \langle \varrho^{-1} \rangle - \omega^2 \langle 1/\varrho c^2 \rangle \quad . \tag{7.3.17}$$

By comparison of (7.3.13 and 17) we see that in the "stretched" coordinate system the effective medium is a liquid with the density $\tilde{\varrho} = \langle \varrho^{-1} \rangle^{-1}$ and sound velocity $\tilde{c} = (\langle \varrho^{-1} \rangle / \langle 1/\varrho c^2 \rangle)^{1/2}$.

In the original coordinate system this effective medium may be considered as a *generalized transversally isotropic fluid* [Ref. 7.34; Chap. 5] with different densities along the horizontal and vertical coordinates. Such a generalized fluid is described by

$$\frac{\partial v_i}{\partial t} = -\frac{1}{\varrho_1} \frac{\partial p}{\partial x_i} \quad , \quad i = 1, 2 \quad ;$$
$$\frac{\partial v_3}{\partial t} = -\frac{1}{\varrho_2} \frac{\partial p}{\partial x_3} \quad ; \quad \frac{\partial v_l}{\partial x_l} + \beta \frac{\partial p}{\partial t} = 0 \quad , \tag{7.3.18}$$

which become the usual equations for sound waves in fluids (1.1.9, 10) if $\varrho_1 = \varrho_2$, $\beta = 1/\varrho_1 c^2$. The parameter β is the compressibility. The matrix $[\kappa]$ for a homogeneous layer in generalized fluid can be found in the usual way:

$$\kappa_{11} = \kappa_{22} = 0 \quad , \quad \kappa_{12} = \varrho_2 \quad , \quad \kappa_{21} = \xi^2/\varrho_1 - \omega^2\beta \quad . \tag{7.3.19}$$

By comparing these relations with (7.3.14) we see that the effective medium is a generalized fluid with parameters $\varrho_1 = \langle \varrho^{-1} \rangle^{-1}$, $\varrho_2 = \langle \varrho \rangle$, $\beta = \langle 1/\varrho c^2 \rangle$.

These results allow us to study also the absorption of waves in finely layered media. For a periodic viscoelastic system (isotropic or transversally isotropic), the effective medium is again described by (7.3.11, 12) where now the elastic constants are complex values. In particular, for a viscous liquid with negligibly small thermal conductivity, when $\lambda = \lambda' = \lambda_0 - i\omega(\zeta - 2\eta/3)$, $\mu = \mu' = \mu'' = -i\omega\eta$, the effective medium is transversally isotropic, viscoelastic body with pure imaginary elastic moduli $\tilde{\mu}$ and $\tilde{\mu}''$ and complex $\tilde{\lambda}$, $\tilde{\lambda}'$ and $\tilde{\mu}'$. In the case of alternating liquid and solid layers we obtain an effective medium of the same kind, according to (7.3.11). The results hold, however, only when the liquid layer thickness h_1 is small not only compared to the sound wavelength λ but also to the wavelength of the viscous wave $\lambda_v = 2\pi(2\eta/\omega\varrho)^{1/2} \ll \lambda$, see (7.1.12). If the viscosity of the liquid is so small that $\lambda_v \ll h_1 \ll \lambda$ and the thickness of the solid layer is small compared to the wavelength of the elastic waves that are propagating in it, then to a first approximation the viscosity of the liquid can be neglected. Such a problem was considered in [Ref. 7.34, Sect. 5.5; Ref. 7.36, Sect. 12.4]. A specific case of this problem is the definition of the effective medium for an elastic finely layered system with, as a boundary condition, free slip between layers [Ref. 7.34, Sect. 5.4). Effective media for periodic systems with different types of anisotropy are found in [7.37, 52].

Other approaches can also be used for the determination of effective media. Among them are the variational method [Ref. 7.38, Chap. 4] and the asymptotic multiscale method [7.39]. These methods are more sophisticated and can also be applied to the general case of a three-dimensional periodic medium. The corresponding methods of averaging are described in [7.38, 39]. By using these methods it is possible to determine effective media even when deviations from periodicity take place. ([Ref. 7.39, Chap. 3] and [7.40]). The theory of a transient wave propagation in a finely layered, anisotropic, viscoelastic media was developed in [7.53, 54].

8. Geometrical Acoustics. WKB Approximation

The importance of geometrical acoustics, or the ray method, in studying sound fields in inhomogeneous media can hardly be exaggerated. Regardless of the physical nature of the waves considered, this approach is also often referred to as geometrical optics or the eikonal approximation. Due to its simplicity and lucidity it is very frequently used in fundamental as well as applied studies. Even beyond its range of direct applicability in most cases geometrical acoustics allows us to qualitatively picture the field structure. Thus, it is of great heuristic value. In this chapter we shall consider waves which are harmonically dependent on horizontal coordinates and time. In regions where a medium is homogeneous, the field reduces to one or two (propagating upwards and downwards) plane waves. For this type of problem the ray approach coincides with the WKB approximation. Geometrical acoustics for the general case of three-dimensionally inhomogeneous, moving and motionless media have been considered in [Ref. 8.65, Chap. 5].

A modern interpretation of the ray approach as an approximate method of a wave theory was initiated by *Debay* [8.1], *Sobolev* [8.2], and *Rytov* [8.3, 4]. Prominent ideas for the WKB method were devised by *Liouville* and *Green*. Important contributions to the development of the method were made by *Wentzel, Kramers,* and *Brillouin* (the abbreviation of whose names gives us the most widely used designation of this method, WKB) as well as by *Rayleigh, Jeffreys* and several others.

Our presentation of geometrical acoustics does not intend to be exhaustive. A more complete and mathematically rigorous account of the subject may be found in the monographs [8.5–10]. A systematic exposition of the fundamentals of the ray method and its numerous applications to various physical problems as well as an extensive bibliography are contained in [8.11].

8.1 The WKB Approximation and Its Range of Validity

8.1.1 Asymptotic Solution of the Wave Equation

A sound field with the harmonic dependence on horizontal coordinates and time given by $\exp(i\boldsymbol{\xi} \cdot \boldsymbol{r} - i\omega t)$, $\boldsymbol{\xi} = (\xi_1, \xi_2, 0)$ obeys the wave equation (1.2.25). We shall assume that density ϱ, velocities of sound c and of flow v_0 are smooth functions of z. We denote $u(z)$ the projection of v_0 on the direction $\boldsymbol{\xi}$. Extracting in (1.2.25) the factor $k_0 = \omega/c_0$ which is proportional to the frequency of the wave, one finds

$$\frac{\partial^2 \Phi}{\partial \zeta^2} + k_0^2 N^2 \Phi = 0 \quad , \quad p = \Phi(\zeta) \exp(i\boldsymbol{\xi} \cdot \boldsymbol{r} - i\omega t) \quad . \tag{8.1.1}$$

Here N is an effective refraction index of a stratified moving fluid. Introducing the quantity $\theta_0 \equiv \arcsin \left[\xi/k_0 \beta(z_0) \right]$ one has

$$N^2 = \left(\frac{\varrho_0}{\varrho \beta^2} \right)^2 \left[n^2 \beta^2 - \left(\frac{\xi}{k_0} \right)^2 \right] \quad , \qquad n = \frac{c_0}{c(z)} \quad , \tag{8.1.2}$$

$$\beta = 1 - \frac{u\xi}{\omega} \quad , \qquad \zeta = \varrho_0^{-1} \int_{z_a}^{z} \varrho \beta^2 dz \quad , \qquad \xi = \frac{k_0 \sin \theta_0}{1 + u_0 c_0^{-1} \sin \theta_0} \quad , \tag{8.1.3}$$

where z_a is a constant; ϱ_0, c_0 and u_0 are the values of the corresponding quantities taken at an arbitrary horizon z_0. Later it will become clear that θ_0 is actually the incidence angle of the wave at $z = z_0$. The quantities N, β, and ζ depend on the incidence angle of the wave, but not on its frequency. At normal incidence upon a medium with constant density ($\varrho \equiv \varrho_0$), (8.1.3) gives $\zeta = z - z_a$ and N coincides with the usual refraction index n. It is worth bearing this simple case in mind to aid understanding of further results.

A sound field depends on frequency via the dimensionless parameter $k_0 L$, where L is a parameter which defines the spatial scale of the variability of the medium. When $L \to \infty$, the medium becomes homogeneous, and the solutions of (8.1.1) are $\exp(\pm ik_0 N\zeta)$. Hence it is expedient to search for the high-frequency solution of (8.1.1) in the form

$$\Phi = \exp \left(ik_0 \int^{\zeta} q \, d\zeta \right) \quad , \tag{8.1.4}$$

where $q \to \pm N$ as $k_0 L \to \infty$. The lower limit of integration is not given as it only affects the normalization of Φ. Substitution of (8.1.4) into (8.1.1) leads to the Riccati equation:

$$q^2 - N^2 = ik_0^{-1} \frac{dq}{d\zeta} \quad . \tag{8.1.5}$$

We shall search for its solution in the form of series expansion in powers of the small quantity $k_0^{-1} = c_0/\omega$:

$$q = \sum_{l=0}^{+\infty} y_l(\zeta) k_0^{-l} \quad . \tag{8.1.6}$$

Equating the coefficients before different powers k_0^{-l} in (8.1.5) one obtains

$$y_0^2 = N^2 \quad ,$$

$$y_m = (2y_0)^{-1} \left(\frac{i d y_{m-1}}{d\zeta} - \sum_{l=1}^{m-1} y_l y_{m-l} \right) \quad , \qquad m = 1, 2, \ldots \quad . \tag{8.1.7}$$

Successively we find

$$y_0 = \pm N \quad , \qquad y_1 = -i \frac{d}{d\zeta} \ln \frac{1}{\sqrt{N}} \quad , \qquad y_2 = \frac{y_0}{2N^{3/2}} \cdot \frac{d^2}{d\zeta^2} \frac{1}{\sqrt{N}} \quad ,$$

$$y_3 = \frac{i}{2} \frac{d}{d\zeta} \frac{y_2}{y_0} \quad , \tag{8.1.8}$$

and so on. Note that when N is real, the functions y_{2m} are real and contribute to the wave phase whereas y_{2m+1} are pure imaginary and determine the wave amplitude. For $N = \pm i|N|$ all y_m take pure imaginary values. Restricting ourselves to the first four terms of the series, we obtain from (8.1.6, 8)

$$\Phi = \frac{1}{\sqrt{N}} \exp\left[-\frac{\varepsilon}{2} \pm ik_0 \int^{\zeta} (1 + \varepsilon) N d\zeta\right] \quad , \tag{8.1.9}$$

where

$$\varepsilon = \frac{1}{2k_0^2 N^{3/2}} \frac{d^2}{d\zeta^2} \frac{1}{\sqrt{N}} = \frac{(\varrho_0/\varrho\beta^2)^2}{2k_0^2 N^{3/2}} \left[\left(\frac{1}{\sqrt{N}}\right)'' - \left(\frac{1}{\sqrt{N}}\right)'(\ln \varrho\beta^2)'\right] \quad . \tag{8.1.10}$$

Here the prime denotes differentiation with respect to z.

Usually (8.1.9) with ε supposed to be zero is taken as the WKB approximation, i.e.,

$$\Phi = \frac{1}{\sqrt{N}} \exp\left(\pm ik_0 \int^{\zeta} N d\zeta\right) = (\varrho_0/\varrho\beta^2)^{-1/2}(n^2\beta^2 - \xi^2/k_0^2)^{-1/4}$$

$$\times \exp\left[\pm ik_0 \int^{z} \sqrt{n^2\beta^2 - \xi^2/k_0^2} dz\right] \quad . \tag{8.1.11}$$

It should be noted that in order to construct and to study high WKB approximations to solutions of (8.1.1), it is convenient to use its relation to the nonlinear differential Milne equation (see [8.12] and references therein) or especially to a linear differential equation of third order, equivalent to the Milne equation [8.13].

8.1.2 WKB Approximation's General Conditions of Use

Equation (8.1.11), constructed from (8.1.8) with the use of y_0 and y_1 only, is close to exact solution if the contributions of subsequent approximations y_2 and y_3 to the wave's amplitude and phase are small. Consequently, by comparing (8.1.9) with (8.1.11) one finds the conditions

$$\varepsilon \ll 1 \quad , \quad F(\zeta_1, \zeta_2) \equiv k_0 \int_{\zeta_1}^{\zeta_2} \varepsilon N d\zeta \ll 1 \quad . \tag{8.1.12}$$

The condition on ε restricts values of the first and the second derivatives of N with respect to ζ. The second condition also contains an interval of ζ values in which (8.1.11) is used. If $|\zeta_2 - \zeta_1| \gtrsim 1/|Nk_0|$, the second condition is more restrictive than the first. Let $N \simeq 1$, $\beta \simeq 1$. Then the inequalities in (8.1.12) may be written as

$$k_0^2 L^2 \gg 1 \quad , \quad |z_2 - z_1| \ll k_0 L^2 \quad . \tag{8.1.13}$$

We see that variations of the medium parameters should be small over distances of the order of the wavelength for the WKB approximation to be applicable. The vertical dimensions of the region where (8.1.11) is valid are limited, but they are large compared to the parameter L defining the space scale of the medium variability.

The second equality in (8.1.12) is not only necessary but also a sufficient condition for the first WKB approximation (8.1.11) to be valid. Let $0 < N^2 < \infty$ in an interval (ζ_1, ζ_2). Consider the *exact* solution $W(\zeta)$ of (8.1.1) which coincides with the approximate solution at some point $\zeta_3 \in (\zeta_1, \zeta_2)$, namely, $W = \Phi$, $dW/d\zeta = d\Phi/d\zeta$ at $\zeta = \zeta_3$. It is supposed, that one of the two signs in the exponent in (8.1.11) is chosen. Let us introduce the following designations:

$$W(\zeta) = \Phi(\zeta)[1 + \delta_1(\zeta)] \quad ,$$

$$\frac{dW}{d\zeta} = \pm i k_0 \sqrt{N} \exp\left(\pm i k_0 \int^{\zeta} N \, d\zeta\right)\left[1 + \delta_2(\zeta) \pm i \frac{dN}{d\zeta} \cdot \frac{1 + \delta_1(\zeta)}{2 k_0 N^2}\right] \quad .(8.1.14)$$

The functions $\delta_{1,2}(\zeta)$ describe the discrepancy between the exact solution and its WKB approximation. An estimate of these functions was obtained by *Olver* [8.14, 15]:

$$|\delta_{1,2}(\zeta)| \le \exp\left[2|F(\zeta, \zeta_3)|\right] - 1 \quad . \tag{8.1.15}$$

An analogous estimate has also been proven in the case when $N^2 < 0$ [Ref. 8.8, Chap. 2] and [Ref. 8.14, 15], see also [8.58] and Sect. 9.1 below. The quantity F in (8.1.12) is proportional to k_0^{-1}. Hence for a fixed angle of incidence and a sufficiently large frequency the WKB approximation is applicable in an arbitrary medium provided that the parameters ϱ, c, v_0 are smooth functions of z and N is not equal to 0 or infinity. In a homogeneous medium $F \equiv 0$ and (8.1.11) is an exact solution of the wave equation.

Under conditions of guided propagation the ratio ξ/ω, for a normal mode of fixed number, depends on frequency, i.e., θ_0 is a function of ω, and the applicability of the WKB approximation is not guaranteed by ω tending to infinity. The above formulation of the conditions, that is, that the medium parameters must change much slower than the vertical variation of the sound field, remains valid. The WKB approximation proves to be good in describing normal modes of high order [Ref. 8.16, Sects. 49, 79]. When the horizontal separation between source and receiver is large, the error in the phase of the normal modes caluclated using the WKB approximation becomes important since this error accumulates with growing distance. In defining the maximum horizontal distance at which the WKB solution is applicable, account of the normal modes interference is important. These issues were considered in ([Ref. 8.16, Sects. 45, 48] and [8.17]). Some interesting qualitative estimates of the distances up to which the ray-acoustic method of calculation yields reasonable results for different sound field features in underwater sound channels in the ocean are made in [8.18].

8.1.3 Regions of Applicability in Vicinities of Turning Points and Horizons of Resonant Interaction

It follows from (8.1.2, 10) that (8.1.11) cannot be applied in the vicinities of horizons where one of the conditions is met:

$$\beta(z_c) = 0 \quad , \tag{8.1.16}$$

$$n(z_t)\beta(z_t) = \pm \left(1 + u_0 c_0^{-1} \sin \theta_0\right)^{-1} \sin \theta_0 \equiv \pm \frac{\xi}{k_0} \quad . \tag{8.1.17}$$

As was pointed out in Sect. 1.2, a resonant interaction between sound and flow occurs in the vicinity of the horizon $z = z_c$. $z = z_t$ is the turning horizon of the sound wave. Here the vertical component of the wave vector $(k^2\beta^2 - \xi^2)^{1/2}$ equals zero. The behavior of the acoustic field in the vicinities of the horizons z_c and z_t will be considered in Chap. 9. Here we shall discuss where one may use WKB approximation when turning points and resonant interaction with a flow take place. It will be assumed that $n(z) \simeq 1$, $\varrho(z)/\varrho_0 \simeq 1$ and that the rates of variation of c, ϱ, and v_0 with z are of the same order of magnitude.

Let there be no points of resonant interaction in the vicinity of the turning point $|z - z_t| \lesssim L$. Then $\beta(z_t) \simeq 1$. Suppose that the WKB approximation is applicable outside a region $|z - z_t| \lesssim d_t$ with $d_t \ll L$. This assumption will be justified by the following calculation. Because L is a scale of the variability of $N^2(z)$, it is reasonable to substitute for $N^2(z)$ in the region $|z - z_t| \lesssim d_t$ the first nonzero term in its series expansion at the point z_t: $N^2(z) = a(z - z_t) + O(d_t^2 L^{-2})$, $|a| \simeq L^{-1}$. [If $|a|$ is small, i.e., $|a| \ll L^{-1}$, but $|d^2N^2/dz^2| \simeq L^{-2}$, then there are two close turning points. The situation in this case is quite similar, but one more term should be retained in the expansion of $N^2(z)$]. Using the N^2 approximation written above, one finds from (8.1.10) that $\varepsilon \simeq (k_0 L N^3)^{-2}$. Then both the inequalities in (8.1.12) lead to the same condition

$$|z - z_t| \gg \sqrt[3]{L/k_0^2} \quad , \quad \text{or} \quad d_t \simeq \sqrt[3]{L/k_0^2} \quad . \tag{8.1.18}$$

[This also ensures that the inequality in (8.2.1) is satisfied, which will be useful later in the physical interpretation of the WKB approximation.]

According to (8.1.18), the vertical dimensions of the region where the ray approach is not applicable are large compared to the wavelength $\lambda = 2\pi/\beta n k_0$, but much smaller than L. With increasing frequency the region narrows as $d_t \sim \omega^{-2/3}$.

One can cast the condition of the applicability of the WKB approximation given by (8.1.18) in the vicinity of the turning horizon in a physically more obvious form by expressing it in terms of the phase integral

$$\varphi(z, z_t) = k_0 \int\limits_{\zeta(z_t)}^{\zeta(z)} N \, d\zeta$$

whose lower limit is this horizon. Namely, the WKB approximation is valid at such horizons for which a phase advance of the wave leaving the horizon z_t is large compared to unity. We shall prove this fact for a more general case than was considered above. Let the effective refractive index N tend to zero as $z \to z_t$ according to the power law $N(z) \approx b(z - z_t)^\alpha L^{-\alpha}$, $\alpha > 0$, $b = \text{const}$, and $\beta(z) \simeq 1$ near z_t. Then for ζ_1 fixed, $\zeta_1 \neq \zeta_t \equiv \zeta(z_t)$, the quantity $F(\zeta_1, \zeta)$ in (8.1.12) as well as the quantity $\varepsilon(\zeta)$ in (8.1.10) tend to infinity as $\zeta \to \zeta_t$. If $(\zeta_1 - \zeta_t)/(\zeta - \zeta_t) > 1$, then in the given vicinity of z_t one has $|\varepsilon(\zeta)| \simeq |F(\zeta_1, \zeta)|^2 \simeq L^\alpha/k_0 |z - z_t|^{1+\alpha} \simeq 1/|\varphi(z, z_t)|$. Consequently, the conditions of the validity of the WKB approximation given in (8.1.12) may be cast in the form

$$\left| k_0 \int\limits_{\zeta(z_t)}^{\zeta(z)} N \, d\zeta \right| = \left| \int\limits_{z_t}^{z} \sqrt{k^2\beta^2 - \xi^2} \, dz \right| \gg 1 \quad . \tag{8.1.19}$$

It turns out, that the inequality in (8.2.1) also holds whenever the condition of (8.1.19) is met.

We now consider the applicability of the WKB approximation in cases where resonant interaction between flow and sound wave occurs. If the incidence angle of the wave is not small ($\xi \simeq k_0$), then there are no turning points close to the point of the resonant interaction, z_c. Suppose once more that the region $|z - z_c| \lesssim d_c$ near the point z_c where the WKB method fails, is narrow compared to L. Then in this region $\beta = b(z - z_c) + O(d_c^2 L^{-2})$, $|b| \simeq L^{-1}$ and $N\beta^2 = (i\varrho_0 \xi/\varrho(z_c)k_0)[1 + O(d_c/L)]$. Now it follows from (8.1.10) that $\varepsilon \simeq k_0^{-2}(z - z_c)^{-2}$. Hence, according to (8.1.12), the conditions that need to be met in order for (8.1.11) to be applicable for $\xi \simeq k_0$ have the form

$$|z - z_c| \gg k_0^{-1} \quad , \quad \text{or} \quad d_c \simeq k_0^{-1} \quad . \tag{8.1.20a}$$

Thus, d_c is much smaller than L and is of the same order of magnitude as the wavelength far from the point z_c or the space period of the field in the horizontal plane.

For $z = z_c$ the vertical component of the wave vector equals $i\xi$. If an angle of incidence isn't small ($\xi \simeq k_0$), then the wave reaches this horizon highly attenuated. On the other hand, (8.1.16) has no solution at all when $\boldsymbol{\xi} \cdot \boldsymbol{v}_0 = 0$. That is why, when considering resonant interaction it is important to study the case where the incidence angle θ_0 is small, but different from zero, and the projection u of the vector $\boldsymbol{v}_0$ on the direction $\boldsymbol{\xi}$ is not small that is, $u/v_0 \simeq 1$. Then $\xi \ll k_0$, and according to (8.1.17) there are two turning points close to z_c. As before the inequalities in (8.1.12) make it possible to analyze the conditions of the applicability of the WKB approximation in this quite particular but interesting case. For small angles of incidence and $|z - z_c| \simeq d_c$ one has $N = (n\varrho_0/\beta\varrho)[1 + O(\xi^2/k_0^2\beta^2)]$. According to (8.1.10), $\varepsilon \simeq (k_0 L \beta^2)^{-2}$. The conditions for the validity of the approximation (8.1.11) are given by

$$|z - z_c| \gg \sqrt{L/k_0} \quad \text{or} \quad d_c \simeq \sqrt{L/k_0} \quad . \tag{8.1.20b}$$

As long as $\xi/k_0 \lesssim |\beta(z_c \pm d_c)|$ was supposed in the derivation of (8.1.20b), this estimate refers to waves with $\xi \lesssim (k_0/L)^{1/2}$. In (8.1.20b) d_c is considerably greater than in (8.1.20a) and $d_c \gg d_t$, but $d_c \ll L$ as before. The distances between z_c and the two closest turning horizons are $|z_t - z_c| \simeq L\xi/k_0 \lesssim d_c$. As expected, there are no turning points in the regions specified by (8.1.20b) where the WKB approximation holds.

Quite similarly to the above derivation of (8.1.19) for the case of a sole turning point, one can show that when $\beta(z)$ tends to zero as $z \to z_c$ according to a power law, the general conditions given in (8.1.12) for the validity of the WKB approximation as well as the consequent conditions (8.1.20a, b) may be written in the intuitive form of (8.1.19), provided that z_c is substituted for z_t.

8.2 Physical Meaning of the Approximate Solutions

We now consider the physical meaning of the solutions of the wave equation obtained by the WKB technique. Expression (8.1.11) represents two noninteracting waves propagating in directions symmetric with respect to the horizontal plane. Thus within the limits of ray acoustics first approximation there is no wave reflection. The quantity in the exponent in (8.1.11) gives the phase advance of a wave propagating between horizons which are the limits of integration. The coefficient standing before the exponent assures satisfaction of energy conservation. Substitution of (8.1.11) into (6.1.7) shows that for $N^2 > 0$ the z-component of a vector of the power flux density, averaged over a period of oscillation, I_z, depends neither on z nor on x and y, quite analogous to the case of plane waves. The choice of the sign in the exponent only affects the sign of I_z. For $N^2 < 0$ one has $I_z = 0$ as in an inhomogeneous plane wave.

The solutions given in (8.1.11) are often referred to as local plane waves with slowly varying amplitudes. In order for this interpretation to be valid, the approximation $\partial \Phi / \partial \zeta \approx \pm i k_0 N \Phi$ should be obeyed. One can obtain the derivative $\partial \Phi / \partial \zeta$ by letting $\delta_1 = \delta_2 = 0$ in (8.1.14). Then one can see that it is possible to consider the amplitude to be slowly varying if

$$\frac{1}{2 k_0 N^2} \frac{dN}{d\zeta} \ll 1 \quad . \tag{8.2.1}$$

In general this inequality is not a necessary condition for the WKB approximation to be applicable in the form of (8.1.11). For instance, if $\varrho \equiv \varrho_0$, $\theta_0 = 0$ and $n = (a + bz)^{-2}$ where a and b are arbitrary constants, $\varepsilon \equiv 0$ according to (8.1.10), and (8.1.11) gives the *exact* solution of the wave equation of (8.1.1)

$$\Phi = \text{const} \frac{1}{\sqrt{n}} \exp \left(\mp i k_0 b^{-1} \sqrt{n} \right) \quad . \tag{8.2.2}$$

Consider those z values for which $n \simeq 1$. The amplitudes of waves given by (8.2.2) are slowly varying compared to their phases only under the condition that $|b| \ll k_0$, consistent with (8.2.1). Thus, it is clear that the WKB approximation (8.1.11) may be close to an exact solution, even when that solution has a rapidly varying amplitude. Special cases similar to (8.2.2) are very rare, however. Usually the inequality (8.2.1) and the first of the conditions given in (8.1.12) reduce to the requirement that $k_0 L \gg 1$ and are satisfied in the region of applicability of the WKB approximation.

8.2.1 Medium at Rest

We now consider the sound field found using the ray approximation in more detail. Suppose initially that a medium is motionless. It is convenient to use a coordinate system with the Ox axis parallel to the vector $\boldsymbol{\xi}$. Then according to (8.1.11), the WKB approximation of the general solution to (8.1.1) is

$$p(x, z) = \sqrt{\varrho}(n^2 - \sin^2 \theta_0)^{-1/4} \left\{ C_1 \exp\left[ik_0 \int_{z_1}^{z} \sqrt{n^2 - \sin^2 \theta_0} dz\right]\right.$$

$$\left. + C_2 \exp\left[-ik_0 \int_{z_1}^{z} \sqrt{n^2 - \sin^2 \theta_0} dz\right]\right\} \exp\left(i\xi x\right) \quad , \tag{8.2.3}$$

where z_1, C_1 and C_2 are arbitrary constants. It is worth noting that density stratification affects only the amplitude of the wave. Consider waves resulting from (8.2.3), when $C_1 \neq 0$, $C_2 = 0$, and $C_1 = 0$, $C_2 \neq 0$. These waves are a generalization of plane waves in layered media. Their fronts (that is, surfaces of constant phase) are specified by the equation

$$\varphi(x, z) = \xi x \pm k_0 \int_{z_1}^{z} \sqrt{n^2 - \sin^2 \theta_0} dz = \text{const} \quad . \tag{8.2.4}$$

In terms of an angle $\theta(z) = \arc \sin \left[\xi/k_0 n(z)\right]$ which obeys Snell's law (2.2.6), the wave vector defined as the gradient of wave phase equals

$$\boldsymbol{q} = \left(\xi, 0, \pm k_0 \sqrt{n^2 - \sin^2 \theta_0}\right) = k_0 n(z) \left(\sin \theta(z), 0, \pm \cos \theta(z)\right). \tag{8.2.5}$$

Hence, $\theta(z)$ is an acute angle between the wave vector and the Oz axis. At a turning point $\theta(z_t) = \frac{\pi}{2}$. Beyond a turning point, where $n < \sin \theta_0$, θ assumes complex values. For real θ the phase velocity of the wave equals $c_{\mathrm{ph}} = \omega q/q^2$ and its magnitude coincides with the local sound speed: $c_{\mathrm{ph}} = c(z)$. It follows from (2.1.11) and (8.2.3) that the power flux density vector is parallel to the vector $\boldsymbol{q}$. According to (8.2.5) the group velocity is $\boldsymbol{c}_{\mathrm{g}} \equiv \partial\omega/\partial\boldsymbol{q} = \partial[qc(z)]/\partial\boldsymbol{q} = \boldsymbol{c}_{\mathrm{ph}}$.

A *ray* is defined as a curve whose tangents at each point are parallel to the power flux density vector. In a medium at rest rays are orthogonal to fronts. A family of parallel rays making the angle $\theta(z)$ with a vertical may be related to each of the considered solutions (8.2.3).

8.2.2 Moving Medium

Equation (8.1.11) shows that stratification of the flow velocity affects the wave amplitude as well as its phase

$$\varphi(x, z) = \xi x \pm k_0 \int_{z_1}^{z} \sqrt{n^2 \beta^2 - \xi^2/k_0^2} dz \quad . \tag{8.2.6}$$

The wave vector $\boldsymbol{q} = \nabla\varphi$ equals

$$\boldsymbol{q} = \left(\xi, 0, \pm \sqrt{k^2 \beta^2 - \xi^2}\right) = k(z)\beta(z)\left(\sin \theta(z), 0, \pm \cos \theta(z)\right) \quad , \tag{8.2.7}$$

and makes an angle

$$\theta(z) = \arc \sin \frac{\xi}{k(z)\beta(z)} \tag{8.2.8}$$

172

with the Oz axis. From (8.2.7), $q^2 = k^2\beta^2 = \omega^2 c^{-2}(1 - \boldsymbol{q} \cdot \boldsymbol{v}_0/\omega)^2$. Hence,

$$\omega = q(z)c(z) + \boldsymbol{q}(z) \cdot \boldsymbol{v}_0(z) \quad . \tag{8.2.9}$$

The dispersion equation (8.2.9), obtained by using the WKB approximation, has the same form as the dispersion equation (1.2.16) pertaining to a homogeneous moving medium, but now $\boldsymbol{q}$ and $\boldsymbol{v}_0$ are allowed to vary from one point to another. Expressions for the phase and group velocities

$$c_{\mathrm{ph}}(z) = \frac{\omega \boldsymbol{q}}{q^2} = \left(c + \frac{\boldsymbol{q} \cdot \boldsymbol{v}_0}{q} \right) \frac{\boldsymbol{q}}{q} \quad , \tag{8.2.10}$$

$$c_{\mathrm{g}}(z) = \frac{\partial \omega}{\partial \boldsymbol{q}} = \boldsymbol{v}_0 + \frac{c\boldsymbol{q}}{q} \quad , \tag{8.2.11}$$

are analogous to (1.2.15, 17). Here $\boldsymbol{v}_0$, c, and $\boldsymbol{q}$ are functions of z. Equation (8.2.10) was obtained for the first time by *Rayleigh* [8.19]. The equality of the speed of the sound energy propagation in a moving medium to the group velocity c_{g} was proven in [8.20]. From (8.2.10, 11) it follows that $c_{\mathrm{g}} \geq c_{\mathrm{ph}}$. The magnitudes and directions of the vectors c_{g} and c_{ph} coincide only under the condition that $\pm \boldsymbol{v}_0 \parallel \boldsymbol{\xi}$. At a turning point both velocities are directed horizontally.

As in the case of a motionless medium, a system of rays may be related to each of the two waves given in (8.1.11). At fixed z all the rays make the same acute angle $\psi(z)$ with the Oz axis. Inside a system an individual ray can be specified, for example, by the coordinates (x_0, y_0) of its point of intersection with a plane $z = z_0$. It is possible to match two arbitrary rays by pure horizontal translation. Because phase and group velocities differ in a moving medium, one should distinguish between the refraction laws of a wave normal $\boldsymbol{\nu} = \boldsymbol{q}/q$ and a vector $\boldsymbol{\tau} = c_{\mathrm{g}}/c_{\mathrm{g}}$ tangent to a ray. The unit vector $\boldsymbol{\nu}(z)$ is normal to the wave front and lies in a vertical plane which is parallel to the vector $\boldsymbol{\xi}$. The angle between the vector $\boldsymbol{\nu}$ and the Oz-axis is given by (8.2.8). Expressing $\boldsymbol{\xi}$ through the value of θ taken at some fixed horizon $z = z_0$, one casts the refraction law for the wave normal as follows [cf. (2.6.8)]:

$$u(z) + \frac{c(z)}{\sin \theta(z)} = u_0 + \frac{c_0}{\sin \theta_0} = \mathrm{const} \quad . \tag{8.2.12}$$

This result was obtained in Chap. 2 for a discretely layered medium.

When the vector $\boldsymbol{v}_0$ is not in the plane of wave incidence, then according to (8.2.11) a ray is not a plane curve. Departing from (8.2.7, 11) it is not difficult to find the projections of the vector $\boldsymbol{\tau}$ onto the coordinate axes. They are

$$\boldsymbol{\tau}(z) = (c^2 + v_0^2 + 2cu \sin \theta)^{-1/2}(c \sin \theta + u, v_{0y}, \pm c \cos \theta) \quad , \tag{8.2.13}$$

where v_{0y} is a $\boldsymbol{v}_0$ projection upon the Oy-axis. Sometimes it is more convenient to define the unit vector $\boldsymbol{\tau}$ by the angles ψ between $\boldsymbol{\tau}$ and the Oz-axis, and α between $\boldsymbol{\xi}$ and $\boldsymbol{\tau}$ projections onto the xy-plane (Fig. 8.1):

$$\boldsymbol{\tau} = (\sin \psi \cos \alpha, \ \sin \psi \sin \alpha, \ \pm \cos \psi) \quad . \tag{8.2.14}$$

From (8.2.13) it follows that

$$\tan \alpha = \frac{v_{0y}}{c \sin \theta + u} \quad ,$$

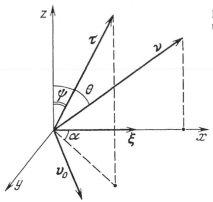

$$\cos\psi = c\cos\theta(c^2 + v_0^2 + 2cu\sin\theta)^{-1/2} \quad . \tag{8.2.15}$$

When the flow velocity tends to zero, the discrepancy between ν and τ vanishes, $\psi \to \theta$, and (8.2.12) reduces to Snell's law given in (2.2.6). The $v_{0y}(z)$ component of the flow velocity doesn't enter (8.1.1) and therefore does not affect the sound field of waves with an $\exp(i\xi x - i\omega t)$ dependence on the horizontal coordinates and time. Although different rays come through a given point (x, y, z) depending on the $v_{0y}(z)$ profile, this does not change the field, since the sound pressure is the same on each ray of the system.

A great number of studies ([8.20–34, 59–64] and others) were devoted to ray acoustics of moving media which are layered or inhomogeneous in three dimensions and its applications to the investigation of sound fields in the ocean and atmosphere. A review of part of this research together with a description of the historical background can be found in [8.35, 36].

8.3 Another Approach to the Ray Acoustics Approximation

In this section we shall use yet another approach, proposed by *Bremmer* [8.37, 38] to solve the wave equation which will enable us to get an obvious interpretation of ray acoustics' high approximations. We shall basically follow [8.37–39], but some generalizations will be made. The difference between this approach and the method described in Sect. 8.1, is the use of convergent expansions instead of the asymptotic series (8.1.6).

The wave equation given in (8.1.1) is equivalent to the system of equations

$$f = \frac{\partial\Phi}{\partial\zeta} \quad , \quad \frac{\partial f}{\partial\zeta} = -k_0^2 N^2 \Phi \quad . \tag{8.3.1}$$

In a medium at rest with $\varrho = \text{const}$ we have $\zeta = z - z_a$, $N = n(z)\cos\theta(z)$. Only the most interesting case, $N^2 > 0$, will be considered. Let us look for solutions of the system given by (8.3.1) in the form

$$\Phi = \frac{1}{\sqrt{N}}\left[\chi_1\exp\left(ik_0\int_0^\zeta N\,d\zeta\right) + \chi_2\exp\left(-ik_0\int_0^\zeta N\,d\zeta\right)\right] \quad , \tag{8.3.2a}$$

$$f = ik_0\sqrt{N}\left[\chi_1 \exp\left(ik_0 \int\limits_0^\zeta N\,d\zeta\right) - \chi_2 \exp\left(-ik_0 \int\limits_0^\zeta N\,d\zeta\right)\right] \quad , \qquad (8.3.2b)$$

where $\chi_{1,2}(\zeta)$ are new unknown functions. Note that for $\chi_{1,2} = $ const, (8.3.2a) is a sum of solutions (8.1.11). Substitution of (8.3.2) into (8.3.1) leads to a system of first-order equations for χ_1 and χ_2

$$\frac{d\chi_1}{d\zeta} = \frac{\chi_2}{2}\left(\frac{\partial}{\partial\zeta}\ln N\right)\exp\left(-2ik_0 \int\limits_0^\zeta N\,d\zeta\right) \quad ,$$

$$\frac{d\chi_2}{d\zeta} = \frac{\chi_1}{2}\left(\frac{\partial}{\partial\zeta}\ln N\right)\exp\left(2ik_0 \int\limits_0^\zeta N\,d\zeta\right) \quad . \qquad (8.3.3)$$

Supposing that $\partial(\ln N)/\partial\zeta$ tends to zero sufficiently rapidly as $|\zeta| \to \infty$, one can write (8.3.3) in the integral form

$$\chi_1(\zeta) = \frac{1}{2}\int\limits_{-\infty}^\zeta \frac{\partial(\ln N)}{\partial\zeta_1} \cdot \chi_2(\zeta_1)\exp\left(-2ik_0 \int\limits_0^{\zeta_1} N\,d\zeta_2\right)d\zeta_1 \quad ,$$

$$\chi_2(\zeta) = \frac{1}{2}\int\limits_{+\infty}^\zeta \frac{\partial(\ln N)}{\partial\zeta_1} \cdot \chi_1(\zeta_1)\exp\left(2ik_0 \int\limits_0^{\zeta_1} N\,d\zeta_2\right)d\zeta_1 \quad . \qquad (8.3.4)$$

It was shown above that the ray acoustic approximation gives better results as $N(\zeta)$ changes more slowly. Therefore, one may consider the factor $\partial(\ln N)/\partial\zeta_1$ in the right-hand side of (8.3.4) to be small and solve the system by a method of successive approximations. In the zeroth approximation, letting the right-hand sides be zero, one finds

$$\chi_1^{(0)} = C_1 \quad , \quad \chi_2^{(0)} = C_2 \quad , \quad C_{1,2} = \text{const} \quad , \qquad (8.3.5)$$

i.e., the ray acoustic result. The term containing $\chi_2^{(0)}$ in (8.3.2) describes a wave propagating towards negative ζ values (a "direct" wave). The term containing $\chi_1^{(0)}$ describes a wave propagating towards positive ζ values (an "inverse" wave). Note that the approximation given by (8.3.5), (8.3.2) differs from (8.1.11) slightly because in the first case f is only approximately equal to $\partial\Phi/\partial\zeta$. The given approximate solutions become equivalent when the inequality (8.2.1) holds.

To simplify further transformations it is convenient to rewrite (8.3.4) in the form

$$\chi_1(\zeta) = \varepsilon \int\limits_{-\infty}^\zeta \lambda_1(\zeta_1)\chi_2(\zeta_1)d\zeta_1 \quad ,$$

$$\chi_2(\zeta) = \varepsilon \int\limits_{+\infty}^\zeta \lambda_2(\zeta_1)\chi_1(\zeta_1)d\zeta_1 \quad , \qquad (8.3.6)$$

where $\varepsilon \simeq (k_0 L)^{-1}$ is a small parameter, $|\lambda_{1,2}| \lesssim 1$. Successive iterations lead to solutions that are series in powers of ε:

$$\chi_j(\zeta) = \sum_{m=0}^{+\infty} \varepsilon^m \chi_j^{(m)}(\zeta) \quad , \quad j = 1, 2 \quad . \tag{8.3.7}$$

Here

$$\chi_1^{(m)}(\zeta) = \int_{-\infty}^{\zeta} \lambda_1(\zeta_1) \chi_2^{(m-1)}(\zeta_1) d\zeta_1 \quad ,$$

$$\chi_2^{(m)}(\zeta) = \int_{+\infty}^{\zeta} \lambda_2(\zeta_1) \chi_1^{(m-1)}(\zeta_1) d\zeta_1 \quad . \tag{8.3.8}$$

When $m = 1$, (8.3.8) shows that at each point of the space an inverse wave of the zeroth approximation initiates a direct wave of the first order $\chi_2^{(1)}$, and vice versa. In other words, a wave of the zeroth approximation is reflected at each horizon due to medium inhomogeneity and in that way generates a wave of the first approximation, which propagates in the reverse direction.

For fixed ζ, a secondary direct wave $\chi_2^{(1)}$ is formed due to reflection of a primary inverse wave $\chi_1^{(0)}$ at all $\zeta_1 > \zeta$. On the contrary, an inverse secondary wave $\chi_1^{(1)}$ is created as a result of reflection of the wave $\chi_2^{(0)}$ at all horizons for which $\zeta_1 < \zeta$. This is illustrated in Fig. 8.2. In this manner generation of waves of arbitrary order m can occur.

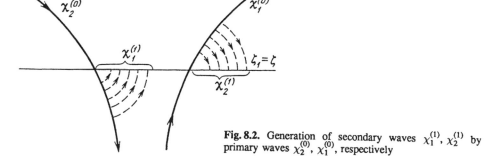

Fig. 8.2. Generation of secondary waves $\chi_1^{(1)}$, $\chi_2^{(1)}$ by primary waves $\chi_2^{(0)}$, $\chi_1^{(0)}$, respectively

By successively using the equalities in (8.3.8), one finds explicit expressions for $\chi_{1,2}^{(m)}$ in terms of m-fold multiple integrals. For $\chi_1^{(m)}$ ($m = 1, 2, \ldots$) these expressions are

$$\chi_1^{(2m)} = C_1 \int_{-\infty}^{\zeta} d\zeta_1 \cdot \lambda_1(\zeta_1) \int_{+\infty}^{\zeta_1} d\zeta_2 \cdot \lambda_2(\zeta_2) \int_{-\infty}^{\zeta_2} d\zeta_3 \cdot \lambda_1(\zeta_3) \cdot \ldots$$

$$\times \int_{+\infty}^{\zeta_{2m-1}} d\zeta_{2m} \cdot \lambda_2(\zeta_{2m}) \quad ,$$

$$\chi_2^{(2m-1)} = C_2 \int\limits_{-\infty}^{\zeta} d\zeta_1 \cdot \lambda_1(\zeta_1) \int\limits_{+\infty}^{\zeta_1} d\zeta_2 \cdot \lambda_2(\zeta_2) \int\limits_{-\infty}^{\zeta_2} d\zeta_3 \cdot \lambda_1(\zeta_3) \cdot \ldots$$

$$\times \int\limits_{-\infty}^{\zeta_{2m-2}} d\zeta_{2m-1} \cdot \lambda_1(\zeta_{2m-1}) \ . \tag{8.3.9}$$

Physically one should treat $\chi_1^{(m)}$ and $\chi_2^{(m)}$ as waves resulting from m-fold reflections in an inhomogeneous medium at the levels $\zeta_1, \zeta_2, \ldots, \zeta_m$, see (8.3.9). To obtain the complete value of $\chi_{1,2}^{(m)}$, integrals are taken over all possible levels of reflection. Figure 8.3 shows the successive generation of waves of different orders.

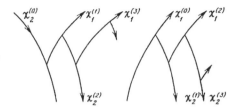

Fig. 8.3. Successive generation of waves of different orders in an inhomogeneous medium

For a large enough frequency ($\omega \sim \varepsilon^{-1}$) the series (8.3.7, 9) are convergent. This was proved in [Ref. 8.9; Chap. 3]. The necessary and sufficient condition for the series to be absolutely summable was stated in [8.40].

So far it has been assumed that the sound wave is monochromatic. When a plane δ-shaped acoustic impulse is incident upon a medium, the type of series considered, i.e., expansion of the sound field into waves with different numbers of reflections, is absolutely summable if the derivative $\partial N/\partial \zeta$ exists and is bounded [8.41]. For a moving fluid with stratified density and sound speed this condition is met when the derivatives $c'(z)$, $\varrho'(z)$, and $v_0'(z)$ just are bounded. In the case of a short impulse better convergence of the series may be explained by the fact that at any arbitrary instant only a limited part of an inhomogeneous medium contributes to the formation of the reflected field. However, another explanation is also possible. Indeed, a typical value of ω for such a pulse is infinity, since any finite frequency range contains an infinitesimal part of the total energy of the pulse. That is why the ray consideration of the reflection of a δ-impulse is applicable to an arbitrary medium with a smooth z-dependence in its parameters.

Bremmer's approach presented above admits an extension to three-dimensional inhomogeneous media [8.66, 67]. It should be noted that one more method of an approximate description of wave reflection by layered media, which to a first approximation also leads to ray acoustics, was given in [8.42]. A version of the WKB method which differs from the versions described above by second and higher approximations was considered in [8.43].

Numerous attempts have been made to modify the WKB method. They were aimed at enlarging its domain of applicability, particularly to describe the field in the vicinity of the turning point. These modifications either use special rather than elementary functions in an *ad hoc* manner [8.44–50] or do not lead to an exact asymptotic representation of the solution [8.51–57]. In many cases the physical meaning of the "modified" solutions is obscure. In our opinion the *reference equation method* is most promising technique for describing high-frequency sound fields in regions where ray acoustics is not valid. The method and several important physical problems which it enables to solve are the subject of the next chapter.

9. The Sound Field in the Case of Turning Horizons and Resonant Interaction with a Flow

An analysis of the sound field in media with smooth, slowly varying parameters (small changes over distances of the order of the wavelength) has been given in the previous chapter. We proceed with this analysis by considering the case where the WKB approximation is not adequate to describe the sound field. We will follow the ideas of *Langer* [9.1,2].

9.1 Reference Equation Method

9.1.1 High-Frequency Solution of the Wave Equation

Let two linearly independent solutions $W_{1,2}$ of the equation

$$\frac{d^2W}{d\eta^2} + k_0^2 M W = 0 \tag{9.1.1}$$

with arbitrary k_0 be known, where $M = M(\eta)$ is independent of k_0. It was shown in Sect. 3.1 that the function

$$f(\zeta) = (d\eta/d\zeta)^{-1/2} W[\eta(\zeta)] \tag{9.1.2}$$

is the exact solution to (8.1.1) with an effective refraction index

$$N(\zeta) = \sqrt{m(\zeta) + (\eta')^2 M[\eta(\zeta)]} \quad ,$$
$$m \equiv (2k_0)^{-2}\{2(\ln \eta')'' - [(\ln \eta')']^2\} \quad , \tag{9.1.3}$$

where η is an arbitrary smooth function of ζ. Here the prime denotes a derivative with respect to ζ. It is impossible to find $\eta(\zeta)$ explicitly from (9.1.3) for arbitrary $N(\zeta)$. For high-frequency waves, however, when $m(\zeta)$ is small an *approximate* solution of (8.1.1) can be found by the change of variables

$$\eta' = \sqrt{N^2(\zeta)/M[\eta(\zeta)]} \quad , \quad \text{or} \quad \int_{\eta_0}^{\eta} M^{1/2}(\eta) d\eta = \int_{\zeta_0}^{\zeta} N(\zeta) d\zeta \quad . \tag{9.1.4}$$

In this case $\eta(\zeta)$ is independent of the frequency. Note that the approximate solution (9.1.2) multiplied by $N^{1/2}$ depends on ζ only via the phase integral

$$\int_{\zeta_0}^{\zeta} N(\zeta_1) d\zeta_1 = \int_{z_0}^{z} \sqrt{n^2\beta^2 - \xi^2/k_0^2} dz \quad .$$

Hence, under the approximation of (9.1.2) and (9.1.4) the density stratification influences only the wave amplitude. As in the ray approximation the function $\varrho^{-1/2}(z)f[\zeta(z)]$ does not depend on the density variation.

The correction term $m(\zeta)$ in (9.1.3) is small if the derivative $(\ln \eta')''$ is bounded. It follows from (9.1.4) that this is the case when the function M is twice continuously differentiable, and the zeros as well as the poles of $N^2(\zeta)$ and $M[\eta(\zeta)]$ coincide. These conditions serve for choice of the *reference function* $M(\eta)$. If $N^2(\zeta)$ and $M(\eta)$ have the same singularities (9.1.1) allows us to obtain the high-frequency asymptotics of the wave field and is then called the *reference equation* with respect to (8.1.1). Each reference equation can describe the asymptotics of the sound field for the whole class of $N(\zeta)$.

In some cases one has to consider (8.1.1) and (9.1.1) with $M(\eta)$ and $N(\zeta)$ also depending on k_0. Such is the problem when, for example, the incidence angle of the wave depends on the frequency. If $M(\eta, k_0)$ and $N(\zeta, k_0)$ tend to finite functions $M_1(\eta)$ and $N_1(\zeta)$ as $k_0 \to \infty$, then M and N in (9.1.4) must be replaced by M_1 and N_1 when we are interested in asymptotic solutions. The differences $M - M_1$ and $N^2 - N_1^2$ make an additional contribution to the discrepancy $m(\zeta, k_0)$ of the coefficients of the wave and reference equations.

9.1.2 An Estimation of the Asymptotic Solution Accuracy

To estimate the proximity of the exact and approximate solutions it is convenient to replace (8.1.1) by an integral equation. Let $G(\zeta, \zeta_1)$ be the Green function of the equation which has known exact solutions, that is

$$\frac{d^2 G(\zeta, \zeta_1)}{d\zeta^2} + k_0^2[N^2(\zeta) + m(\zeta)]G(\zeta, \zeta_1) = -\delta(\zeta - \zeta_1) \quad . \tag{9.1.5}$$

Adding $k_0^2 m \Phi$ to both parts of (8.1.1) and considering the right-hand part as a heterogeneity we obtain

$$\Phi(\zeta) = (\eta')^{-1/2} W(\eta(\zeta)) - k_0^2 \int_{-\infty}^{+\infty} d\zeta_1 m(\zeta_1) \Phi(\zeta_1) G(\zeta, \zeta_1) \quad . \tag{9.1.6}$$

It is advisable to convert this integral equation into one with a variable upper integration limit. The Green's function is defined by (9.1.5) with an accuracy up to an arbitrary solution of the homogeneous equation. We choose this solution from the requirement that $G \equiv 0$ at $\zeta < \zeta_1$. Integrating both parts of (9.1.5) over ζ from $\zeta_1 - \varepsilon$ up to $\zeta_1 + \varepsilon$ and assuming $\varepsilon \to 0$ yields

$$G(\zeta, \zeta_1) = 0 \quad , \quad \left.\frac{\partial G}{\partial \zeta}\right|_{\zeta = \zeta_1} = -1 \quad . \tag{9.1.7}$$

At $\zeta > \zeta_1$ we have $G = (\eta')^{-1/2}\{A_1 W_1[\eta(\zeta)] + A_2 W_2[(\eta(\zeta)]\}$. The unknown coefficients $A_{1,2}$ can be found from (9.1.7). Then,

$$G(\zeta, \zeta_1) = [\eta'(\zeta)\eta'(\zeta_1)]^{-1/2} g(\eta, \eta_1) \quad ,$$
$$g(\eta, \eta_1) = [W_1(\eta_1)W_2(\eta) - W_1(\eta)W_2(\eta_1)]w^{-1} \quad . \tag{9.1.8}$$

where $\zeta > \zeta_1$, $\eta = \eta(\zeta)$, $\eta_1 = \eta(\zeta_1)$, and

$$w \equiv W_2 \frac{dW_1}{d\eta} - W_1 \frac{dW_2}{d\eta} = \text{const}$$

is the Wronskian.

Let us require that the approximate solution $f(\zeta)$ (9.1.2) be equal to the exact one $\Phi(\zeta)$ together with their first derivatives at some point ζ_0:

$$\Phi(\zeta_0) = a_1 = f(\zeta_0) \quad , \quad \Phi'(\zeta_0) = a_2 = f'(\zeta_0) \quad . \tag{9.1.9}$$

Taking into account that G is zero at $\zeta_1 > \zeta$ and that the integral in (9.1.6) is also zero at $\zeta = \zeta_0$ we obtain

$$\Phi(\zeta) = f(\zeta) - k_0^2 \int_{\zeta_0}^{\zeta} d\zeta_1 m(\zeta_1)\Phi(\zeta_1)G(\zeta,\zeta_1) \quad . \tag{9.1.10a}$$

By differentiation of (9.1.10a) with respect to ζ and taking into account (9.1.5, 7) it is easy to verify that $\Phi(\zeta)$ satisfies the wave equation given in (8.1.1) as well as the initial conditions (9.1.9). By using (9.1.8) for the Green's function, (9.1.10a) can be written as

$$\varphi(\zeta) = \sqrt{\eta'(\zeta)}f(\zeta) - k_0^2 \int_{\zeta_0}^{\zeta} d\zeta_1 \frac{m(\zeta_1)}{\eta'(\zeta_1)}\varphi(\zeta_1)g(\eta,\eta_1) \quad , \tag{9.1.10b}$$

where $\varphi(\zeta) \equiv (\eta')^{1/2}\Phi(\zeta)$.

Equations (9.1.10) are linear Volterra integral equations [Ref. 9.3, Chap. 4] for $\Phi(\zeta)$ and $\varphi(\zeta)$. Solutions of these equations can be found by the method of successive approximations. To obtain the iteration formula for (9.1.10b) we assume as the zeroth approximation $\varphi^{(0)}(\zeta) = (\eta')^{1/2}f(\zeta)$, then

$$\varphi^{(l)}(\zeta) = \varphi^{(0)}(\zeta) - k_0^2 \int_{\zeta_0}^{\zeta} d\zeta_1 \frac{m(\zeta_1)}{\eta'(\zeta_1)}\varphi^{(l-1)}(\zeta_1)g(\eta,\eta_1) \quad , \quad l = 1, 2, \ldots \quad . \tag{9.1.11}$$

The lth iteration can be represented as a sum of iterated integrals

$$\varphi^{(l)}(\zeta) = \sum_{\nu=0}^{l} I_\nu(\zeta), I_0 \equiv \varphi^{(0)} ,$$

$$I_\nu = (-k_0^2)^\nu \int_{\zeta_0}^{\zeta} du_1 \frac{m(u_1)}{\eta'(u_1)}\varphi^{(0)}(u_1)g(\eta,\eta_1) \int_{\zeta_0}^{u_1} du_2 \frac{m(u_2)}{\eta'(u_2)}$$

$$\times \varphi^{(0)}(u_2)g(\eta_1,\eta_2)\ldots \int_{\zeta_0}^{u_{\nu-1}} du_\nu \frac{m(u_\nu)}{\eta'(u_\nu)}\varphi^{(0)}(u_\nu)g(\eta_{\nu-1},\eta_\nu) \quad ,$$

$$\nu = 1, 2, \ldots \quad . \tag{9.1.12}$$

Here $u_0 \equiv \zeta$, $\eta_\nu = \eta(u_\nu)$.

Let the function g obtained by using the solutions of the reference equation be bounded: $|g(\eta_1, \eta_2)| \leq D$, $D = \text{const}$. Then at $\zeta \geq \zeta_0$

$$|I_\nu(\zeta)| \leq (k_0^2 D)^\nu \int_{\zeta_0}^{\zeta} du_1 \left| \frac{m(u_1)\varphi^{(0)}(u_1)}{\eta'(u_1)} \right| \cdot \dots$$

$$\times \int_{\zeta_0}^{u_{\nu-1}} du_\nu \left| \frac{m(u_\nu)\varphi^{(0)}(u_\nu)}{\eta'(u_\nu)} \right| = \frac{[k_0^2 DF(\zeta, \zeta_0)]^\nu}{\nu!} \quad , \tag{9.1.13}$$

where

$$F(\zeta, \zeta_0) = \int_{\zeta_0}^{\zeta} du_1 \left| \frac{m(u_1)\varphi^{(0)}(u_1)}{\eta'(u_1)} \right| \quad . \tag{9.1.14}$$

If $\zeta < \zeta_0$, the function $F(\zeta, \zeta_0)$ in (9.1.13) must be replaced by $F(\zeta_0, \zeta)$. It follows from (9.1.13) that the iteration sequence (9.1.11) converges uniformly in any finite interval of ζ to the solution of the integral equation (9.1.10), $\varphi = \sum_{\nu=0}^{+\infty} I_\nu(\zeta)$ and

$$|\varphi(\zeta) - \varphi^{(l)}(\zeta)| \leq \sum_{\nu=l+1}^{\infty} [k_0^2 D|F(\zeta, \zeta_0)|]^\nu / \nu! = \exp Q - \sum_{\nu=0}^{l} Q^\nu / \nu! \quad ,$$

$$Q \equiv k_0^2 D|F(\zeta, \zeta_0)| \quad . \tag{9.1.15}$$

In particular, by assuming that $l = 0$ one obtains an upper bound for the deviation of the asymptotic solution of the wave equation (9.1.2) from the exact solution

$$|\Phi(\zeta) - f(\zeta)| \leq |\eta'|^{-1/2}(\exp Q - 1) \quad . \tag{9.1.16}$$

The rate of change of the function $W(\eta)$ in (9.1.1) increases with increasing k_0 and $g(\eta, \eta_1) \to 0$ as $k_0 \to \infty$ according to (9.1.8). In particular, if $M = A\eta^\alpha$ and $\alpha \geq 1$ we have $W'(\eta)/W(\eta) \sim k_0^{2/(\alpha+2)}$ and $D \sim k_0^{-2/(\alpha+2)}$. It follows from (9.1.3, 14) that $k_0^2 F$ is independent of k_0. Hence, the parameter Q is small if $k_0 \to \infty$, the iteration series in (9.1.11) converges rapidly and the difference between the asymptotic solution (9.1.2) and the exact one tends to zero. A sufficient condition for the validity of the asymptotic solution is

$$k_0^2 D|F(\zeta, \zeta_0)| \ll 1 \quad . \tag{9.1.17}$$

Specifically, when $M(\eta)$ is the power function of η, f and Φ differ from each other only by the factor $1 + O[(k_0 L)^{-2/(\alpha+2)}]$, where L is the characteristic spatial dimension of the medium variability. Subsequent terms of the expansion of $\Phi(\zeta)$ in powers of $1/k_0$ are given by (9.1.12). They can also be obtained by looking for the solution to (8.1.1) in the form $\Phi = W[\eta(\zeta)]q_1(\zeta) + k_0^{-\gamma} W'[\eta(\zeta)]q_2(\zeta)$, where the parameter $\gamma > 0$ is determined by the reference equation (9.1.1) and $q_{1,2}(\zeta)$ are expandable in a series of *integral powers* of $1/k_0$ [9.1, 4]. In another approach [9.5] the asymptotic solution is found in the form of (9.1.2) and $\eta(\zeta)$ is represented as the iterative solution to (9.1.3), where (9.1.4) is the zeroth approximation. For the reader who is interested in further details of the reference equation method the mathematical literature [9.2, 6–11] can be recommended.

9.1.3 The Simplest Example

Let us consider an example. Assume that the effective wave number $N(\zeta)$ has no zeros and does not go to infinity anywhere in the interval of ζ under consideration. Then the constant $M(\eta) \equiv 1$ can be chosen as a reference function. The asymptotic solution to the wave equation will be, see (9.1.2,4),

$$f(\zeta) = N^{-1/2}(\zeta)\left[B_1 \exp\left(ik_0 \int_{\zeta_0}^{\zeta} N d\zeta_1 \right) + B_2 \exp\left(-ik_0 \int_{\zeta_0}^{\zeta} N d\zeta_1 \right) \right] \quad,$$
$$(9.1.18)$$

which coincides with the WKB solution (8.1.11). The difference in the coefficients of the wave and reference equation is

$$m(\zeta) = (2k_0)^{-2}\{2(\ln N)'' - [(\ln N)']^2\} = -2N^2(\zeta)\varepsilon(\zeta) \quad, \tag{9.1.19}$$

where $\varepsilon(\zeta)$ is determined from (8.1.10). This elucidates why requirement of ε to be small, which appeared in Chap. 8 during the comparison of the first approximation with second and third approximations in the WKB method, is enough for neglecting the higher-order approximations. The Green's function in (9.1.8) can be written as

$$G(\zeta_1, \zeta_2) = [N(\zeta_1)N(\zeta_2)]^{-1/2} g(\eta_1, \eta_2), \quad \eta_{1,2}$$

$$= \int_{\zeta_0}^{\zeta_{1,2}} N d\zeta, \quad g(\eta_1, \eta_2) = \frac{1}{k_0} \sin k_0(\eta_1 - \eta_2) \quad. \tag{9.1.20}$$

Let $N^2(\zeta) > 0$. Then we can assume that $D = k_0^{-1}$ in (9.1.17). A sufficient condition for the validity of the asymptotic formula in (9.1.18) is now

$$k_0 \int_{\zeta_0}^{\zeta} |N(\zeta_1)\varepsilon(\zeta_1)| d\zeta_1 \ll 1 \quad. \tag{9.1.21}$$

It is easy to show that in this case the estimate given in (9.1.16) is identical to (8.1.15).

9.2 Sound Field in the Vicinity of a Turning Point

Next more complex is the case where $N^2(\zeta)$ is a bounded function and has only one zero in the given interval of ζ. Let this point be the origin of the coordinate ζ and assume besides that $[N^2(0)]' \neq 0$. From (9.1.4) one can see that $1/\eta'$ will be bounded only if $M = 0$ at $\eta = \eta(0)$. The simplest reference function which has this property is $M = a[\eta - \eta(0)]$, $a = $ const. By assuming $\eta(0) = 0$, we get from (9.1.4)

$$\sqrt{a} \int_0^{\eta} \eta^{1/2} d\eta = \int_0^{\zeta} N(\zeta) d\zeta \quad, \quad \text{or} \quad \eta = \left(\frac{3\varphi(\zeta)}{2\sqrt{a}} \right)^{2/3} \quad, \quad \text{where} \tag{9.2.1a}$$

$$\varphi(\zeta) = \int_0^{\zeta} N d\zeta \quad. \tag{9.2.1b}$$

Assume for definiteness that $a > 0$; $N^2 > 0$ at $\zeta > 0$ and $N^2 < 0$ at $\zeta < 0$. The function η is a three-valued function of ζ. It is convenient to choose the regular branch at which η is real when ζ is real. We assume $\arg \varphi(\zeta) \in [0, 2\pi)$ for this purpose. Then we have $\varphi \geq 0$ and $\eta \geq 0$ when $\zeta \geq 0$. If $\zeta < 0$ we have $N = i|N|$, $\varphi = |\varphi| \exp(3\pi i/2)$ and $\eta = |\eta| \exp(i\pi)$. [Unless otherwise stated we let the quantity $|w|^\alpha \exp(i\alpha \arg w)$ be the power w^α of the complex number $w = |w| \exp(i \arg w)$.] Hence, the change of variables (9.2.1) can be written as

$$\eta(\zeta) = \left(\frac{3|\varphi(\zeta)|}{2\sqrt{a}} \right)^{2/3} \operatorname{sgn} \zeta \quad . \tag{9.2.2}$$

Here $\operatorname{sgn} \zeta = 1$ if $\zeta > 0$ and -1 if $\zeta < 0$.

In the case under consideration the reference equation (9.1.1) can be easily reduced to the Airy equation, see (3.5.2–4) at $\xi^2 = k_0^2$. Its solutions are Airy functions of the argument $-(k_0^2 a)^{1/3} \eta$. Properties of these functions were described in Sect. 3.5. Now, using (9.1.2) and (9.2.2) we find the asymptotic solution to the wave equation:

$$
f(\zeta) = \left| \frac{3 k_0 \varphi}{2 N^3} \right|^{1/6} \left[B_1 u \left(- \left| \frac{3 k_0 \varphi(\zeta)}{2} \right|^{2/3} \operatorname{sgn} \zeta \right) \right.
$$
$$
\left. + B_2 v \left(- \left| \frac{3 k_0 \varphi(\zeta)}{2} \right|^{2/3} \operatorname{sgn} \zeta \right) \right] \quad . \tag{9.2.3}
$$

If N^2 is positive at $\zeta < 0$ and negative at $\zeta > 0$, the corresponding asymptotics differ from (9.2.3) only in the sign of the argument of the Airy function.

Calculation of the correction term $m(\zeta)$ by using (9.1.3), (9.2.1) shows that it is limited at any ζ including turning point $\zeta = 0$. Hence (9.2.3) is the uniform (i.e., valid at any ζ) asymptotic solution to the wave equation. The function g in (9.1.8) is

$$g(\eta_1, \eta_2) = \frac{u(\eta_2) v(\eta_1) - u(\eta_1) v(\eta_2)}{(-a k_0^2)^{1/3}} \quad . \tag{9.2.4}$$

Here, we have taken into account that the Wronskian of the Airy functions $u(\eta)$ and $v(\eta)$ is equal to 1, see (3.5.9). Therefore, we have $D \sim k_0^{-2/3}$ in (9.1.15) and the relative error in the asymptotic formula (9.2.3) can not be greater than $O[(k_0 L)^{-2/3}]$. Moreover, it can be shown that the stricter estimate of the error is valid; (9.2.3) differs from the exact solution by the factor $1 + O[(k_0 L)^{-1}]$. An elaborate study of the accuracy of the asymptotic formula of (9.2.3) was given in [9.12, 13].

The argument of the Airy functions in (9.2.3) becomes zero at a turning point and is small in its vicinity. In (9.2.1) $N(\zeta)$ can be replaced by $(a_1 \zeta)^{1/2}[1 + O(\zeta/L)]$ if $|\zeta| \ll L$, where $a_1 = dN^2/d\zeta|_{\zeta=0} \simeq L^{-1}$. Then we obtain from (9.2.3)

$$
f(\zeta) \approx |k_0/a_1|^{1/6} \{ B_1 u[-(k_0^2 a_1)^{1/3} \zeta]
$$
$$
+ B_2 v[-(k_0^2 a_1)^{1/3} \zeta] \} \quad , \quad |\zeta| \ll L \quad . \tag{9.2.5}
$$

This expression could also be obtained from the simple requirement that the coefficients of the wave and reference equations that is $k_0^2 a_1 \zeta [1 + O(\zeta/L)]$ and $k_0^2 a \eta$ be nearly equal. In another limiting case when $|\zeta| \gg (L/k_0^2)^{1/3}$ the argument of the Airy functions is large and these functions can be replaced by their asymptotic expressions (3.5.13, 14). Taking only the main terms we obtain

$$f \approx N^{-1/2}[B_1 \cos (k_0|\varphi| + \tfrac{\pi}{4}) + B_2 \sin (k_0|\varphi| + \tfrac{\pi}{4})] \quad ,$$
$$N^2 > 0 \quad , \quad |\zeta| \gg (L/k_0^2)^{1/3} \quad , \tag{9.2.6}$$

$$f \approx |N|^{-1/2}[B_1 \exp (k_0|\varphi|) + 0.5 B_2 \exp (-k_0|\varphi|)] \quad ,$$
$$N^2 < 0 \quad , \quad |\zeta| \gg (L/k_0^2)^{1/3} \quad . \tag{9.2.7}$$

These expressions can be easily identified as the WKB approximation given by (8.1.11). Note also that the condition required for (9.2.6, 7) to be valid agrees with the inequality (8.1.18).

Equations (9.2.5–7) are *local* asymptotic expressions for the wave field. The regions where they are valid overlap since $(L/k_0^2)^{1/3} \ll L$. Together they represent the high-frequency approximation for Φ for any ζ. They are much simpler than (9.2.3) and valid for any sign of a_1 and are rather convenient for numerical calculations since they do not require calculation of Airy functions with large arguments. Local asymptotics do not include uncertainties of the kind 0/0 (as, for example, φ/N^3 in (9.2.3) as $N \rightarrow 0$) which can be troublesome in numerical calculations. On the other hand, uniform asymptotic expressions are more convenient for analytical computations since (9.2.3) is valid at any ζ.

One can see from (9.2.3) and (9.2.5) that near the turning point $\zeta = 0$ there exists a region $|\zeta| \lesssim (L/k_0^2)^{1/3}$, that is small, as compared to L, where the wave amplitude is frequency dependent. The width of this region decreases as $\omega^{-2/3}$ with increasing frequency. Simultaneously, the wave amplitude increases as $\omega^{1/6}$ and approaches the value determined from ray acoustics as $\omega \rightarrow \infty$. As for the sound pressure dependence on the vertical coordinate, one can use the WKB expressions

$$f(\zeta) = N^{-1/2}\{C_1 \exp [ik_0\varphi(\zeta)]$$
$$+ C_2 \exp [-ik_0\varphi(\zeta)]\} \quad , \quad N^2 > 0 \quad , \tag{9.2.8a}$$

$$f(\zeta) = N^{-1/2}[C_3 \exp (k_0|\varphi(\zeta)|)$$
$$+ C_4 \exp (-k_0|\varphi(\zeta)|)] \quad , \quad N^2 < 0 \quad , \tag{9.2.8b}$$

if one does not need to know the field in the vicinity of the turning point. By comparing the last expressions with (9.2.6 and 7) we can obtain the relations between $C_{1,2}$ that is, the amplitudes in the insonified region, and $C_{3,4}$ the amplitudes in the shadow. In the first region the waves are propagating, in the second they are inhomogeneous. If $N^2 > 0$ for $\zeta > 0$ [$\varphi(\zeta)$ is positive in the insonified region] these relations are

$$C_3 = C_1 + iC_2 \quad , \quad C_4 = (iC_1 + C_2)/2 \quad . \tag{9.2.9}$$

If $N^2 > 0$ for $\zeta < 0$ [$\varphi(\zeta) < 0$ in the insonified region], C_1 and C_2 must be interchanged in (9.2.9). In the same manner the high-frequency asymptotic expression for the field with more than one turning point but with distances between neighbouring points much larger than $(L/k_0^2)^{1/3}$ can be obtained. In the vicinity of each such point and far from the points the field will be described by an expression such as (9.2.5) and by the WKB expressions (9.2.6 and 9.2.7), respectively.

The number of local asymptotics that must be used can be reduced by more than half by the use of (9.2.3) for the description of the field near each of the turning points. These asymptotic equations are matched in the common region of applicability between the turning points.

We have seen in Chap. 8 that in the framework of the ray approximation the wave propagates without reflection if turning points are absent. Now consider reflection at the turning point. Let the wave be incident from above upon the turning horizon $\zeta = 0$. (It is assumed that $N^2 < 0$ for $\zeta < 0$.) Far from the turning point the field is described by (9.2.8). We suppose that $|\varphi| \to \infty$ as $\zeta \to -\infty$.[1] Then the field will be finite only when $C_3 = 0$. From (9.2.9) we get then $C_1 = -iC_4$, $C_2 = C_4$. In (9.2.8a) the term proportional to C_1 is a reflected wave whereas the term proportional to C_2 is the incident one. For the ratio of these terms that is the reflection coefficient, we get

$$V = C_1 C_2^{-1} \exp\left[2ik_0\varphi(\zeta)\right] = \exp\left(2ik_0 \int_0^\zeta N \, d\zeta - i\frac{\pi}{2}\right) . \qquad (9.2.10)$$

We see that $|V| = 1$ which means total reflection at the turning point. The integral in the exponent in (9.2.10) is the geometrical phase increment when the wave propagates from the horizon ζ to 0 and back. The term $-i\frac{\pi}{2}$ is the $\frac{\pi}{2}$ loss of the phase at the turning point. This phenomenon is not predicted by the ray approximation.

9.3 Reflection from a "Potential Barrier"

9.3.1 Uniform Asymptotics of the Sound Field

Let the refraction coefficient $n(z)$ have a minimum $n = n_m$ at some point z_m and the medim be at rest. Then for waves with a horizontal wave vector component $\xi > k_0 n_m$, the effective refraction coefficient becomes zero at two points $z_{1,2}$: $z_1 < z_m < z_2$ (Fig. 9.1). In the region $z_1 < z < z_2$ we have $N^2 < 0$ and the waves are inhomogeneous there. This region is a sort of barrier for the propagation of waves from the half-space $z < z_1$ into the half-space $z > z_2$ and back. By analogy with quantum mechanics we will use the expression "reflection from a potential barrier". The wave has no turning points if $\xi < k_0 n_m$. However, the WKB approximation is not valid near the top of the barrier $z = z_m$ when ξ is close to $k_0 n_m$. In this case, a noticeable reflection called the "over-barrier reflection" occurs. These effects were studied by a number of researchers, see, for example, [9.14–19],

---

[1] Only if N tends to zero sufficiently rapidly when $\zeta \to -\infty$, will the value $|\varphi(-\infty)|$ be finite. This is analogous to the case when the second turning point exists at infinity and the problem becomes very close to that of reflection from a potential barrier considered in the next section.

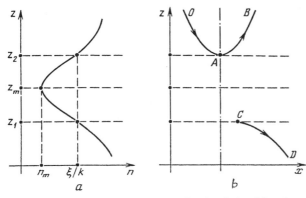

Fig. 9.1. Vertical dependence of the refraction index (**a**) and geometry of the rays when two turning horizons are present (**b**) in reflection from a potential barrier in a motionless medium

[Ref. 9.20, Sects. 23, 50], [Ref. 9.21, Chap. 3]. Analogous phenomena take place in moving media. Then besides $n(z)$ the shape and height of the potential barrier are also determined by the flow velocity profile $v_0(z)$. In this section we suppose that the dependence of n and v_0 on z in the neighborhood of z_m is analytical and that $\beta(z) > 0$.

If two turning points are far enough from each other, the high-frequency asymptotic expression for the field can be obtained by using the results of the previous section. It becomes meaningless, however, when $|z_1 - z_2| \to 0$. In this case, a reference function with two zeros must be used. The simplest reference equation is

$$\frac{d^2W}{d\eta^2} + k_0^2 a^2(\eta^2 - \eta_0^2)W = 0 \quad . \tag{9.3.1}$$

In Sect. 3.2 we had an analogous equation, see (3.2.25) with $\beta_3 = 0$. Linear-independent solutions to (9.3.1) are the parabolic cylinder functions, also called Weber functions,

$$W = D_{-1/2+i\alpha}\left(\pm\eta\exp\left(-i\frac{\pi}{4}\right)\sqrt{2k_0a}\right) \quad , \quad \alpha = -k_0a\eta_0^2/2 \quad , \tag{9.3.2}$$

described and tabulated in [Ref. 9.22, Chap. 19] and [9.23, 24]. We assume that $a > 0$.[2]

In the case under consideration the change of variables in (9.1.4) becomes

$$\int_{\eta_0}^{\eta} a\sqrt{\eta^2 - \eta_0^2}\,d\eta \equiv \frac{a\eta_0^2}{2}\left\{\frac{\eta}{\eta_0}\sqrt{\frac{\eta^2}{\eta_0^2} - 1} - \ln\left[\frac{\eta}{\eta_0} + +\sqrt{\frac{\eta^2}{\eta_0^2} - 1}\right]\right\}$$

$$= \int_{\zeta_2}^{\zeta} N(\zeta)\,d\zeta \quad . \tag{9.3.3}$$

We have denoted $\zeta_{1,2} = \zeta(z_{1,2})$, hence $N(\zeta_{1,2}) = 0$. Specific choice of the lower

---

[2] In the case when $N^2 > 0$ at $z_1 < z < z_2$ and $N^2 < 0$ outside of this region (i.e., $N^2(\zeta)$ has a maximum) one can again use (9.3.1) assuming $a = \pm i|a|$.

integration limits in (9.3.3) ensures that one of zeros of the reference function $M[\eta(\zeta)]$ will coincide with a zero of $N(\zeta)$. The coincidence of the two other zeros $[\eta(\zeta_1) = -\eta_0]$ can be ensured by suitable choice of η_0. Namely, η_0 should satisfy the equation

$$\int_{\zeta_1}^{\zeta_2} N(\zeta)d\zeta = \int_{-\eta_0}^{\eta_0} a\sqrt{\eta^2 - \eta_0^2}d\eta \equiv \frac{i\pi}{2}a\eta_0^2 \quad . \tag{9.3.4}$$

According to (9.3.2, 4) the parameter α is

$$\alpha = \frac{i}{\pi}k_0 \int_{\zeta_1}^{\zeta_2} N d\zeta = -\frac{1}{\pi}\int_{z_1}^{z_2} \sqrt{\xi^2 - k_0^2 n^2 \beta^2}dz \quad . \tag{9.3.5}$$

It is easy to verify that the change of variables given in (9.3.3) under the condition given by (9.3.5) ensures that the correction term $m(\zeta)$ will be bounded at any ζ.

In the case of reflection from a potential barrier ($\zeta_{1,2}$ are real), in the region $\zeta_1 < \zeta < \zeta_2$ we have $N = i|N|$ and $\eta_0^2 > 0$ according to (9.3.4) and also $\alpha < 0$. In contrast, in the case of over-barrier propagation $\eta_0^2 < 0$ and $\alpha > 0$ and the minimum value of $N^2(\zeta)$ at the real axes $N^2(\zeta_m) \equiv N_m^2$ is positive. However, the shape of lines of equal values of the analytical function in the vicinity of a saddle point ζ_m where $(N^2)' = 0$ is such that ζ_m lies on the line (in the complex plane ζ) at which N^2 appears to be real and less than N_m^2 [Ref. 9.25, Sect. 45]. The integration in (9.3.4, 5) is along this line between the complex turning points $\zeta_1 = \zeta_2^*$. We assume that $N_m^2 \ll 1$. [If $N_m^2 \simeq 1$, $N^2(\zeta)$ may have no zeros in the complex plane.] Then the turning points are close to ζ_m and the following expressions can be easily obtained

$$\zeta_{1,2} \approx \zeta_m \mp i\sqrt{2N_m^2/[N^2(\zeta_m)]''} \quad .$$

Obviously $|\zeta_1 - \zeta_2| \ll L$. It follows from (9.3.5) that $\alpha \ll k_0 L$, but can be much larger than unity.

The uniform asymptotic expression for the vertical dependence of the sound field in the case of two turning points will be according to (9.1.2):

$$f(\zeta) = \left[\frac{N^2(\zeta)}{\eta^2(\zeta) - \eta_0^2}\right]^{-1/4} \{A_1 D_{-1/2+i\alpha}[\exp(-i\tfrac{\pi}{4})\eta\sqrt{2k_0a}]$$

$$+ A_2 D_{-1/2+i\alpha}[\exp(3i\tfrac{\pi}{4})\eta\sqrt{2k_0a}]\} \quad . \tag{9.3.6}$$

It can be shown that $f(\zeta)$ differs from the exact solution only by the factor $1 + O[(k_0L)^{-1}]$. Replacement of the parabolic cylinder functions in (9.3.6) by a corresponding power series or by asymptotic expressions ([Ref. 9.22, Chap. 19], [9.26]) yields various local asymptotics. In particular, when $\alpha < 0$ and $|\alpha| \gg 1$ from (9.3.6) we get two asymptotic expressions as in (9.2.3) with a common region of validity between the turning points ζ_1 and ζ_2.

9.3.2 Relation to the WKB Approximation

Let us more thoroughly consider the transformation of (9.3.6) into the WKB expressions (8.1.11). For this purpose we must use the representation of the functions W (9.3.2) at large η. For $|u/\nu| \gg 1$ we have the asymptotic expressions

$$D_\nu(u) = \exp(-u^2/4)u^\nu[1 + O(u^{-2})] \quad ;$$

$$D_\nu(u) = \exp(-u^2/4)u^\nu[1 + O(u^{-2})]$$

$$- \sqrt{2\pi}\,\Gamma^{-1}(-\nu)\exp(u^2/4 \pm i\pi\nu)u^{-1-\nu}[1 + O(u^{-2})] \quad . \qquad (9.3.7)$$

The best approximation is achieved when the first equation for $D_\nu(u)$ is used at $\arg u \in (-\frac{\pi}{2}, \frac{\pi}{2})$, the second one, with the upper sign, when $\arg u \in (\frac{\pi}{2}, \pi]$ and with the lower sign if $\arg u \in (-\pi, -\frac{\pi}{2})$ [Ref. 9.23, Sect. 5]. The so-called Darvin expansions ([Ref. 9.22, Chap. 19], [9.23]) have a much broader region of applicability if $|\nu| \gg 1$. We write the main terms for one of the parabolic cylinder functions in (9.3.6) at $|\eta/\eta_0| > 1$:

$$D_{-1/2+i\alpha}[\exp(3\pi i/4)\eta\sqrt{2k_0 a}]$$

$$\sim B(\eta^2 - \eta_0^2)^{-1/4}\exp\left\{i\alpha\frac{\eta}{\eta_0}\sqrt{\frac{\eta^2}{\eta_0^2} - 1} - i\alpha\ln\left[-\frac{\eta}{\eta_0} - \sqrt{\frac{\eta^2}{\eta_0^2} - 1}\right]\right\} \quad ,$$

$$B = \frac{(-\alpha)^{i\alpha/2}}{(2k_0 a)^{1/4}}\exp\left[\frac{\pi\alpha}{4} + i\left(\frac{\pi}{8} - \frac{\alpha}{2}\right)\right] \quad ;$$

$$\arg[\eta\exp(3\pi i/4)] \in \left(-\frac{\pi}{2}, \frac{\pi}{2}\right) \quad , \qquad\qquad (9.3.8)$$

$$D_{-1/2+i\alpha}[\exp(3\pi i/4)\eta\sqrt{2k_0 a}]$$

$$\sim \sum_{j=1}^{2} E_j(\eta^2 - \eta_0^2)^{-1/4}\exp\left\{(-1)^j i\alpha\frac{\eta}{\eta_0}\sqrt{\frac{\eta^2}{\eta_0^2} - 1}\right.$$

$$\left. -(-1)^j i\alpha\ln\left[\frac{\eta}{\eta_0} + \sqrt{\frac{\eta^2}{\eta_0^2} - 1}\right]\right\} \quad ,$$

$$E_1 = \frac{(-\alpha)^{i\alpha/2}}{(2k_0 a)^{1/4}}\exp\left[-\frac{3\pi\alpha}{4} - i\left(\frac{3\pi}{8} + \alpha\right)\right] \quad ,$$

$$E_2 = \frac{\sqrt{2\pi}(-\alpha)^{-i\alpha/2}}{(2k_0 a)^{1/4}\Gamma(1/2 - i\alpha)}\exp\left[-\frac{\pi\alpha}{4} + i\left(\frac{\pi}{8} + \frac{\alpha}{2}\right)\right] \quad ;$$

$$\arg[\eta\exp(3\pi i/4)] \in \left(-\pi, -\frac{\pi}{2}\right) \cup \left(\frac{\pi}{2}, \pi\right] \quad . \qquad (9.3.9)$$

As a matter of fact, the Darvin expansions are solutions to the reference equation given by (9.3.1) in the WKB approximation when they are normalized in such a way that they coincide with (9.3.7) as $|\eta| \to \infty$. Let $k_0/a \gg 1$. Calculating $\varepsilon(\eta)$ from (8.1.10) where $N^2(\eta) = a^2(\eta^2 - \eta_0^2)$ and taking into account the inequalities in (8.1.12) after simple operations we find the region of validity of (9.3.8, 9) for real η

$$|\eta| \gg (k_0 a)^{-1/2} \quad , \quad \text{if} \quad |\alpha| \lesssim 1 \quad , \tag{9.3.10}$$

$$|\eta \pm \eta_0| \gg (k_0 a)^{-1/2} |\alpha|^{-1/6} \quad , \quad \text{if} \quad \alpha < 0, \quad |\alpha| \gg 1 \quad . \tag{9.3.11}$$

When α is a large positive number (9.3.8) and (9.3.9) can be used at any real η.

If we use only the main terms of the Darvin expansion for the parabolic cylinder function in (9.3.6) and take into account (9.3.3), then (9.3.6) turns into the usual WKB asymptotics of the solution of the wave equation. Hence, the inequalities in (9.3.10, 11) determine the region of the validity of the WKB approximation when there are two turning points. The physical meaning of these conditions is rather simple. When the turning points are close to one another ($|\alpha| \lesssim 1$) the region in the vicinity of the top of the barrier with the dimension of the order of $(L/k_0)^{1/2}$ must be excluded in the case of reflection from the barrier as well as in the case of over-barrier propagation. When the turning points are far apart from each other the WKB approximation is invalid only in the vicinities of the horizons $\zeta_{1,2}$. By (9.3.3) both conditions (9.3.10, 11) can be written in the same and very simple phase-integral form:

$$\left| k_0 \int_{\zeta_{1,2}}^{\zeta} N(\zeta) d\zeta \right| \gg 1 \quad . \tag{9.3.12}$$

9.3.3 Reflection and Transmission Coefficients

We now proceed to the calculation of the plane wave reflection coefficient from a potential barrier. Let the wave be incident from $\zeta = +\infty$. Behind the barrier (as $\zeta \to -\infty$) only the transmitted wave exists. According to (9.3.3), $\eta \to -\infty$ when $\zeta \to -\infty$. It follows from (9.3.7) that, as $\eta \to -\infty$, the term proportional to A_1 in (9.3.6) gives waves propagating in both directions whereas the term proportional to A_2 yields only a wave described by $\exp(ik_0 a^2 \eta^2 / 2)$ and propagating to $\zeta = -\infty$. Hence $A_1 = 0$. According to (9.3.3, 6, 8 and 9) we now obtain for the points far from the barrier's top

$$f(\zeta) = A_2 N^{-1/2}(\zeta) B \exp\left(-ik_0 \int_{\zeta_1}^{\zeta} N \, d\zeta\right) \quad , \quad \zeta < \zeta_1 \quad , \tag{9.3.13a}$$

$$f(\zeta) = A_2 N^{-1/2}(\zeta) \left[E_1 \exp\left(ik_0 \int_{\zeta_2}^{\zeta} N \, d\zeta\right) \right.$$

$$\left. + E_2 \exp\left(-ik_0 \int_{\zeta_2}^{\zeta} N \, d\zeta\right) \right] \quad , \quad \zeta > \zeta_2 \quad . \tag{9.3.13b}$$

Expression (9.3.13b) is a superposition of incident and reflected waves, and (9.3.13a) is the transmitted wave. The quantities $V = E_1/E_2$ and $W = B/E_2$ are the reflection and transmission coefficients. The amplitudes and phases of reflected and transmitted waves far from the layer (ζ_1, ζ_2) can be found by multiplying the complex amplitude of the incident wave by V and W, respectively, and taking into account the phase advance calculated by the geometrical acoustic approximation. Using the quantities B and $E_{1,2}$ from (9.3.8) and (9.3.9) and also (3.4.22) for the modulus of a Gamma function of complex argument we find

$$V = [1 + \exp(2\pi\alpha)]^{-1/2} \exp[-i(\chi + \tfrac{\pi}{4})] \quad ,$$
$$W = [1 + \exp(-2\pi\alpha)]^{-1/2} \exp(-i\chi) \quad ,$$
$$\chi(\alpha) = \alpha - \alpha \ln|\alpha| - \mathrm{Im}\{\ln \Gamma(\tfrac{1}{2} - i\alpha)\} \quad . \tag{9.3.14}$$

Note that $\chi \to 0$ as $\alpha \to 0$ and $|\alpha| \to \infty$. It was shown in [9.27] that $|\chi(\alpha)| \leq 0.095\tfrac{\pi}{2}$. One can get an idea of the accuracy of the asymptotic results of (9.3.14) by comparing them to exact solutions. Reference 9.27, for example, makes this comparison for an Epstein layer (Sect. 3.4). It turns out that the approximate solutions given by (9.3.14) are close to the exact ones given in (3.4.21, 29) if $b\lambda_0 \lesssim 1$.

The energy reflection and transmission coefficients are:

$$|V|^2 = [1 + \exp(2\pi\alpha)]^{-1} \quad , \quad |W|^2 = [1 + \exp(-2\pi\alpha)]^{-1} \quad . \tag{9.3.15}$$

In quantum mechanics this result is referred to as the Kemble approximation. Note that $|V(\alpha)|^2 = |W(-\alpha)|^2$. In the case of over-barrier propagation ($\alpha > 0$), $|V|^2$ becomes exponentially small when α increases whereas $|W|^2$ tends to unity. In the limit we obtain the same result as in geometrical acoustics. When reflection from the barrier ($\alpha < 0$) is considered, then as $|\alpha| \to \infty$ one obtains the same result as in the case of a single turning point. For ξ value corresponding to $\alpha = 0$, two turning points merge at the top of the barrier and $|V|^2 = |W|^2 = \tfrac{1}{2}$, i.e., half of the energy is transmitted and half is reflected. In this case, the rays corresponding to the incident wave become horizontal at the horizon z_m and produce rays corresponding to reflected and transmitted waves. The first turn and go to $z = +\infty$, and the second to $z = -\infty$.

In the case of reflection from a barrier, the rays exist in the half-spaces $z > z_2$ and $z < z_1$. The relationship between rays of incident and transmitted waves was studied by *Murphy* [9.27]. In considering the reflection of a bounded beam he found the following result (Fig. 9.1b). When the layer is sufficiently thick the reflected ray *OAB* is just like a ray in the case of one turning point (including a phase delay of $\tfrac{\pi}{2}$) but, naturally, its amplitude differs from that of the incident ray. The ray jumps over from A to C without a phase delay, but with a decrease in the amplitude and with horizontal displacement by a distance depending on the wavelength.

In the framework of geometrical acoustics there exist only incident and reflected rays. Penetration of the sound energy through a barrier ("tunneling") is a diffraction effect which disappears at $k_0 \to \infty$. Rays in the region $z < z_1$ are an example of the so-called *diffracted rays*. Such rays can also be treated [9.28] by the methods of *complex geometrical acoustics* [9.29] where rays are considered as curves in complex space.

9.4 Amplification of Sound in an Inhomogeneous Flow

9.4.1 Reference Problem

We now consider the reflection of a sound wave from a medium with continuously stratified flow. In this case, the reflection coefficient could be greater than unity. First we shall discuss this phenomenon in a model case where an exact solution of the problem can be found. Suppose that medium density and the sound velocity are constant whereas the flow velocity is linearly dependent on the vertical coordinate: $v_0 = (az, 0, 0)$, $a > 0$. Let us have at $z = +\infty$ an incident wave with harmonic dependence on t and horizontal coordinates: $\exp(i\xi \cdot r - i\omega t)$, $\xi = (-\xi, 0, 0)$. We have also $\beta(z) \equiv 1 - \xi \cdot v_0/\omega = 1 - z/z_c$, where $z_c = -\omega/\xi a$ is the horizon at which the velocity of the phase propagation along the horizontal plane ck/ξ equals the flow velocity. The latter is larger than the sound velocity if $k > \xi$. Note that the z-component of the wave vector is zero at the horizons $z_{1,2} = z_c(1 \pm \xi/k)$ (Fig. 9.2).

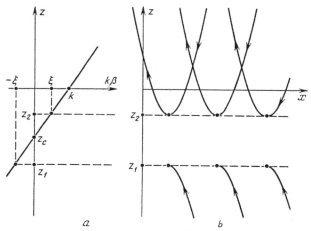

Fig. 9.2. Sound reflection from a medium with a linear flow profile. **a** Vertical dependence of $k\beta$. **b** Rays corresponding to incident, reflected, and transmitted waves. Turning points are designated by $z_{1,2}$, z_c is a horizon of resonance interaction

The sound pressure depends on z via the function $\Phi(z)$ satisfying (1.2.21) which, in the case under the consideration, is written

$$\frac{\partial^2 \Psi}{\partial z^2} + [\xi^2 a^2 c^{-2}(z - z_c)^2 - \xi^2 - 2(z - z_c)^{-2}]\Psi = 0 \quad ,$$

$$\Psi = (z - z_c)^{-1}\Phi(z) \quad . \tag{9.4.1}$$

The general solution of this equation can be expressed in terms of the Whittaker function $W_{l,-3/4}(\eta)$ (Sect. 3.7.2) which can also be written in terms of parabolic cylinder functions [Ref. 9.22; Chap. 13]:

$$\Phi(z) = A_1 W(\alpha, y) + A_2 W(\alpha, -y) \quad , \tag{9.4.2a}$$

$$y = \exp\left(\frac{i\pi}{4}\right) q(z - z_c) \quad , \quad q = \sqrt{2\frac{\xi a}{c}} \quad , \quad \alpha = \frac{\xi^2}{q^2} \quad ; \tag{9.4.2b}$$

$$W(\alpha, y) \equiv 2^{i\alpha}\sqrt{y}\,W_{i\alpha/2,-3/4}(y^2/2)$$
$$= y D_{\nu+1}(y) + D_\nu(y) \quad , \quad \nu = -\tfrac{3}{2} + i\alpha \quad . \tag{9.4.3}$$

Taking into account (9.3.1,2) one can easily prove that the function $y^{-1}W(\alpha, \pm y)$ satisfies (9.4.1). Later we shall see that α is related to the transparency of the potential barrier $z_1 < z < z_2$. One can verify that according to (9.3.7) the sound field (9.4.2) far from the horizon $z = z_c$ (when $|y| \gg |\nu|$) is the superposition of waves propagating in directions which are symmetrical with respect to the horizontal plane

$$\Phi(z) = \{[A_1 + i\exp(-\pi\alpha)A_2]y^{\nu+2}b^{-1} + \sqrt{2\pi}(\nu + 2)\Gamma^{-1}(-\nu)$$
$$\times A_1(-y)^{-1-\nu}b\}[1 + O(y^{-2})] \quad , \quad z < z_c \quad ; \tag{9.4.4}$$

$$\Phi(z) = \{[A_1 - i\exp(\pi\alpha)A_2]y^{\nu+2}b^{-1} + \sqrt{2\pi}(\nu + 2)\Gamma^{-1}(-\nu)$$
$$\times A_2 y^{-1-\nu}b\}[1 + O(y^{-2})] \quad , \quad z > z_c \quad . \tag{9.4.5}$$

Here $b \equiv \exp[iq^2(z - z_c)^2/4]$. The amplitudes of both terms in braces in (9.4.4, 5) are proportional to $\sqrt{|y|}$. For $z > z_c$ one has $\beta > 0$. In this case the phase of the first term in (9.4.5) increases when z decreases and thus this term is the incident wave, whereas the second term represents the reflected wave. The ratio of the coefficients in front of b and b^{-1} is, obviously, the reflection coefficient V. The function $|\Gamma(-\nu)|$ can be calculated by (3.4.22), hence

$$|V| = \frac{\sqrt{1 + \exp(-2\pi\alpha)}}{|1 + i\exp(-\pi\alpha)A_1/A_2|} \quad . \tag{9.4.6}$$

The ratio A_1/A_2 is determined from the conditions for $\Phi(z)$ as $z \to -\infty$. Under the condition $\beta(z) > 0$ one should require that the coefficient before $y^{\nu+2}b^{-1}$ (the wave whose phase increases when z increases) be zero. However, we have $\beta < 0$ for $z < z_c$. In this case as $z \to -\infty$, the sound field can not contain a wave with a negative z-component in the phase velocity.

To prove this we shall first consider, for simplicity, the reflection of a plane wave incident from a medium at rest upon a homogeneous liquid half-space where the flow velocity is $v_0 = (-u, 0, 0) = $ const. For $z < 0$, one has two plane waves as solutions to the wave equation:

$$p = a_1 \exp[-i(\xi x + \mu z + \omega t)] + a_2 \exp[-i(\xi x - \mu z + \omega t)] \quad ,$$

where $\mu = (k^2\beta^2 - \xi^2)^{1/2}$, $\beta = 1 - \xi u/\omega$. Now it is reasonable to use the coordinate system $\tilde{x} = x + ut$, $\tilde{y} = y$, $\tilde{z} = z$ which moves together with the liquid. We have in this system

$$p = a_1 \exp[-i(\xi\tilde{x} + \mu\tilde{z} + \omega\beta t)] + a_2 \exp[-i(\xi\tilde{x} - \mu\tilde{z} + \omega\beta t)] \quad .$$

The frequency is "negative" in the moving system if $\beta < 0$. Therefore, the sign of the z-component of the phase and group velocities is opposite to that of the z-component of the wave vector. The wave vector of the wave with the amplitude a_1 points in the direction of negative z, whereas the perturbation propagates upward. Presence of this wave contradicts the causality principle, hence, this wave must be eliminated. (In fact, this rule for the correct choice of the solutions was already used in Sect. 2.6 by assuming that the sign of the z-component of the wave vector in the refracted wave is opposite to that of β). For an observer in the coordinate system at rest and $\beta < 0$, the refracted wave's phase propagates upward. This does not contradict the causality principle since z-component of the group velocity is negative according to (1.2.16), (1.2.17).

As above, in a continuously stratified medium and when $\beta < 0$ the acoustical perturbation propagates downward only in the wave where the z-component of the wave vector is positive [see (8.2.9) and (8.2.11)]. Hence one must assume that $A_1 = 0$ in (9.4.4). For the modulus of the transmission coefficient, i.e., the ratio of the amplitudes of transmitted and incident waves, we have from (9.4.4, 5)

$$|W| = \frac{|A_2 - iA_1 \exp(\pi\alpha)|}{|A_2 \exp(\pi\alpha) + iA_1|} \, . \tag{9.4.7}$$

Since $A_1 = 0$ from (9.4.6, 7) we get: $|V|^2 = 1 + \exp(-2\pi\alpha)$, $|W|^2 = \exp(-2\pi\alpha)$. Hence, the wave is amplified after reflection from a stratified flow. Let the vertical component of the power-density flux I_z in the incident wave be unity. In the reflected and transmitted waves these quantities are $|V|^2$ and $|W|^2$, respectively. It is clear from the equality $|V|^2 = |W|^2 + 1$ that sound amplification in the process of reflection is due to an influx of energy from $z = -\infty$. The value $|V|^2 - 1 = \exp(-2\pi\alpha)$ gives the energy exchange between the sound field and flow. It increases with increasing $\vec{\nabla} v_0$ and decreasing incidence angle of the wave.

We have considered above the case when $-\xi \| v_0$ and the wave is incident from the region where $\beta > 0$. Let us now omit these restrictions. Let ξ make the angle φ with the x-axis. The direction of ξ influences the sound field via β. If $\cos \varphi < 0$, one must replace a by $-a \cos \varphi$ in the expressions for z_c and q. All other formulas do not change. In particular, we have $\alpha = \xi^2/q^2 = -\xi c/2a \cos \varphi$. In the case where $\cos \varphi > 0$ the sound wave is incident from the region where $\beta < 0$. Equations (9.4.1–5) hold if one replaces a by $a \cos \varphi$. According to (9.4.5), the amplitude of the reflected wave is proportional to $A_1 - iA_2 \exp(\pi\alpha)$ whereas that of the incident wave is proportional to A_2. Since $\beta(z) > 0$, the coefficient before b^{-1} must tend to zero when $z \to -\infty$. As a result we again obtain for the modulus of the reflection coefficient $|V|^2 = 1 + \exp(-2\pi\alpha) = 1 + |W|^2$. It is also obvious that the flow does not affect the sound field and that there is no reflection if $\xi \cdot v_0 = 0$. Note also that at fixed φ the reflection coefficient is a discontinuous function of ξ: $|V|^2 = 2$ as $\xi \to +0$ and $|V| = 0$ at normal incidence. This is due to the fact that for any $\xi \neq 0$ there are turning horizons and a horizon of resonant interaction, but not if $\xi = 0$. If the flow velocity $v_0(z)$ is finite at any z, then we would have the reflection coefficient as a continuous function of ξ.

9.4.2 General Flow. Well Separated Horizon of Resonant Interaction and Turning Points

Let us now consider reflection of high-frequency sound from the medium with $c(z)$, $\varrho(z)$, and $v_0(z)$ as continuous arbitrary functions. We suppose that the sound velocity and density tend to constant values when $z \to +\infty$ and that $v_0(+\infty) = 0$. In addition, we also suppose that $k(z)\beta(z)$ increases monotonically from $k(-\infty)\beta(-\infty) < -\xi < 0$ up to $k(+\infty)\beta(+\infty) = k(+\infty)$ and that $\beta'(z) \neq 0$. Then the sound wave has one horizon $z = z_c$ where $\beta(z_c) = 0$ and two turning horizons z_1 and z_2 so that $z_1 < z_c < z_2$. The wave is inhomogeneous in the region $z_1 < z < z_2$ and propagating for $z > z_2$ and $z < z_1$.

We shall use the wave equation in the form given in (1.2.21)

$$\Psi'' + \left\{ k^2\beta^2 - \xi^2 + \frac{(\varrho\beta^2)''}{2\varrho\beta^2} - 3\left[\frac{(\varrho\beta^2)'}{2\varrho\beta^2} \right]^2 \right\} \Psi = 0 \quad , \quad \Psi = \frac{\Phi(z)}{\beta\varrho^{1/2}} \quad . \qquad (9.4.8)$$

Note that at $z = z_c$ the derivatives $\partial^l \beta / \partial z^l$ and analogous derivatives of c, ϱ, and v_0 are of the order of magnitude of L^{-l}, since $\xi \cdot v_0(z_c)/\omega = 1$. It follows from the equalities $k\beta|_{z=z_{1,2}} = \mp \xi$ that $|z_{1,2} - z_c| = O(\xi L/k)$. In the region $|z - z_c| \ll L$ the functions $c(z)$, $\varrho(z)$, and $v_0(z)$ can be represented by their Taylor series. Then (9.4.8) becomes

$$\Psi'' + [k^2\beta^2 - \xi^2 - 2(z - z_c)^{-2} - \kappa\xi(z - z_c)^{-1} + O(L^{-2})]\Psi = 0 \quad ,$$

$$\kappa \equiv \xi^{-1}(\beta''/\beta' + \varrho'/\varrho)_{z=z_c} \quad . \qquad (9.4.9)$$

If one takes into account the fluid viscosity, then, as $z \to z_c$, the coefficient before Ψ in (9.4.9) behaves in such a manner [9.6] that z_c must be replaced by the complex value $z_c - i\delta \, \mathrm{sgn}\, \beta'(z_c)$ where $\delta \to +0$. In the following it will be assumed that $\xi L \gg 1$. (The physical meaning of this condition is that the Mach number $M = v_0(z_c)/c(z_c)$ at $z = z_c$ is small compared to $kL \gg 1$. Both $M \simeq 1$ and $M \gg 1$ are permissible). Then in comparison to ξ^2 the terms of order $O(L^{-2})$ in the coefficient before Ψ are negligible. Note also that $\kappa \simeq (\xi L)^{-1} \ll 1$. This parameter is the ratio of the characteristic scales of the variability of the sound field and the medium in the vicinity of $z = z_c$. The term $\kappa\xi(z - z_c)^{-1}$ in the coefficient in (9.4.9) is relatively small at any z but it can not be neglected. It will be shown later that this term is important to the solutions of the equation and has interesting physical consequences.

We do not know the reference equation corresponding to the wave equation (9.4.8) [that is, one having two turning points and the singularity as in (9.4.8)] which is solvable in terms of a known special function. This prevents us from obtaining the uniform asymptotics for the sound field, and we shall therefore use the set of local ones. Here, the values of the parameters $\alpha_{1,2}$ are important:

$$\alpha_j = \frac{2i}{\pi}(-1)^{j+1}\varphi(z_j, z_c) \quad , \quad j = 1, 2 \quad \text{where}$$

$$\varphi(u, v) \equiv \int_v^u \sqrt{k^2\beta^2 - \xi^2}\, dz \qquad (9.4.10)$$

is the **phase integral**. Note that $\alpha_j > 0$ and $\alpha_1 \simeq \alpha_2 \simeq \xi^2 L/k$ to within an order of magnitude.

First consider the case $\alpha_{1,2} \gg 1$. It can be shown (Sect. 8.1) that in this case the WKB approximation is valid everywhere except in narrow neighbourhoods near the horizons $z = z_c$, $z = z_{1,2}$. As a reference equation in the region $|z - z_c| \ll |z_{1,2} - z_c|$ we use the Whittaker equation (3.2.1) and the substitution $\eta(z) = -2i\varphi(z, z_c)$. Note that $\eta(z) \approx 2\xi(z - z_c)$. According to (3.1.9) and (3.2.1) the function

$$f(z) = (i\eta'/2)^{-1/2}\{B_1 W_{l,m}[\eta(z)]$$
$$+ B_2 W_{-l,m}[-\eta(z)]\} \quad , \quad B_{1,2} = \text{const} \tag{9.4.11}$$

is the exact solution of the equation

$$f'' + \left\{(k^2\beta^2 - \xi^2)\left(1 - \frac{4l}{\eta} + \frac{1 - 4m^2}{\eta^2}\right)\right.$$
$$\left. + \left(\frac{1}{2}\ln \eta'\right)'' - \left[\left(\frac{1}{2}\ln \eta'\right)'\right]^2\right\}f = 0 \quad , \tag{9.4.12}$$

where $W_{\pm l,m}$ are Whittaker functions.

It can be easily shown that when $m = \frac{3}{2}$ and $l = -\kappa/2$ the discrepancy between the coefficients of (9.4.9, 12) is $O(L^{-2} + \xi^2\alpha^{-2})$ and negligible compared to ξ^2. Hence (9.4.11) is the main term in the asymptotic expansion (with respect to the parameter $\xi L \gg 1$) of $\Psi(z)$ in the region under consideration.

Whittaker functions can be replaced by their asymptotics if $|\eta| \gg 1$. At the point $\eta = 0$ the functions $W_{l,m}(\pm\eta)$ have a logarithmic singularity. Therefore the coefficients in the asymptotic expansions, found in [9.30], depend on $\arg\eta$. In the problem under consideration, taking into account what was said about the displacement of z_c from the real axis z, we can see that $\arg\eta$ changes from 0 for $z > z_c$ up to π for $z < z_c$. Then we obtain from (9.4.11), using the results of [9.30]:

$$f(z) \approx (k^2\beta^2 - \xi^2)^{-1/4}\left\{B_1 \exp[i\varphi(z, z_c)]\right.$$
$$\left. + B_2\left(1 - \frac{i\pi}{2}\kappa\right)\exp[-i\varphi(z, z_c)]\right\} \quad , \quad z > z_c \quad , \tag{9.4.13}$$

$$f(z) \approx (k^2\beta^2 - \xi^2)^{-1/4}\{(B_1 + i\tfrac{\pi}{2}\kappa B_2)\exp[i\varphi(z, z_c)]$$
$$+ [B_2 - i\tfrac{\pi}{2}\kappa(B_1 + B_2)]\exp[-i\varphi(z, z_c)]\} \quad , \quad z < z_c \quad , \tag{9.4.14}$$

within an accuracy to the factor $1 + O(\kappa^2 + |\eta|^{-1})$. We see that in the region $\xi^{-1} \ll |z - z_c| \ll |z_{1,2} - z_c|$ the sound field is of WKB type.

It can be shown that for $|z - z_c| \gtrsim |z_{1,2} - z_c|$ the terms $(z - z_c)^{-2}$ and $\kappa\xi(z - z_c)^{-1}$ in the coefficient of the wave equation are of the order of $\xi^2 O(\alpha^{-2})$ and can be neglected. Hence, the sound field in the neighborhood of turning horizons is described in terms of Airy functions by the usual asymptotic expressions (Sect. 9.2), which transform into the WKB solutions:

$$f(z) = (k^2\beta^2 - \xi^2)^{-1/4}\{D_1 \exp[i\varphi(z, z_1)]$$
$$+ D_2 \exp[-i\varphi(z, z_1)]\} \quad , \quad z < z_1 \quad , \tag{9.4.15}$$

$$f(z) = (k^2\beta^2 - \xi^2)^{-1/4}\{D_3 \exp[i\varphi(z, z_2)]$$
$$+ D_4 \exp[-i\varphi(z, z_2)]\} \quad , \quad z > z_2 \quad , \tag{9.4.16}$$

if $|z - z_{1,2}| \gg (Lk_0^{-2})^{1/3}$.

The relation between the coefficients in the WKB solutions (9.4.13 and 16), (9.4.14 and 15) above and below the turning horizons is given by (9.2.9). Taking into account that $\varphi(z, z_c) = \varphi(z, z_1) - i\pi\alpha_1/2 = \varphi(z, z_2) + i\pi\alpha_2/2$, we obtain

$$D_1 = (B_1 + i\pi\kappa B_2/2)\exp(\pi\alpha_1/2)$$
$$- [iB_2/2 + \pi\kappa(B_1 + B_2)/4]\exp(-\pi\alpha_1/2) \quad ,$$

$$D_2 = [B_2/2 - i\pi\kappa(B_1 + B_2)/4]\exp(-\pi\alpha_1/2)$$
$$- (iB_1 - \pi\kappa B_2/2)\exp(\pi\alpha_1/2) \quad ,$$

$$D_3 = 0.5B_1 \exp(-\pi\alpha_2/2) - iB_2(1 - i\pi\kappa/2)\exp(\pi\alpha_2/2) \quad ,$$

$$D_4 = B_2(1 - i\pi\kappa/2)\exp(\pi\alpha_2/2) - 0.5iB_1 \exp(-\pi\alpha_2/2) \quad . \tag{9.4.17}$$

We have seen above that according to the casuality principle $D_2 = 0$. Then after simple operations we obtain for the reflection and transmission coefficients

$$|V|^2 \equiv \left|\frac{D_3}{D_4}\right|^2 = 1 + \exp[-\pi(\alpha_1 + \alpha_2)] + \pi\kappa \exp(-\pi\alpha_2) \quad , \tag{9.4.18}$$

$$|W|^2 \equiv \left|\frac{D_1}{D_4}\right|^2 = \exp[-\pi(\alpha_1 + \alpha_2)] \quad . \tag{9.4.19}$$

Here we retain only the main terms of the expansion of $|V|^2$ and $|W|^2$ in powers of the small quantities κ and $\exp(-\pi\alpha_{1,2})$. By knowing D_j and using (9.2.5–8) one can find the coefficients of the Airy functions for asymptotics in the neighbourhood of the turning horizons.

9.4.3 General Flow. Arbitrary Separated Horizon of Resonant Interaction and Turning Points

When $\alpha_{1,2} \lesssim 1$ there exists no region between z_1 and z_2 where the WKB approximation is valid and the method described above can not be used for obtaining the asymptotic expressions for the field. Since $\alpha_{1,2} \simeq \xi^2 L/k$ we have $\xi/k \ll 1$ for $\alpha_{1,2} \lesssim 1$ due to the assumption $\xi L \gg 1$. We note also that the inequalities $\xi/k \ll 1$ and $\alpha_{1,2} \gg 1$ can hold simultaneously. Hence, analysis of the cases where $\xi/k \ll 1$ and $\alpha_{1,2} \gg 1$ covers all the possible cases in this problem.

Thus, we now assume that $\xi/k \ll 1$ (near-normal incidence). As a reference equation we use (9.4.1) with the following redefinition of the independent variable: $q(z - z_c) = \eta$. We choose the variable substitution in the reference equation in such a way that coefficients before terms proportional to ω^2 are the same in the reference equation and in (9.4.8) [see (9.1.4)]:

$$\int_0^\eta \sqrt{\eta^2/4 - \alpha}\, d\eta = \varphi(z, z_c) \quad , \quad \text{or} \quad \eta = \sqrt{2\alpha}\, \sin \tau \quad , \quad \text{where}$$

$$\tau + \frac{1}{2}\sin 2\tau = \varphi(z, z_c)/i\alpha \quad . \tag{9.4.20}$$

Specific choice of the lower limit of integration over η ensures that the poles at $z = z_c$ and $\eta(z_c) = 0$ in the wave and reference equations are coincident. Coincidence of the turning points at $z = z_1$ and $\eta(z_1) = -2\alpha^{1/2}$ can be achieved by a suitable choice of α. Taking into account (9.4.10), we obtain from (9.4.20) $\alpha = \alpha_1$. By analogy the turning points $z = z_2$ and $\eta(z_2) = 2\alpha^{1/2}$ will coincide if $\alpha = \alpha_2$. Since generally $\alpha_1 \neq \alpha_2$, we have to use different variable substitutions $\eta_{2,1}(z)$ for $z > z_c$ and for $z < z_c$ with different values of α.

The discrepancy between the values of the coefficients in the wave and reference equations is finite for any z if $\kappa = 0$. It is possible to prove by using (9.4.20) that the order of magnitude of this discrepancy is not bigger than that of $k/\xi L^2$ and tends to zero when $L \to \infty$. Hence, the main term of the high-frequency asymptotics for $\Psi(z)$ will be

$$f(z) = (\eta_1^2 \eta_1')^{-1/2}\{E_1 W[\alpha_1, \exp(i\tfrac{\pi}{4})\eta_1(z)]$$
$$+ E_2 W[\alpha_1, \exp(-3i\tfrac{\pi}{4})\eta_1(z)]\} \quad , \quad z < z_c \quad , \tag{9.4.21}$$

$$f(z) = (\eta_2^2 \eta_2')^{-1/2}\{E_3 W[\alpha_2, \exp(i\tfrac{\pi}{4})\eta_2(z)]$$
$$+ E_4 W[\alpha_2, \exp(-3i\tfrac{\pi}{4})\eta_2(z)]\} \quad , \quad z > z_c \quad , \tag{9.4.22}$$

where E_j are arbitrary constants.

The discrepancy includes the singular term $-\kappa\xi(z - z_c)^{-1}$ at $z \to z_c$ if $\kappa \neq 0$. In this case, the asymptotic character of the solutions (9.4.21, 22) can be guaranteed only in the region not very close to z_c, in spite of $\kappa \to 0$ as $L \to \infty$. It is possible to obtain a condition under which (9.4.21, 22) are asymptotic solutions of the wave equation by requiring that $\kappa\xi(z - z_c)^{-1}$ be small compared to ξ^2 in the coefficient of (9.4.9). From this requirement we find: $\xi|z - z_c| \gg 1/\xi L$.

In the vicinity of the horizon $z = z_c$ it is reasonable to supplement (9.4.21 and 22) with the series representation of $\Psi(z)$ in powers of $z - z_c$. Neglecting the term $O(L^{-2})$ as compared with ξ^2 in (9.4.9), we obtain solutions [Ref. 9.31, Part 1, Sect. 25.7]:

$$\Psi(z) = F_1 g_1(u) + F_2 g_2(u) \quad , \quad u = \xi(z - z_c) \quad ,$$
$$g_1 = u^2(1 + \kappa u/4 + u^2/10 + \ldots) \quad ,$$
$$g_2 = (\kappa/3)(1 - \kappa^2/4)g_1(u)\ln u - u^{-1}(1 - \kappa u/2 - u^2/2 + \ldots) \quad , \tag{9.4.23}$$

where $F_{1,2} = \text{const}$. The solution g_2 has a singularity at $u = 0$ due to the singularity in the wave equation's coefficient at this point. The sound pressure $p \sim \beta \varrho^{1/2}\Psi$, however, as well as the derivatives $\partial p/\partial z$ and $\partial^2 p/\partial z^2$ remain finite. If one accounts for the viscosity, $\text{Im}\{z_c\} < 0$, therefore, $\arg u$ increases by π when the transition from $z > z_c$ to $z < z_c$ takes place. Taking this into account, we obtain

$$g_1(\kappa, -u) = g_1(-\kappa, u) \quad ;$$

$$g_2(\kappa, -u) = -g_2(-\kappa, u) + \frac{i\pi\kappa}{3}\left(1 - \frac{\kappa^2}{4}\right)g_1(-\kappa, u) \quad . \tag{9.4.24}$$

The solution (9.4.23) has common regions of applicability with (9.4.21, 22), namely, $1/\xi L \ll \xi |z - z_c| \ll 1$. To find the relationship between the coefficients $F_{1,2}$ and $E_{1,2}$, $E_{3,4}$ we represent the parabolic cylinder functions as an expansion in powers of their arguments [Ref. 9.22, Chap. 19]. Then for $f(z)$ we obtain in (9.4.22)

$$f(z) = [\eta_2'(z_c)]^{-3/2}\frac{\xi}{u}W(\alpha_2, 0)\left[(E_3 + E_4)\left(1 - \frac{u^2}{2} + \dots\right)\right.$$

$$\left. + \frac{R(\alpha_2)}{3}(E_3 - E_4)(u^3 + \dots)\right] \tag{9.4.25}$$

where

$$R(\alpha) = \pi\sqrt{\alpha/2}[\cosh(\pi\alpha)]^{-1}\exp(\pi\alpha/2)[1 - i\exp(-\pi\alpha)]$$

$$\times [1 + (2\alpha)^{-2}]|\Gamma(3/4 - i\alpha/2)|^{-2} \quad . \tag{9.4.26}$$

Let $|\kappa| \ll \xi |z - z_c| \ll 1$. Then in $g_{1,2}$ (9.4.23) we need to retain only the first terms of the series since $\kappa u^3 \ln u$ and κu are negligible compared to u^2. Comparing (9.4.23 and 25) further we find that

$$g_{1,2}(u) \sim \eta^{-1}\{W[\alpha, \eta\exp(i\tfrac{\pi}{4})] \mp W[\alpha, \eta\exp(-3i\tfrac{\pi}{4})]\} \quad .$$

Matching of the solutions at $z > z_c$ is written as

$$\xi[\eta_2'(z_c)]^{-3/2}W(\alpha_2, 0)(E_3 + E_4) = -F_2 \quad ,$$

$$(E_3 - E_4)R(\alpha_2)/3(E_3 + E_4) = -F_1/F_2 \quad . \tag{9.4.27}$$

Analogously, taking into account (9.4.24) one obtains at $z < z_c$

$$\frac{\xi W(\alpha_1, 0)}{[\eta_1'(z_c)]^{3/2}}(E_1 + E_2) = -F_2 \quad ,$$

$$\frac{R(\alpha_1)(E_1 - E_2)}{3(E_1 + E_2)} = -\frac{F_1}{F_2} - \frac{i\pi\kappa}{3}\left(1 - \frac{\kappa^2}{4}\right) \quad . \tag{9.4.28}$$

Using (9.4.21, 22) one can find the reflection and transmission coefficients in the same manner as for the reference problem with a linear flow profile $v_0(z)$. According to the casuality principle we have $E_1 = 0$. As a result we obtain the expression for the reflection coefficient which coincides with (9.4.6). The only difference is that instead of A_1/A_2 and α we now have E_3/E_4 and α_2, respectively. The modulus of the transmission coefficient is

$$|W| = \frac{\exp[-\pi(\alpha_1 + 3\alpha_2)/4]|E_2|}{|E_4 + iE_3\exp(-\pi\alpha_2)|} \quad . \tag{9.4.29}$$

Equations (9.4.27, 28) allow us to express $F_{1,2}$ and $E_{3,4}$ in terms of E_2. In particular,

$$\frac{E_3}{E_4} = \frac{R(\alpha_2) - R(\alpha_1) + i\pi\kappa(1 - \kappa^2/4)}{R(\alpha_2) + R(\alpha_1) - i\pi\kappa(1 - \kappa^2/4)} \quad . \tag{9.4.30}$$

This expression can be simplified due to the smallness of the ratio ξ/k. In the integrals for $\alpha_{1,2}$ in (9.4.10) we shall use the new variable $s = k\beta/\xi$. Then

$$\alpha_j = (-1)^j \frac{2\xi}{\pi} \int_0^{(-1)^j} \sqrt{1 - s^2}\frac{dz}{ds}\,ds \quad . \tag{9.4.31}$$

By expanding s in a series in powers of $z - z_c$, inverting this series, and substituting it into (9.4.31) we find

$$\alpha_j = \frac{\xi^2}{2k\beta'}\left[1 - (-1)^j\frac{4\xi}{3\pi k\beta'}\left(2\frac{k'}{k} + \frac{\beta''}{\beta'}\right) + O\left(\frac{\xi^2}{k^2}\right)\right] \quad , \tag{9.4.32}$$

where k, k', β', and β'' are taken at $z = z_c$. Note that $|\alpha_2 - \alpha_1| \simeq \xi\alpha_2/k$ and $|E_3/E_4| \ll 1$. It follows from the asymptotics of the Γ-function of an argument with a large modulus [Ref. 9.22, Chap. 6] that $R(\alpha) = 1 + O(\alpha^{-2})$. Taking this into account and retaining only the terms to the first power of $(\alpha_2 - \alpha_1)/\alpha_2$ and κ we obtain by using (9.4.26, 30)

$$\mathrm{Im}\left\{\frac{E_3}{E_4}\right\} = \frac{1}{2}\left\{\frac{\exp(-\pi\alpha_1) - \exp(-\pi\alpha_2)}{1 + \exp(-2\pi\alpha_2)}\right.$$

$$\left. + \frac{\pi\kappa}{|R(\alpha_2)|\sqrt{1 + \exp(-2\pi\alpha_2)}}\right\} \quad . \tag{9.4.33}$$

With the same accuracy we obtain by using (9.4.6, 29)

$$|V|^2 = [1 + \exp(-2\pi\alpha_2)][1 + 2\exp(-\pi\alpha_2)\,\mathrm{Im}\{E_3/E_4\}]$$
$$= 1 + \exp[-\pi(\alpha_1 + \alpha_2)] + \pi\kappa|R(\alpha_2)|^{-1}$$
$$\times \exp(-\pi\alpha_2)[1 + \exp(-2\pi\alpha_2)]^{1/2} \quad . \tag{9.4.34}$$

$$|W|^2 = \frac{\exp[-\pi(\alpha_1 + \alpha_2)/2]}{2\cosh(\pi\alpha_2)}\left(\frac{\alpha_2}{\alpha_1}\right)^{3/2}$$

$$\times \left|\frac{R(\alpha_1) + R(\alpha_2)}{2R(\alpha_2)}\frac{\Gamma(5/4 - i\alpha_1/2)}{\Gamma(5/4 - i\alpha_2/2)}\right|^2 |V|^2 \quad . \tag{9.4.35}$$

When $\alpha_{1,2} \gg 1$, (9.4.34 and 35) reduce to (9.4.18 and 19) obtained without assuming that ξ/k is small. Hence (9.4.34, 35) give the values of the energy reflection and transmission coefficients for any values of parameters in the problem under consideration. If $\kappa = 0$ and $\alpha_1 = \alpha_2$ one has $|V|^2 = 1 + |W|^2 = 1 + \exp(-2\pi\alpha_1)$ as in the case of the reference problem with a linear profile $v_0(z)$.

Note that the vertical structure of the sound field can be described without the help of Whittaker functions. Indeed, as we have seen above, the validity regions of the asymptotics (9.4.21, 22) and the power expansion (9.4.23) cover the entire medium at any ξ/k. Those asymptotics which include parabolic cylinder functions allow us to obtain the main term of the sound-pressure expansion in powers of $1/kL$ even at $z = z_c$; calculation of the derivative $\partial p/\partial z$ in the region $|z - z_c| \lesssim 1/\xi^2 L$ requires (9.4.23), however.

9.4.4 Discussion of the Results

Let us discuss the physical meaning of the obtained results. If the value κ in (9.4.9) is positive, the sound wave becomes amplified in the process of reflection. This is due to two reasons. First, there exists an acoustical energy influx from $z = -\infty$. This phenomenon also occurs in the case of reflection from a homogeneous moving medium (Sect. 2.6) because the sound wave which penetrates into the region $z < z_c$ possesses negative energy [9.37], that is, the flow energy is larger in the absence of the wave. The second reason for wave amplification is related to the processes which take place in the vicinity of the horizon $z = z_c$. This process is represented in (9.4.34) by the term which is proportional to κ. This mechanism of sound amplification is due to a resonant interaction between the oscillations of particles in the sound wave and their motion in the flow. Since the flow is assumed to be steady the resonance occurs at zero frequency (in the reference system moving together with the fluid particles). This is quite analogous to the Landau amplification (or attenuation, for other particle distribution functions) of plasma oscillations [9.32, 33].

At the horizon $z = z_c$, the projection of the particles velocity upon the direction ξ is equal to the velocity of the wave trail. That is why in the coordinate system moving together with the particles the wave frequency is zero. Hence, the energy acquired by the sound from the flow is not zero when averaged over the wave period. Due to viscosity, even when it is infinitely small[3], fluid particles which overtake the wave decelerate and transfer the energy to the wave. In contrast, particles which lag behind the wave take energy from it and accelerate. As a result, if the wave amplitude is nearly constant at $z \approx z_c$, the sound is amplified when the number of particles overtaking the wave is greater than the number lagging behind it. The opposite would result in the case of negative viscosity. This explains why V depends on the direction of the going around the singular point $z = z_c$ in the wave equation[4].

In contrast to the first mechanism of sound amplification, resonant interaction is sensitive to density stratification, see (9.4.9). Under the conditions $\varrho = $ const and $v_0 \| \xi$ the resonant interaction leads to energy transfer from the flow to the sound wave if $v_0''(z_c) < 0$. In contrast, we have energy transfer from the sound to the flow if $v_0''(z_c) > 0$, $\varrho'(z_c) = 0$. When $\alpha_{1,2} \lesssim 1$ the main mechanism for energy exchange between sound and flow is the first one. In this case we have $|V| > 1$ independent of the sign of κ. With increasing angle of incidence the width of the potential barrier (the region $z_1 < z < z_2$ where the sound wave is inhomogeneous) also increases whereas the amplitude of the transmitted wave decreases exponentially. Here the resonant interaction becomes dominant; the sign of $|V^2| - 1$ is the same as the sign of κ.

Note that under some conditions a third mechanism of sound amplification through a change in the sign of viscous dissipation in a moving medium manifests itself [9.32, 34].

---

[3]   Infinitely small viscosity can be the cause of finite effects in monochromatic waves because the smaller viscosity is, the longer the time after which oscillations become steady-state.

[4]   For a quantitative analysis of wave-flow energy exchange along these lines, see [9.38]. When $\rho'(z_c) \neq 0$, an account of wave-intensity variation at $z \approx z_c$ turns out to be necessary. It can be achieved [9.38] by considering the sound wave as a gas of particles (phonons) with z-dependent concentration. The results of such an analysis are consistent with those obtained above in a different way.

In the previous discussion we have supposed that the sound wave has two turning points. If the projection of the flow velocity $u(z)$ on the direction ξ obeys the inequality $u(z) < c(z) + \omega/\xi$, then $-\xi < k\beta < 0$ at $z < z_c$ and the wave has only one turning horizon $z = z_2 > z_c$. In the absence of resonant interaction the reflection will be total ($|V| = 1$) since the wave is inhomogeneous at $z < z_2$. In the given case, the sound field may be found by transfer to the limit $\alpha_1 \to +\infty$ in the asymptotic expressions obtained earlier for the case of two turning points. In particular, we find from (9.4.18) for the reflection coefficient

$$|V|^2 = 1 + \pi\kappa \exp\left(-\pi\alpha_2\right) \ . \tag{9.4.36}$$

A problem of practical interest is sound reflection by a jet of finite width when $v_0 \to 0$ as $|z| \to \infty$. Suppose that $u(z)$ has only one maximum $u(z_m) = u_m$ and that the product $k(z)\beta(z)$ increases for $z > z_m$ and decreases for $z < z_m$. If $\beta(z_m) > 0$, it is straightforward to find the acoustic field by using results of preceeding sections. Let $-\xi < k(z_m)\beta(z_m) < 0$. The sound wave has two horizons of resonance interaction $z_{c1,c2}$ and two turning horizons $z_{1,2}$ (Fig. 9.3a). We shall consider only the simplest case, when $|\varphi(z_1, z_{c1})| \gg 1$, $|\varphi(z_{c1}, z_{c2})| \gg 1$, $|\varphi(z_{c2}, z_2)| \gg 1$, where the phase integral φ is defined in (9.4.10). Under such conditions one may use local asymptotics of the type given in (9.2.5) in the vicinities of turning points $z = z_{1,2}$ and of the type given in (9.4.11) in vicinities of horizons $z = z_{c1,c2}$. Everywhere outside these vicinities the WKB approximation is valid. By requiring that there be no wave coming from $z = -\infty$ for $z < z_1$ and then successively calculating the coefficients of the WKB asymptotics in their regions of validity from the lower to upper layers by using (9.2.9) and (9.4.17), we find

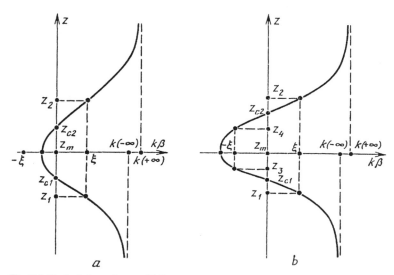

a b

Fig. 9.3. Vertical dependence of $k\beta$ at sound wave reflection from a bounded jet. Turning horizons $z = z_j$, $j = 1, 2, 3, 4$ and resonant interaction horizons $z = z_{cj}$, $j = 1, 2$ in two cases: the wave is inhomogeneous in the vicinity of the horizon z_m (a); there is a region of transparency $z_3 < z < z_4$ inside the jet in which the sound waves are propagating (b)

$$|V|^2 = 1 - \exp\left(-2|\varphi(z_2, z_1)|\right) + \pi\kappa(z_{c2})\exp\left(-2|\varphi(z_{c2}, z_2)|\right)$$
$$\quad - \pi\kappa(z_{c1})\exp\left(-2|\varphi(z_{c1}, z_2)|\right) \ , \tag{9.4.37a}$$

$$|W|^2 = \exp\left(-2|\varphi(z_2, z_1)|\right) \ ,$$
$$\kappa(z) \equiv [\beta''(z)/\beta'(z) + \varrho'(z)/\varrho(z)]/\xi \ . \tag{9.4.37b}$$

Here, only the leading terms in the expansions of $|V|^2$ and $|W|^2$ in powers of the small parameters $\exp(-|\varphi|)$ and κ are preserved.

Due to exponential damping of the wave in the region $z_1 < z < z_2$, it is the upper horizon of resonant interaction ($z = z_{c2}$) which mainly contributes to the energy exchange between the sound field and flow. When resonant interaction is absent ($\kappa = 0$) we have the same expressions for the moduli of the reflection and transmission coefficient as in the problem of reflection from a potential barrier with $\beta(z) > 0$, see (9.3.15), where $-\pi\alpha = |\varphi(z_2, z_1)| \gg 1$. The case where $k(z_m)\beta(z_m) < -\xi$ (Fig. 9.3b), when the wave has four turning points [9.34] as well as other problems where the turning and resonant interaction horizons are far apart may be studied by the same method.

Consider, for example, sound amplification upon reflection of a plane wave incident from a homogeneous medium at rest on the half-space $z < 0$ with exponential stratification of flow velocity. The velocity profile is given by $v_0(z) = (u(z), 0, 0)$, $u(z) = u_m[1 - \exp(z/L)]$. (Sound speed and density are supposed to be uniform over the entire space). Then $\beta(z) = 1 - \xi k^{-1} M[1 - \exp(z/L)]$, where $M \equiv u_m c^{-1} \cos\varphi$ and φ is angle between vector $\boldsymbol{\xi}$ and Ox axis. We imply that $\cos\varphi > 0$ since certainly $|V| < 1$ for $\cos\varphi \leq 0$. If $(1 + M)^{-1} < \xi/k \leq M^{-1}$, the wave has a single turning point and there is no resonant interaction horizon. In this case total reflection occurs; $|V| = 1$. If $M^{-1} < \xi/k < (M - 1)^{-1}$, there is a horizon of resonant interaction at $z = z_c$ and a turning point at $z = z_2$. $|V|^2$ is given by (9.4.36). If $\xi/k > (M - 1)^{-1}$, there are two turning points z_1 and z_2 with a horizon of resonant interaction lying between them. In this case $|V|^2$ should be taken from (9.4.34). One can find values $\alpha_{1,2}(\xi)$ for the given exponential velocity profile analytically.

Results of the calculation of the energy reflection coefficient $|V|^2$ at $kL = 500/\pi$ and three different values of the Mach number M are presented in Fig. 9.4 for $\xi/k > (1 + M)^{-1}$. When $M = 50$, the main contribution to sound amplification in the region of maximum $|V|^2$ values is due to energy influx from $z = -\infty$. There is a sharp maximum in $|V|^2$ close to the incident angle at the minimum of $\alpha_1 + \alpha_2$. When $M = 5$, wave amplification is mainly (and for $M = 2$ exclusively) due to resonant interaction with the flow. $|V|^2$ has a maximum at angles of incidence where α_2 is close to its minimum. The dependence of $|V|^2 - 1$ on the Mach number of the flow u_m/c and on the azimuth φ is very strong. Thus, $|V|^2_{\max} = 1.47$ if $M = 50$, but $|V|^2 - 1 < 10^{-79}$ for all (real) angles of incidence if $M = 2$. In reality, due to volumetric sound absorption described by the imaginary part of the wave number k, there is no wave amplification at all if $M - 1 \lesssim 1$.

Our presentation in Sect. 9.4 is based essentially on [9.40]. An extension of the theory and further discussion of the results can be found in [9.38]. Previously,

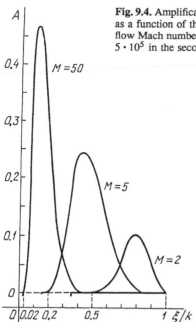

Fig. 9.4. Amplification of a sound wave by an inhomogeneous flow $A = |V|^2 - 1$ as a function of the angle of incidence $\theta = \arcsin(\xi/k)$ at different values of the flow Mach number: $M = 50$, $M = 5$ and $M = 2$. For clarity, A is multiplied by $5 \cdot 10^5$ in the second case and by 10^{78} in the third

sound amplification upon reflection from a supersonic flow (in addition to the case of discretely layered media discussed in Sect. 2.6) has been considered in the literature for a flow with a thin (compared to the wavelength) transition layer [9.35, 39] or with a $v_0(z)$ profile close to linear [9.36]. In the latter resonant interaction was not taken into account. A profound analysis of sound amplification in homogeneous media with flow of constant direction was given in [9.34]. In that study it is assumed that the WKB approximation is valid between the turning points and the resonant interaction horizon, that is, in our notation, $\alpha_{1,2} \gg 1$.

10. Sound Reflection from a Medium with Arbitrarily Varying Parameters

As is seen from Chap. 3, exact solutions of the problem of plane sound wave reflection from a layered medium exist only for a few cases. Although the study of these cases is quite valuable and reveals a number of important regularities, it does not eliminate the problem of the study of reflection of acoustic waves from layers in which the medium parameters have an arbitrary dependence on the coordinate z. It is also important to note that in real geophysical situations the parameters of the medium do not remain constant but undergo both systematic and fluctuational changes as time passes. It is necessary to know how these changes affect the reflection coefficient. As shown in Sect. 6.3, even small variations of the medium parameters can have a significant effect on the reflection coefficient. A number of results for a general layered inhomogeneous medium were obtained in [10.1–18, 56–58]. The exposition of this chapter is based on works of the present authors [10.19–22].

Let the density of the medium, the sound speed, and the flow velocity be given by the functions $\varrho(z)$, $c(z)$, and $v_0(z)$. It is assumed that as $z \to +\infty$ and $z \to -\infty$ the medium parameters approach constant values equal to ϱ_1, c_1, v_{01} and ϱ_2, c_2, v_{02}, respectively. We assume that there are no horizons of resonant interaction of the sound with flow in the medium. For brevity, the common factor $\exp(i\boldsymbol{\xi} \cdot \boldsymbol{r} - i\omega t)$ will be omitted in all the expressions for the acoustic field.

10.1 Differential Equations for Reflection Coefficient and Impedance of a Sound Wave

10.1.1 Riccati Equation

Let a plane wave be specified for $z = +\infty$, propagating in the direction of negative z (the incident wave). The wave equation (1.2.25) can be satisfied in the general case only under the assumption that there also exists a reflected wave at $z = +\infty$. Our problem will be the estimation of the ratio of complex amplitudes of the reflected and incident waves, i.e., the reflection coefficient V in modulus and phase. We shall not follow the usual procedure, according to which it is necessary to find solutions of the wave equation and then to calculate the reflection coefficient by using them. Instead we shall obtain an equation in which V enters directly as a function of the vertical coordinate. The dependence of the reflection coefficient on $\boldsymbol{\xi}$ enters this equation parametrically.

We take the wave equation for the sound field in an inhomogeneous moving medium in the form (8.1.1–3). The vertical dependence of the acoustic pressure, Φ, and of the z-component of particle displacement in the wave, f, are related through (8.3.1). To satisfy (8.3.1), we introduce the concept of the "incident" (i) and "reflected" (r) waves for any z by defining them in the following fashion:

$$\Phi^{(i)} = \Phi_1(\zeta) \quad , \quad f^{(i)} = -ik_0 N \Phi_1(\zeta) \quad ;$$
$$\Phi^{(r)} = \Phi_2(\zeta) \quad , \quad f^{(r)} = ik_0 N \Phi_2(\zeta) \quad , \tag{10.1.1}$$

where $\Phi_{1,2}$ are two new unknown functions. Note that the relationship between Φ and f in each of the waves is the same as in the ray-theory approximation when the incident and reflected waves propagate without interacting with one another, cf. (10.1.1) with (8.3.2) under condition (8.3.5).

We define $V \equiv \Phi_2(\zeta)/\Phi_1(\zeta)$ to be the reflection coefficient for an arbitrary horizon. In spite of the arbitrariness of this definition we shall not yet introduce any approximations. Substituting the sums $f = ik_0 N \Phi_1 (V - 1)$, $\Phi = \Phi_1 (1 + V)$ of the incident and reflected waves (10.1.1) into (8.3.1), we obtain the equations

$$ik_0 N \Phi_1 (V - 1) = \frac{\partial}{\partial \zeta} [\Phi_1 (1 + V)] \quad ,$$

$$\frac{\partial}{\partial \zeta} [N \Phi_1 (V - 1)] = ik_0 N^2 (1 + V) \Phi_1 \quad . \tag{10.1.2}$$

Multiplying the first of the equations by $(V - 1)/\Phi_1$ and the second one by $(V + 1)/\Phi_1$, and then adding one to the other, we obtain a *Riccati equation* for the reflection coefficient:

$$\frac{\partial V}{\partial \zeta} = 2ik_0 NV + \gamma (1 - V^2) \quad , \quad \gamma \equiv 0.5 \frac{\partial (\ln N)}{\partial \zeta} \quad . \tag{10.1.3}$$

If z is taken as the vertical coordinate, according to (8.1.2, 3), Eq. (10.1.3) becomes

$$\frac{\partial V}{\partial z} = 2ik_0 \sqrt{n^2 \beta^2 - \xi^2/k_0^2} V + \gamma_1 (1 - V^2) \quad ,$$

$$\gamma_1 \equiv \frac{1}{4} \frac{\partial}{\partial z} \ln \frac{n^2 \beta^2 - \xi^2/k_0^2}{\varrho^2 \beta^4} \quad . \tag{10.1.4}$$

As a boundary condition which is necessary for the determination of the unique solution of (10.1.3) or (10.1.4), one can take

$$\lim_{z \to -\infty} V = 0 \quad , \tag{10.1.5}$$

since there is no reflected wave as $z \to -\infty$ (behind the inhomogeneous layer). A specification of V for a certain horizon can also serve as a boundary condition. Thus, if the medium is homogeneous for $z < z_1$ then $V(z_1) = 0$. If there is an absolutely rigid or a pressure-release boundary at $z = z_1$, then $V(z_1) = \pm 1$.

We can also obtain the Riccati equation for the impedance. The definition of the sound wave impedance in a moving medium was given in Sect. 2.6:

$$Z \equiv -i\omega \varrho_0 p (\partial p/\partial \zeta)^{-1} = -i\omega \varrho_0 \Phi (\partial \Phi/\partial \zeta)^{-1} \quad . \tag{10.1.6}$$

Differentiating (10.1.6) with respect to ζ one obtains

$$\frac{\partial Z}{\partial \zeta} = -i\omega \varrho_0 \left[1 - \Phi \frac{\partial^2 \Phi}{\partial \zeta^2} \left(\frac{\partial \Phi}{\partial \zeta} \right)^{-2} \right] \quad ,$$

which after account of (8.1.1) gives the desired differential equation

$$\frac{\partial Z}{\partial \zeta} = i\omega \varrho_0 [(k_0 N/\omega \varrho_0)^2 Z^2 - 1] \quad . \tag{10.1.7}$$

When the impedance is considered to be a function of the coordinate z, (10.1.7) becomes

$$\frac{\partial Z}{\partial z} = i\omega \varrho \beta^2 [(k_0/\omega \varrho \beta^2)^2 (n^2 \beta^2 - \xi^2/k_0^2) Z^2 - 1] \quad . \tag{10.1.8}$$

If the medium is homogeneous for $z < z_1$, then the boundary condition for Z is that $Z(z_1)$ and the impedance of the plane wave propagating towards negative z values must be equal. It is also not difficult to specify the boundary condition for Z in other cases.

With the help of (10.1.6), we can express Φ in terms of Z as

$$\Phi(\zeta) = \exp\left[-i\omega \varrho_0 \int_{\zeta_0}^{\zeta} Z^{-1}(\zeta_1) d\zeta_1 \right] \quad , \tag{10.1.9}$$

where the arbitrary quantity ζ_0 determines normalization of Φ. Thus, after the impedance $Z(\zeta)$ is found, the pressure in the entire medium can be calculated by direct integration. The reflection coefficient can be found from $Z(\zeta)$ according to

$$V(\zeta) = \frac{Z(\zeta) - \omega \varrho_0/k_0 N}{Z(\zeta) + \omega \varrho_0/k_0 N} \quad , \tag{10.1.10}$$

which follows from the definitions of Z and V.

We see that the second-order linear differential equation (8.1.1) and the first-order nonlinear equations (10.1.4, 7) are equivalent, that is, by knowing the solution of one of the equations, we can construct the solutions of the two other ones. In a number of problems the Riccati equation is the most convenient tool for finding approximate analytical and numerical solutions. References [10.23, 24] give examples of the use of the Riccati equation in numerical calculations of acoustic fields in fluids. A matrix analogue of (10.1.8) is used in calculating the elastic wave fields in solids with piecewise-continuous stratified parameters [10.25, 26] whereas an operator analogue plays an important role in mathematical modeling of the wave propagation in range-dependent waveguides [10.59, 60]. A far-reaching generalization of the transformation of (8.1.1) to the Riccati equation is the *invariant imbedding method* which reduces solution of the boundary value problem for the wave equation to integration of nonlinear differential equations of the first order [10.27, 28, 61, 62]. This method proves to be especially effective in studying statistical problems [10.27, 28].

10.1.2 Two Properties of the Reflection Coefficient in Inhomogeneous Media

We now return to (10.1.3, 4) for the reflection coefficient V. By $V(z)$ we mean the ratio of the complex (i.e., including phases) amplitudes of the direct and returning waves. Thus, for example, if in a homogeneous medium there were an absolutely reflecting plane at some $z = z_0$ for which $V = 1$, then at normal incidence our reflection coefficient would be equal to $V(z) = \exp[2ik_0 n(z - z_0)]$. This expression is obtained immediately after integration of (10.1.4) for $\gamma_1 = 0$ and the boundary condition $V(z_0) = 1$. It is not difficult to see that if the point, for which the reflection coefficient from an inhomogeneous layer is obtained, is moved from $z = z_1$ to $z = z_2$ when both z_1 and z_2 lie outside the region where significant reflection takes place, then the following relationship between the values of the reflection coefficient at these two points is valid:

$$V(z_2) = V(z_1) \exp\left[2ik_0 \int_{z_1}^{z_2} \sqrt{n^2\beta^2 - \xi^2/k_0^2} \, dz\right] . \tag{10.1.11}$$

According to (10.1.3), the derivative $\partial V/\partial \zeta$ is bounded unless the medium parameters are discontinuous. Let $N = N_1$ at $z = z_0 + 0$ and $N = N_2 \neq N_1$ at $z = z_0 - 0$. Dividing both sides of (10.1.3) by $1 - V^2$ and then integrating over ζ, one finds

$$\int_{\zeta(z_0)-\varepsilon}^{\zeta(z_0)+\varepsilon} \frac{dV}{1 - V^2} = 2ik_0 \int_{\zeta(z_0)-\varepsilon}^{\zeta(z_0)+\varepsilon} \frac{NV \, d\zeta}{1 - V^2} + \frac{1}{2} \int_{\zeta(z_0)-\varepsilon}^{\zeta(z_0)+\varepsilon} d\ln N .$$

Note that at $V \neq 1$ the integrand in the first item in the right-hand side is bounded. In the limit $\varepsilon \to 0$ we then find

$$\frac{V_+ + 1}{V_+ - 1} = \frac{N_1(V_- + 1)}{N_2(V_- - 1)} , \tag{10.1.12}$$

which relates the values of the reflection coefficient above (V_+) and below (V_-) the discontinuity. In particular, if the medium is homogeneous for $z < z_0$, then $V_- = 0$ and we obtain for V_+ the Fresnel expression $V_+ = (N_1 - N_2)/(N_1 + N_2)$. By using (8.1.2, 3), and (2.6.7) it is straightforward to show that V_+ coincides with the expression (2.6.14) for the plane wave reflection coefficient at an interface of two moving homogeneous half-spaces.

10.1.3 On Separation of the Wave Field into Direct and Inverse Waves

Note that the wave equation (8.1.1) from which we started determines only the total value of the field. The separation of the field into a sum of incident and reflected waves, as was done above, has a certain degree of arbitrariness. Only for the cases of a homogeneous medium or a medium with slowly changing properties (as far as the leading terms of the high-frequency asymptotic expressions of the field are considered) can the sound field be unambiguously separated into waves traveling in one direction or the other.

In an inhomogeneous medium an expression of the form $A(z) \exp[i\varphi(z)]$ is usually called a traveling wave, where A is the wave amplitude and $\varphi(z)$ is the phase. However, this expression can also represent a standing wave if only $A(z)$ is not a constant or a slowly changing function. In oder to establish this we consider the following example, set forth by *Schelkunoff* [10.29].

The function

$$\Phi(z) = \cos bz + \varepsilon \exp(ibz) \tag{10.1.13}$$

with $\varepsilon \ll 1$ describes a wave which is essentially a standing wave, since the first term is dominant. However, the same expression can be represented in the form

$$\Phi(z) = A(z) \exp\left[i\varphi(z)\right] \quad , \qquad \text{where} \tag{10.1.14}$$

$$A(z) = \sqrt{\cos^2 bz + 2\varepsilon \cos bz + \varepsilon^2} \quad ,$$
$$\varphi(z) = \arctan\left[\varepsilon(1 + \varepsilon)^{-1} \tan bz\right] \quad , \tag{10.1.15}$$

which can then be considered as a traveling wave.

The field in an inhomogeneous medium can always be represented in the general case in the form given in (10.1.14), but it is not possible to separate this expression uniquely into the sum of an incident and a reflected wave. Moreover, in general, such a separation would have no physical significance. Nevertheless, new recipes for separating the total field into direct and inverse waves do appear in the literature at times. A critical analysis of one such recipe was presented in [10.30].

10.2 Reflection from a Thin Inhomogeneous Layer

10.2.1 Reduction of the Problem to an Integral Equation

The reflection coefficient of an inhomogeneous layer for which the product of its width and the vertical component of the incident wave vector is small compared to unity, can be found without any assumptions about the stratification in the elastic parameters. In [10.19, 20] a method of successive approximations was proposed to solve this problem in the case of a fluid at rest. Later the method was generalized to the case of reflection from a solid inhomogeneous layer [10.31]. But one can prove convergence of the method only for incidence angles that are neither close to $\frac{\pi}{2}$ nor to the critical angle of total reflection. Here we shall describe another approach [10.21, 22] to calculation of the sound field in a thin layer which is suitable at all incidence angles of the wave.

Let there be a layer with the arbitrary piecewise-smooth parameters $\varrho(z)$, $c(z)$, and $v_0(z)$ between two homogeneous fluid half-spaces with the parameters ϱ_1, c_1, v_{01} (for $z > 0$) and ϱ_2, c_2, v_{02} (for $z < -H$). To simplify transformations we shall suppose $v_{01} = 0$. This assumption does not limit the generality of the solution since one can reduce the flow velocity to zero at any horizon by transition to a uniformly moving reference system.

Suppose that in the upper half-space there is an incident plane wave with horizontal wave vector $\boldsymbol{\xi}$. Then in the lower half-space the sound field is a plane wave propagating towards negative z. It is convenient to choose a normalization factor such that the amplitude of this wave equals unity. According to (8.1.1) the sound pressure in the lower medium is given by

$$p = \exp\left[-ik_0(\zeta + \zeta_0)N_2 + i\boldsymbol{\xi} \cdot \boldsymbol{r}\right] \quad , \qquad \zeta \le -\zeta_0 \quad , \tag{10.2.1}$$

where the following designations are used

$$\zeta(z) = \varrho_2^{-1} \int_0^z \varrho(z_1)\beta^2(z_1)dz_1 \quad , \qquad \zeta_0 \equiv -\zeta(-H) > 0 \quad , \tag{10.2.2a}$$

209

$$k_0^2 N_2^2 = \beta_2^{-4}(k_2^2\beta_2^2 - \xi^2) \quad , \qquad \beta_2 = 1 - \boldsymbol{\xi}\cdot\boldsymbol{v}_{02}/\omega \quad . \tag{10.2.2b}$$

Inside the heterogeneous layer the sound pressure obeys the equation

$$\frac{\partial^2 \Phi}{\partial \zeta^2} + (k^2\beta^2 - \xi^2)\left(\frac{\varrho_2}{\varrho\beta^2}\right)^2 \Phi = 0 \quad , \qquad -\zeta_0 \le \zeta \le 0 \quad . \tag{10.2.3}$$

The quantities Φ and $\partial\Phi/\partial\zeta$ should be continuous at the layer boundary $\zeta = -\zeta_0$. With account of (10.2.1), these conditions give

$$\Phi(-\zeta_0) = 1 \quad , \qquad \frac{\partial\Phi(-\zeta_0)}{\partial\zeta} = -ik_0 N_2 \quad . \tag{10.2.4}$$

In the upper half-space

$$\Phi(\zeta) = A\exp(ik_0 N_1\zeta) + B\exp(-ik_0 N_1\zeta) \quad , \qquad k_0 N_1 = \sqrt{k_1^2 - \xi^2}\,\varrho_2/\varrho_1 \quad ,$$
$$\zeta \ge 0 \quad . \tag{10.2.5}$$

When the field inside the layer is known, the reflection and transmission coefficients, $V(\xi)$ and $W(\xi)$, can be readily found from the conditions of continuity of Φ and $\partial\Phi/\partial\zeta$ at $\zeta = 0$:

$$V(\xi) = \frac{A}{B} = \frac{k_0 N_1\Phi(0) - i\partial\Phi(0)/\partial\zeta}{k_0 N_1\Phi(0) + i\partial\Phi(0)/\partial\zeta} \quad , \tag{10.2.6}$$

$$W(\xi) = \frac{1}{B} = \frac{2k_0 N_1}{k_0 N_1\Phi(0) + i\partial\Phi(0)/\partial\zeta} \quad . \tag{10.2.7}$$

Equations (10.2.3, 4) can be considered as defining the intitial value problem for the function $\Phi(\zeta)$ inside the layer. This problem is equivalent to the integral equation

$$\Phi(\zeta) = 1 - ik_0 N_2 \cdot (\zeta + \zeta_0) - k_0^2 \int_{-\zeta_0}^{\zeta} (\zeta - u)N^2\Phi(u)du \quad . \tag{10.2.8}$$

Indeed, substitution of $\zeta = -\zeta_0$ into (10.2.8) gives the boundary conditions of (10.2.4), and double differentiation of both parts of (10.2.8) with respect to ζ reduces the integral equation to (10.2.3).

10.2.2 Iterative Solution of the Integral Equation

Equation (10.2.8) is a Volterra integral equation of the second kind. The theory for such equations is well developed, see for instance, [10.32]. They possess a valuable property, namely, successive approximations always converge to the solution of the equation. To be exact: the iterative sequence

$$\Phi^{(0)}(\zeta) = 1 - ik_0(\zeta + \zeta_0)N_2 \quad ,$$

$$\Phi^{(l)}(\zeta) = \Phi^{(0)}(\zeta) - k_0^2 \int_{-\zeta_0}^{\zeta} (\zeta - u)N^2\Phi^{(l-1)}(u)du \quad , \qquad l = 1, 2, \dots \tag{10.2.9}$$

converges absolutely and uniformly (with respect to ζ) to the solution of (10.2.8). The following estimate holds [Ref. 10.32; Sect. 17] [see also (9.1.15)] for the discrepancy between the lth iteration and the exact solution $\Phi(\zeta)$:

$$\Delta^{(l)} \equiv \max_{-\zeta_0 \le \zeta \le 0} |\Phi^{(l)}(\zeta) - \Phi(\zeta)| \le \sqrt{1+Q} \sum_{s=l+1}^{+\infty} \frac{Q^s}{s!} \quad , \tag{10.2.10}$$

where

$$Q = \max_{-H < z < 0} \left|(k^2\beta^2 - \xi^2)(\varrho\beta^2)^{-2}\left(\int_{-H}^{0} \varrho\beta^2 dz\right)^2\right| \quad . \tag{10.2.11}$$

By using a specific form of the kernel of the integral operator in (10.2.8) [namely, that the kernel is proportional to $(\zeta - u)$], one can prove that the stricter estimate of the error is

$$\Delta^{(l)} \le \sqrt{1+Q} \sum_{s=l+1}^{+\infty} \frac{Q^s}{(2s)!} \quad . \tag{10.2.12}$$

It follows from (10.2.12) that in order to obtain the necessary accuracy for $\Phi(\zeta)$, the number of required iterations is of the order of $Q^{1/2}$ if $Q \gg 1$. For thick layers the successive approximations method of (10.2.9) is, therefore, not effective. On the other hand, when $Q \ll 1$ the sequence of iterations converges rapidly.

10.2.3 Physical Consequences

Let us suppose that the inhomogeneous layer is thin compared to the wavelength of the incident wave, i.e., $k_1 H \ll 1$ and the quantities $k(z)$, k_1, and k_2, $\varrho(z)$, ϱ_1, and ϱ_2, as well as $\beta(z)$, 1, and β_2 are of the same order of magnitude. Then $Q \simeq (k_1 H)^2 \ll 1$ according to (10.2.11). For thin layers, l iterations enable us to find the field to within accuracy of $O[(k_1 H)^{2l+2}]$. After one iteration as defined by (10.2.9) and simple transformations, by using formula (10.2.6) we can find the reflection coefficient:

$$V(\xi) =$$
$$\frac{N_1 - N_2 - ik_0(N_1 N_2 \zeta_0 - s_0) + k_0^2(N_1 s_1 + N_2 s_1 + N_2 \zeta_0 s_0) + O\left(k_1^3 H^3\right)}{N_1 + N_2 - ik_0(N_1 N_2 \zeta_0 + s_0) + k_0^2(N_1 s_1 - N_2 s_1 - N_2 \zeta_0 s_0) + O\left(k_1^3 H^3\right)} \quad .$$
$$\tag{10.2.13}$$

Here

$$s_0 \equiv \int_{-\zeta_0}^{0} N^2(u)du = \frac{\varrho_2}{k_0^2} \int_{-H}^{0} \left(k^2 - \frac{\xi^2}{\beta^2}\right)\frac{dz}{\varrho} \quad ,$$

$$\tag{10.2.14}$$

$$s_1 \equiv \int_{-\zeta_0}^{0} uN^2(u)du = \frac{1}{k_0^2} \int_{-H}^{0} \left(k^2 - \frac{\xi^2}{\beta^2}\right)\frac{dz}{\varrho} \int_{0}^{z} \varrho(z_1)\beta^2(z_1)\,dz_1 \quad .$$

To within linear terms in the layer thickness, we obtain from (10.2.13) the simpler expression:

$$V(\xi) = V_0(\xi) + \frac{i[1 - V_0^2(\xi)]}{2k_0 N_2}\left[\int_{-H}^{0} \frac{\varrho_2}{\varrho}\left(k^2 - \frac{\xi^2}{\beta^2}\right)dz - k_0^2 N_2^2 \int_{-H}^{0} \frac{\varrho\beta^2}{\varrho_2}dz\right]$$

$$+ O(k_1^2 H^2) \quad , \tag{10.2.15}$$

where

$$V_0(\xi) = \frac{N_1 - N_2}{N_1 + N_2} \tag{10.2.16}$$

is the reflection coefficient when the layer thickness equals zero; that is, the usual Fresnel reflection coefficient at the interface of two homogeneous moving media with parameters corresponding to the upper and lower half-spaces. For an unmoving medium with constant density the result given in (10.2.15) was obtained by a number of authors [10.6; 10.33; 10.18, Chap. 3]. In the opposite case of a homogeneous moving medium analogous results were derived in [10.34] for several types of flow velocity profiles $v_0(z)$.

Keeping the second-order terms in the expansion of (10.2.13) in powers of $k_1 H$, in the case of a motionless medium we obtain:

$$V(\xi) = V_0(\xi) + [1 - V_0^2(\xi)]\frac{k_1^2 H}{2(k_2^2 - \xi^2)^{1/2}}\left\{i\left(M_1 + \frac{k_2^2 - \xi^2}{k_1^2}M_3\right)\right.$$

$$+ H\left[\sqrt{k_2^2 - \xi^2}\left(2M_2 + \frac{k_2^2 - \xi^2}{k_1^2}M_4\right)\right.$$

$$\left.\left. - \frac{k_1^2\varrho_1(M_1 + (k_2^2 - \xi^2)k_1^{-2}M_3)^2}{\varrho_2\sqrt{k_1^2 - \xi^2} + \varrho_1\sqrt{k_2^2 - \xi^2}}\right]\right\} + O(k_1^3 H^3) \quad . \tag{10.2.17}$$

Here we explicitly extract the dependence of V on ξ by using the designations

$$M_1 = (k_1^2 H)^{-1}\varrho_2 \int_{-H}^{0} dz[k^2(z) - k_2^2]/\varrho(z) \quad ,$$

$$M_2 = (k_1 H)^{-2} \int_{-H}^{0} dz\left[\int_{0}^{z}\varrho(u)du\right][k^2(z) - k_2^2]/\varrho(z) \quad ,$$

$$M_3 = \frac{1}{H}\int_{-H}^{0} dz\frac{\varrho_2^2 - \varrho^2(z)}{\varrho_2\varrho(z)} \quad ,$$

$$M_4 = \frac{1}{H^2}\left[\int_{-H}^{0}\frac{\varrho(z)}{\varrho_2}dz\right]^2 + \frac{2}{H^2}\int_{-H}^{0}\frac{dz}{\varrho(z)}\left[\int_{0}^{z}\varrho(u)du\right] \quad . \tag{10.2.18}$$

Note that the reflection coefficient depends only on some integral characteristics of the density, the sound speed, and the flow velocity in the layer. Thus, the wave effectively averages the medium parameters over distances that are small compared to its own vertical scale $2\pi(k_1^2 - \xi^2)^{-1/2}$. If there is no dissipation and no total reflection, V_0 is a pure real quantity, but the first-order correction in $k_1 H$ to V_0 is pure imaginary. This affects only the phase of $V(\xi)$, i.e., the presence of the layer has the same influence on reflection as a vertical shift of the reflecting interface.

In the case of a motionless medium, according to (10.2.17), the inhomogeneous layer is specified acoustically by four parameters which are independent of frequency and angle of incidence of the wave. In the linear approximation (with respect to $k_1 H$), one has only two parameters, namely, M_1 and M_3. In the case of a moving medium one should calculate the first integral in (10.2.15) separately for each ξ. Hence, the angular dependence of the reflection coefficient may be considerably more complex than in the case of a motionless medium. As to the frequency dependence of V under fixed angle of incidence, in the linear approximation (with respect to $k_1 H$) it is specified by the sole integral parameter [see (10.2.15), where $\xi \sim \omega$ and β is independent of frequency]. The power expansions (10.2.15, 17) in layer thickness are generally valid for *arbitrary* angles of incidence, i.e., arbitrary ξ. The only exception occurs in the specific case when $N_2 = 0$ at $\xi = k_1$, that is, in the limit of grazing incidence. In a motionless fluid, this is the case when the sound velocities in the homogeneous halfspaces $z > 0$ and $z < -H$ are equal, $c = c_1$. In a moving fluid, it happens for one or two values of the angle between the vectors ν_{02} and ξ provided $|c_1 - c_2| \leq \nu_{02}$, according to (10.2.2b). In the specific case considered, some coefficients in (10.2.15, 17) become singular as N_1 and N_2 tend to zero simultaneously. Note that, unlike (10.2.15, 17), Eq. (10.2.13) still holds in this case unless the three quantities, $s_0(\xi)$, $N_1(\xi)$, and $N_2(\xi)$, all turn to zero simultaneously for some $\xi = \xi_0$. (Clearly, the latter is a very particular, degenerate case.) Then third- and higher-order terms in $k_1 H$ become important in both nominator and denominator of (10.2.6) for $\xi \approx \xi_0$. A uniformly valid expression for the reflection coefficient, covering even the degenerate case, follows from (10.2.6) when further iterations (10.2.9) are used to calculate $\Phi(0)$ and $\partial\Phi(0)/\partial\zeta$.

This ends our discussion of (10.2.15, 17). A more detailed analysis was presented in [10.35, 63].

In the final part of this section we shall use the integral equation (10.2.8) to prove the statements made in Sect. 6.1 on the analytical dependence of the reflection and transmission coefficients as well as the sound pressure on the parameter ξ in fluids at rest. In this application it is important that the iterations in (10.2.9) converge for arbitrary layer thickness. Consider a region $|\xi| < \xi_0$ at the complex ξ plane (more precisely, at the Riemann surface). We substitute for the quantity Q in the estimate of (10.2.10) a larger quantity Q_0, where

$$Q_0 = \max_{-H < z < 0} \left[(|k^2| + \xi_0^2)\varrho^{-2} \left(\int_{-H}^{0} \varrho\, dz \right)^2 \right] . \tag{10.2.19}$$

Then it follows from (10.2.10) that the sequence $\Phi^{(l)}(\xi, \zeta)$ converges to the solution $\Phi(\xi, \zeta)$ absolutely and uniformly with respect to ξ belonging to the circle $|\xi| < \xi_0$.

The function $\Phi^{(0)}$ depends on ξ analytically. The function $\Phi^{(l)}$ is defined in (10.2.9) through the integral of $\Phi^{(l-1)}$ and, hence, is also an analytic function

[Ref. 10.36, Sect. 16]. Then $\Phi(\xi,\zeta)$ depends on ξ analytically as the limit of a uniformly convergent sequence. Because the value of ξ_0 was taken to be arbitrary, Φ is an analytical function of ξ for all finite ξ. The analyticity of the reflection and transmission coefficients then follows from (10.2.6,7).

10.3 Method of Successive Approximations for Weakly Reflecting Layers

10.3.1 Integral Equation for the Reflection Coefficient

Equation (10.1.3) can be written in the form

$$\frac{\partial}{\partial \zeta}\left[V \exp\left(-2ik_0 \int_{\zeta_0}^{\zeta} N d\zeta_1\right)\right] = \gamma(1 - V^2) \exp\left(-2ik_0 \int_{\zeta_0}^{\zeta} N d\zeta_1\right). \qquad (10.3.1)$$

The lower limit of integration ζ_0 in the exponential is arbitrary. Equation (10.3.1) with the boundary condition of (10.1.5) is equivalent to the integral equation

$$\exp\left(-2ik_0 \int_{\zeta_0}^{\zeta} N d\zeta_1\right) V(\zeta) = \int_{-\infty}^{\zeta} \gamma(1 - V^2) \exp\left(-2ik_0 \int_{\zeta_0}^{\zeta_2} N d\zeta_1\right) d\zeta_2 \ .$$

$$(10.3.2)$$

We shall solve this equation by again using the method of successive approximations. By assuming γ to be small and neglecting the right side in the zeroth approximation, we get $V^{(0)}(\zeta) = 0$. This corresponds to the (first) approximation of geometrical acoustics in which, as we have seen in Chap. 8, the wave is propagating in the medium without reflection.

Substituting $V = 0$ in the right-hand side of (10.3.2) and denoting

$$s(\zeta) = 2k_0 \int_{\zeta_0}^{\zeta} N d\zeta_1 \ , \qquad (10.3.3)$$

we get in the first- and higher-order approximations [10.20]

$$V^{(1)}(\zeta) = \exp[is(\zeta)] \int_{-\infty}^{\zeta} \gamma(\zeta_1) \exp[-is(\zeta_1)] d\zeta_1 \ ,$$

$$V^{(m+1)}(\zeta) = V^{(1)}(\zeta) - \exp[is(\zeta)] \int_{-\infty}^{\zeta} \gamma(\zeta_1)[V^{(m)}(\zeta_1)]^2 \exp[-is(\zeta_1)] d\zeta_1 \ ,$$

$$(10.3.4)$$

where $m = 1, 2, \ldots$ In fact, the integration in (10.3.4) is fulfilled only inside the inhomogeneous layer. When it is convenient, one can easily transform integrals over ζ in (10.3.3, 4) into integrals over z by using the identity $\varrho_0 d\zeta = \varrho\beta^2 dz$ from

(8.1.3). If the function $\gamma(\zeta)$ is bounded at all points, which according to (10.1.3) and (8.1.2) means that there are no discontinuities in the variation of the medium parameters nor turning points, nor horizons of resonance interaction (Sect. 8.1), then we obtain a converging sequence for $V(\zeta)$ as $m \to \infty$. This follows directly from the general convergence criteria in the method of successive approximations (see, for example, [Ref. 10.37; Part I, Chap. 1, Sect. 1] or [Ref. 10.38, Chap. 3]). The obtained sequence of approximations converges more rapidly the smaller the absolute value of the square of the reflection coefficient $|V|^2$. In particular, due to the smallness of γ a rapid convergence occurs for media with a large space scale of variability L. A generalization of this method for the case of reflection by elastic layers was presented in [10.39].

10.3.2 The Born Approximation

Consider a medium in which the effective refraction index differs slightly from a constant value $N_1 = (n_1^2 \beta_1^2 - \xi^2/k_0^2)^{1/2} \varrho_0/\varrho_1 \beta_1^2$, that is

$$N^2 = N_1^2[1 + \varepsilon(\zeta)] \quad , \quad |\varepsilon| \ll 1 \quad . \tag{10.3.5}$$

Variations in the density, the sound speed, or the flow velocity or in all three parameters can contribute to ε. Let the point ζ lie above the inhomogeneous layer. Then the upper limit of integration in (10.3.4) may be replaced by $+\infty$ without changing the values of the integrals. By integrating by parts the first approximation in (10.3.4) one finds

$$V^{(1)}(\zeta) = \frac{ik_0}{2} \exp[is(\zeta)] \int_{-\infty}^{+\infty} \ln(N^2/N_1^2) \exp[-is(\zeta_1)]N \, d\zeta_1 \quad . \tag{10.3.6}$$

In the case of small inhomogeneities $\ln(N^2/N_1^2) = O(\varepsilon)$. By using instead of ζ the integration variable z which is independent of the medium parameters and taking into account (10.3.3), for small ε we find from (10.3.6)

$$V(z) = \frac{ik_0}{2} \sqrt{n_1^2 \beta_1^2 - \xi^2/k_0^2} \int_{-\infty}^{+\infty} \varepsilon(z_1) \exp[2ik_0(z - z_1)]$$

$$\times \sqrt{n_1^2 \beta_1^2 - \xi^2/k_0^2}] dz_1 + O(\varepsilon^2) \quad . \tag{10.3.7}$$

The difference $V^{(2)} - V^{(1)}$ is proportional to ε^3, see (10.3.4). That is why the difference between $V^{(1)}$ and the limit of the iterative sequence may be neglected at sufficiently small ε. Thus, the main part of the inaccuracy in (10.3.7) is due to omitting the terms $O(\varepsilon^2)$ in (10.3.6).

The result given in (10.3.7) is called the *Born* (or *Rayleigh*) *approximation* for the reflection coefficient [Ref. 10.40; Chap. 3, Sect. 5], [Ref. 10.41, Chap. 4]. In this approximation the plane wave reflection coefficient is proportional to the spectrum density $\bar{\varepsilon}(\kappa)$ of the perturbation $\varepsilon(z)$, where

$$\bar{\varepsilon}(\kappa) \equiv \frac{1}{2\pi} \int_{-\infty}^{+\infty} \varepsilon(z_1) \exp(-i\kappa z_1) dz_1 \quad , \tag{10.3.8}$$

taken at κ, which is equal to twice the vertical component of the incident wave vector:

$$V(z) = i\pi\mu_1 \exp(2i\mu_1 z)\bar{\varepsilon}(2\mu_1) \quad , \quad \mu_1 \equiv \sqrt{k_0^2 n_1^2 \beta_1^2 - \xi^2} \quad . \tag{10.3.9}$$

The conditions under which the Born approximation is valid are rather restrictive. Not only must the medium perturbation at a point ($|\varepsilon| \ll 1$) be small, but also the reflection coefficient and the wave's phase perturbation $s(z) - 2\mu_1(z - z_0)$ throughout the medium. The two latter requirements imply restrictions on the inhomogeneous layer width.

Physically, the effects of the medium's heterogeneity on the incident (but not on the reflected) wave are omitted in the Born approximation. On the other hand one can show [Ref. 10.42, Sect. 25.2] that in the iterative sequence (10.3.4) the first approximation corresponds to taking into account single reflections in an inhomogeneous medium, the next approximations correspond to double reflections, triple reflections, etc. (cf. Sect. 8.3). There are no assumptions about the incident and reflected wave phases.

10.3.3 Numerical Example

As an example of the calculation of the reflection coefficient by the method of successive approximations, we consider the normal incidence of a wave on a layer $-H < z < 0$, in which the sound speed changes linearly:

$$c = c_1(1 - Az/H) \quad . \tag{10.3.10}$$

The density is assumed to be constant throughout the medium. In the half-spaces $z > 0$ and $z < -H$ the sound speed equals c_1 and $c_2 = c_1(1 + A)$, respectively. Note that the gradient of the sound speed is discontinuous at the layer boundaries. In the case considered the exact solution for the reflection coefficient is given by (3.6.23). We compare it to the results of the second approximation in both methods.

Figure 10.1a gives the results of the calculations of the modulus of the reflection coefficient $\varrho = |V|$ for $A = -\frac{1}{2}$. Along the ordinate ϱ is plotted on a linear scale, and $k_1 H$ is plotted along the abscissa on a logarithmic scale. Curve 1 is calculated from the exact formula. Curve 2 is determined by the method of successive approximations for weakly reflecting layers by (10.3.4). These curves are very close for all $k_1 H$ (the discrepancy between the curves is smaller than 1%). By its intrinsic nature the method of (10.3.4) is very suitable to find the positions of the zeros of the reflection coefficient. The approximate value of $k_1 H$ for which the reflection coefficient vanishes is within 0.02% the correct value. The ϱ values calculated from (10.2.17) are rather accurate for small $k_1 H$. But with increasing $k_1 H$ the accuracy drops rapidly and is about 1% at $k_1 H = \frac{\pi}{4}$. For larger $k_1 H$ the approximation (10.2.17) obtained under the assumption that $k_1 H \ll 1$ gives unsatisfactory results.

Figure 10.1.b corresponds to the case where $A = 6$, i.e., the sound speed gradient is considerably greater. As it is to be expected, (10.2.17) works well for small $k_1 H$ in this case, too. When $k_1 H = \pi$, the discrepancy between the exact and approximate solutions is 0.7%. In contrast, the method of (10.3.4) gives ϱ values within 1%

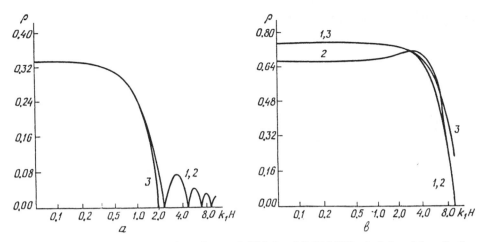

Fig. 10.1 Exact (*curve 1*) and approximate [*curves 2* (10.3.4) and *3* (10.2.17)] calculation of the reflection coefficient for a layer with the linear sound speed profile of (10.3.10) (a) $A = -\frac{1}{2}$ and (b) $A = 6$

error only when $k_1 H > 8.25$ that is, when the modulus of the reflection coefficient becomes sufficiently small.

The methods of successive approximations given by (10.2.9) and (10.3.4) are to some extent mutually complementary. The first one is capable of handling large gradients and discontinuities in the medium parameters, but is suitable only when the phase advance of the wave in the layer is small. The second method allows for large phase changes in the wave in inhomogeneous media, but poorly describes reflection at interfaces inside the layer. Apparently, it should be possible to obtain a more widely applicable version of this method by using the Riccati equation of (10.1.7) for the impedance which is a continuous function of the vertical coordinate instead of the Riccati equation given by (10.1.3) for $V(\zeta)$, which is discontinuous at horizons where $N(\zeta)$ is discontinuous.

In addition to the approaches described above, other methods have been used to calculate the reflection coefficient. In particular, a method using an integral equation for $\Phi(z)$, i.e., a vertical dependence of the acoustic pressure, was presented in [Ref. 10.41; Chap. 3, Sect. 5] more generally and with more details than in Sect. 10.2. One can obtain an estimate of the reflection coefficient of a medium with piecewise-continuous parameters by averaging them over regions of continuity and then matching solutions of the wave equation (corresponding to the averaged medium) at the interface [10.43]. This gives an exact value for the reflection coefficient of a discretely layered medium, and in the general case is suitable for a set of thin or weakly inhomogeneous layers. *Variational* approaches to a calculation of the reflection coefficient were considered in [Ref. 10.18, Chap. 4].

10.4 Reflection at Interfaces in Continuously Layered Media

The WKB approximation and the results of the reference equation method give a rather complete description of wave propagation in a medium whose parameters are sufficiently smooth functions of z and vary only slightly over distances of the order of the wavelength. There are many physical problems in which the properties of the media satisfy these requirements in separate layers, but the density, sound speed, and flow velocity or the derivatives of these parameters are discontinuous at the layer interfaces. In ocean acoustics the surface of the sea bottom is as an example of such an interface. The media which lie on either side of this interface are often considered to be continuously layered, but at the surface itself the sound speed and the density are discontinuous. In seismology, Moho and core-mantle interfaces in the Earth are boundaries of this type. Wave propagation in such *smoothly layered media with interfaces* is the topic of the present section.

In numerical simulation of wave propagation, effective algorithms are usually based on approximating the given medium by a set of layers for which exact or approximate analytical solutions of the wave equation are known [10.44–46] and [Ref. 10.47, Chaps. 7, 9; Ref. 10.64, Chaps. 4, 5]. Hence, artificial interfaces between the layers may arise. If the medium is approximated by a set of the homogeneous layers then the density of the medium or speeds of sound and flow are discontinuous at the interfaces. The same accuracy of the approximation can be achieved with significantly fewer layers by using linear or other, more smooth approximations, when the derivatives of the medium parameters with respect to z, but not the parameters themselves, are discontinuous at the interfaces. Interfaces of this kind are called *weak*. The origin of this term will become clear later. The same term also applies to interfaces at which the parameters themselves are discontinuous but where the relative values of the discontinuities are small compared to unity.

10.4.1 General Approach

In principle, calculating the reflection coefficient from a smoothly layered medium with interfaces is straightforward. As in Sect. 2.5, let there be $n - 1$ fluid layers between two homogeneous fluid half-spaces which are labeled by the indexes 1 and $n + 1$. The density $\varrho_j(z)$ and the speeds of sound $c_j(z)$ and velocities of flow $v_{0j}(z)$ are smooth functions, defined in the intervals $z_{j-1} < z < z_j$, $j = 2, \ldots, n$. The horizon $z = z_j$ is the upper boundary of the jth layer. Consider reflection of the plane wave incident from above on the boundary $z = z_n$ of the upper layer. We designate $\boldsymbol{\xi}$ the horizontal wave vector and $u_j(z)$ the projections of the vectors $\boldsymbol{v}_{0j}(z)$ on the direction $\boldsymbol{\xi}$. The effective refraction index in each layer is given by (8.1.2, 3) in which it is convenient to take $\varrho_0 \equiv \varrho_1$, $z_0 \equiv z_1$. We also denote $\zeta_j \equiv \zeta(z_j)$ as the layer boundaries in terms of the vertical coordinate ζ. The general solution of the wave equation inside each layer is given by the equations in Chaps. 8 and 9:

$$\Phi_j(\zeta) = A_j^{(1)} \Phi_j^{(1)}(\zeta) + A_j^{(2)} \Phi_j^{(2)}(\zeta) \quad , \quad A_j^{(1,2)} = \text{const} \quad . \tag{10.4.1}$$

If there are no turning points and no horizons of resonance interaction then $\Phi_j^{(1,2)}$ are the WKB solutions; if there is only one turning point, then $\Phi_j^{(1,2)}$ can be expressed in terms of the Airy functions, etc. In general, the linearly independent solutions $\Phi_j^{(1,2)}$ are an asymptotic series in powers of k_0^{-1}. In the lower medium ($z < z_1$) only the wave leaving the boundary, given by

$$\Phi_1(\zeta) = W \exp(-ik_0 N_1 \zeta) \tag{10.4.2}$$

is present.

To find the reflection coefficient we use the same recursive scheme for calculating input impedances which was applied in Chap. 2 to the case of discretely layered medium. Let $Z_{in}^{(j)}$ be the wave impedance at $\zeta = \zeta_j$. This quantity is the input impedance of the set of $(j - 1)$ layers lying on the half-space. According to (10.1.6),

$$Z(\zeta) = -i\omega\varrho_1 \left(\Phi_j^{(1)} + \frac{A_j^{(2)}}{A_j^{(1)}} \Phi_j^{(1)} \right) \bigg/ \left(\frac{\partial \Phi_j^{(1)}}{\partial \zeta} + \frac{A_j^{(2)}}{A_j^{(1)}} \frac{\partial \Phi_j^{(2)}}{\partial \zeta} \right) ,$$

$$\zeta_{j-1} \leq \zeta \leq \zeta_j . \tag{10.4.3}$$

At $\zeta = \zeta_{j-1}$ we have $Z = Z_{in}^{(j-1)}$, and $Z = Z_{in}^{(j)}$ at $\zeta = \zeta_j$. Using (10.4.3) to express the quotient $A_j^{(2)}/A_j^{(1)}$ in terms of $Z_{in}^{(j-1)}$ and then substituting the quotient back into (10.4.3), after simple transformations we find the following relation between impedances at the upper and lower boundaries of the jth layer:

$$Z_{in}^{(j)} = i\omega\varrho_1(a_1 Z_{in}^{(j-1)} + a_2 i\omega\varrho_1)/(a_3 i\omega\varrho_1 + a_4 Z_{in}^{(j-1)}) , \tag{10.4.4}$$

where

$$a_1 = \Phi_j^{(1)}(\zeta_j) \frac{\partial \Phi_j^{(2)}(\zeta_{j-1})}{\partial \zeta} - \Phi_j^{(2)}(\zeta_j) \frac{\partial \Phi_j^{(1)}(\zeta_{j-1})}{\partial \zeta} ,$$

$$a_2 = \Phi_j^{(1)}(\zeta_j) \Phi_j^{(2)}(\zeta_{j-1}) - \Phi_j^{(1)}(\zeta_{j-1}) \Phi_j^{(2)}(\zeta_j) ,$$

$$a_3 = \Phi_j^{(1)}(\zeta_{j-1}) \frac{\partial \Phi_j^{(2)}(\zeta_j)}{\partial \zeta} - \Phi_j^{(2)}(\zeta_{j-1}) \frac{\partial \Phi_j^{(1)}(\zeta_j)}{\partial \zeta} ,$$

$$a_4 = \frac{\partial \Phi_j^{(1)}(\zeta_{j-1})}{\partial \zeta} \frac{\partial \Phi_j^{(2)}(\zeta_j)}{\partial \zeta} - \frac{\partial \Phi_j^{(1)}(\zeta_j)}{\partial \zeta} \frac{\partial \Phi_j^{(2)}(\zeta_{j-1})}{\partial \zeta} . \tag{10.4.5}$$

Taking $\Phi_j^{(1,2)}(\zeta) = \exp[\pm ik_0(\zeta - \zeta_{j-1})N_j]$, one can check that for homogeneous layes (10.4.4, 5) lead to the known result of (2.5.3).

According to (10.1.6) and (10.4.2), the input impedance of the lower half-space equals

$$Z_{in}^{(1)} = \frac{\omega\varrho_1}{k_0 N_1} . \tag{10.4.6}$$

By successively using (10.4.4) one finds $Z_{in}^{(n)}$. Then the reflection coefficient is given by (2.5.4), where

$$Z_{n+1} = \frac{\omega\varrho_1}{k_0 N_{n+1}} \tag{10.4.7}$$

is the impedance of the upper half-space. The transmission coefficient may be calculated with the help of (2.5.11).

Generalization of this analysis to the case where the half-spaces $\zeta > \zeta_n$ and $\zeta < \zeta_1 = 0$ are smoothly inhomogeneous is straightforward. Note that if there are no turning points and $\beta \equiv 1 - \xi u/\omega > 0$ in the given half-spaces, one may use the above results with the understanding that N_1 and N_{n+1} in (10.4.6, 7) are $N_1(\zeta_1)$ and $N_{n+1}(\zeta_n)$.

A scattering matrix calculation for smoothly layered elastic media with interfaces is analogous to that used above for fluids and is a generalization of the approach used in Chap. 4 to the case of discretely layered media. This problem is studied in [Ref. 10.47, Chap. 9].

10.4.2 Sound Reflection at Weak Interface

We now consider the case of a sole interface in a smoothly layered fluid in more detail. Let the interface be situated at $\zeta = 0$, and, for simplicity, assume that there are no turning points close to the interface. Then at both sides of the interface the solutions $\Phi_j^{(1,2)}$ in (10.4.1) may be taken in the WKB approximation. According to (8.1.9)

$$\Phi_j^{(1,2)}(\zeta) = N_j^{-1/2} \exp\left[\pm i k_0 \int_0^\zeta (1 + \varepsilon_j) N_j d\zeta\right] \quad , \quad j = 1, 2 \quad , \tag{10.4.8}$$

are the linearly independent solutions of the wave equation. Equation (10.4.8) gives $\Phi_j^{(1,2)}$ to within an accuracy of a factor of $1 + O[(k_0 L)^{-2}]$, where L is a typical space scale of the medium variability. The quantity $\varepsilon_j = O[(k_0 L)^{-2}]$ is defined by (8.1.10) with $N(\zeta) = N_j(\zeta)$. By using (10.1.6) and (10.4.1) to calculate the wave impedance for $\zeta > 0$ and $\zeta < 0$, and then equating the resulting expressions at the interface, one finds the following relation for four amplitude coefficients $A_j^{(1,2)}$:

$$\frac{A_1^{(1)} - A_1^{(2)}}{A_1^{(1)} + A_1^{(2)}} i k_0 N_1 (1 + \varepsilon_1) - \frac{1}{2N_1} \frac{\partial N_1}{\partial \zeta}$$

$$= \frac{A_2^{(1)} - A_2^{(2)}}{A_2^{(1)} + A_2^{(2)}} i k_0 N_2 (1 + \varepsilon_2) - \frac{1}{2N_2} \frac{\partial N_2}{\partial \zeta} . \tag{10.4.9}$$

Here and in (10.4.11, 12, 19, 20) the quantities $\varepsilon_{1,2}$ and $N_{1,2}$, and the derivatives of $N_{1,2}$ are taken at $\zeta = 0$. The second relation between the amplitude coefficients follows from the condition of sound pressure continuity at the interface

$$A_1^{(1)} + A_1^{(2)} = A_2^{(1)} + A_2^{(2)} . \tag{10.4.10}$$

Let the coefficients $A_2^{(1,2)}$ be known. Then one finds from (10.4.9, 10)

$$A_1^{(1,2)} = \frac{1}{2}\left(A_2^{(1)} + A_2^{(2)}\right) \pm \left[\left(A_2^{(1)} - A_2^{(2)}\right)\frac{N_2}{2N_1} - \left(A_2^{(1)} + A_2^{(2)}\right)\right.$$

$$\left. \times \frac{1}{8 i k_0 N_1}\left(\frac{1}{N_2^2}\frac{\partial N_2^2}{\partial \zeta} - \frac{1}{N_1^2}\frac{\partial N_1^2}{\partial \zeta}\right)\right] . \tag{10.4.11}$$

These expressions are valid up to terms on the order of $(k_0 L)^{-2}$.

When the wave is incident from above $A_2^{(1)} = 0$ because of the condition of limiting absorption as $\zeta \to -\infty$. If one considers, by convention, the wave $\Phi_1^{(2)}$ for which the phase increases with decreasing ζ as the incident one and the wave $\Phi_1^{(1)}$ as the reflected one, then the ratio $A_1^{(1)}/A_1^{(2)}$ is the reflection coefficient V at $\zeta = 0$. This definition of the reflection coefficient is perfectly rigorous if the half-space $\zeta > \zeta_1$ is homogeneous. It follows from (10.4.9) that

$$V = \frac{[N_1(1 + \varepsilon_1) - N_2(1 + \varepsilon_2) - (N_2^{-2}\partial N_2^2/\partial\zeta - N_1^{-2}\partial N_1^2/\partial\zeta)/4ik_0]}{[N_1(1 + \varepsilon_1) + N_2(1 + \varepsilon_2) + (N_2^{-2}\partial N_2^2/\partial\zeta - N_1^{-2}\partial N_1^2/\partial\zeta)/4ik_0]} .$$

$$(10.4.12)$$

It is worth noting that retaining in (10.4.8) more terms of the expansion (8.1.6) in powers of k_0^{-1} leads to the corrections $O[(k_0 L)^{-2}]$ in (10.4.9) and $O[(k_0 L)^{-3}]$ in (10.4.12). Hence (10.4.12) gives the reflection coefficient up to terms of the order of $(k_0 L)^{-2}$ inclusive. We now consider three cases.

At first let $N_1 \neq N_2$. Then up to the small corrections $O[(k_0 L)^{-1}]$ we have $V = (N_1 - N_2)/(N_1 + N_2)$, that is, the reflection coefficient coincides with the Fresnel coefficient. In other words, the local plane waves of (10.4.8) are reflected at the usual interfaces in the same way as plane waves at the boundary of homogeneous media. If there is a turning point $\zeta(z_t) \equiv \zeta_t > 0$ in the upper half-space, then $N_1(\zeta) = i|N_1(\zeta)|$ for $\zeta < \zeta_t$ and the amplitude of the wave at the interface $\zeta = 0$ is small. Consider the effect on the sound field of a reflecting boundary situated behind a turning point. If the boundary is close to the turning point, the expressions containing the Airy functions should be used for $\Phi^{(1,2)}(\zeta)$ (Chap. 9). Here we study the simpler case, where inequality (8.1.18) holds fo the horizons z_1 and z_t. Then for $\zeta_t > \zeta \geq 0$ the vertical dependence of the sound field may be cast in the form

$$\Phi(\zeta) = |N|^{-1/2}\left[B_1 \exp\left(-k_0 \int_\zeta^{\zeta_t} |N|d\zeta\right) + B_2 \exp\left(k_0 \int_\zeta^{\zeta_t} |N|d\zeta\right)\right] ,$$

$$(10.4.13)$$

where the exponentially damped wave with amplitude B_1 is the "incident" one and that with amplitude B_2, which would be absent were it not for the boundary, is the "reflected" wave. Above the turning point ($\zeta > \zeta_t$) we have

$$\Phi(\zeta) = N^{-1/2}\left[B_3 \exp\left(ik_0 \int_{\zeta_t}^\zeta N\,d\zeta\right) + B_4 \exp\left(-ik_0 \int_{\zeta_t}^\zeta N\,d\zeta\right)\right]. \quad (10.4.14)$$

According to (9.2.9), the amplitude coefficients above and below the turning horizon are related through

$$B_3 = \exp\left(-i\tfrac{\pi}{4}\right)(B_1 + 0.5iB_2) \quad, \quad B_4 = \exp\left(i\tfrac{\pi}{4}\right)(B_1 - 0.5iB_2) . \quad (10.4.15)$$

The reflection coefficient at the horizon $\zeta = 0$ was found earlier:

$$V(0) \equiv B_2 \exp\left(k_0 \int_0^{\zeta_t} |N|d\zeta\right) \bigg/ B_1 \exp\left(-k_0 \int_0^{\zeta_t} |N|d\zeta\right) = \frac{N_1 - N_2}{N_1 + N_2} .$$

$$(10.4.16)$$

Expressing the ratio B_3/B_4 in terms of B_1/B_2 with the help of (10.4.15), we find for the reflection coefficient at the horizon $\zeta > \zeta_t$

$$V(\zeta) = \frac{B_3}{B_4}\exp\left(2ik_0\int_{\zeta_t}^{\zeta} N\,d\zeta\right)$$

$$= \frac{1+i\alpha}{1-i\alpha}\exp\left(2ik_0\int_{\zeta_t}^{\zeta} N\,d\zeta - i\frac{\pi}{2}\right) \quad , \tag{10.4.17}$$

$$\alpha = \frac{1}{2}\cdot\frac{N_1-N_2}{N_1+N_2}\exp\left(-2k_0\int_{0}^{\zeta_t} |N|\,d\zeta\right) \quad . \tag{10.4.18}$$

We have $|\alpha| \ll 1$ when the WKB approximation is valid. The difference between the reflection coefficients by a turning point of (9.2.10) and V in (10.4.17) is exponentially small. The difference vanishes if $N_1 = N_2$ or $\zeta_t \to +\infty$ as could be expected. If N_2 is real at $\zeta < 0$, i.e., the wave transmits energy to $\zeta = -\infty$, then $\mathrm{Im}\{\alpha\} > 0$ and it follows from (10.4.17) that $|V(\zeta)| < 1$.

We now turn to the second case. For weak interfaces, terms in (10.4.12) that are small compared to unity are significant. If $N_1 = N_2$ but $\partial N_1/\partial\zeta \neq \partial N_2/\partial\zeta$ at $\zeta = 0$ (i.e., c, ϱ, and u are continuous at the interface, but the gradient of one or more of these parameters is discontinuous) then

$$V \approx \frac{\partial N_1^2/\partial\zeta - \partial N_2^2/\partial\zeta}{8ik_0 N_1^3} \quad . \tag{10.4.19}$$

In this case the reflection coefficient is proportional to the change in the gradient of the squared effective refraction index at the interface and is inversely proportional to the frequency. The reflection coefficient is small, as under the conditions of the validity of the WKB approximation $k_0 L \gg 1$, $N_1^2 \gg (k_0 L)^{-2/3}$, see (8.1.18). Now we see that it is weakly reflecting boundaries which are called weak interfaces. Note that at $\beta \equiv 1$ (10.4.19), when applied to the case of an interface between a homogeneous medium and a half-space whose refraction index squared is linearly dependent on z, reduces to the result (3.5.34) obtained from the exact solution of the problem.

In the third case, when $N_1 = N_2$, $\partial N_1/\partial\zeta = \partial N_2/\partial\zeta$, but $\partial^2 N_1/\partial\zeta^2 \neq \partial^2 N_2/\partial\zeta^2$ at the interface, after accounting for (8.1.10) one finds from (10.4.12)

$$V \approx \frac{\partial^2 N_2^2/\partial\zeta^2 - \partial^2 N_1^2/\partial\zeta^2}{16k_0^2 N_1^4} \quad . \tag{10.4.20}$$

The reflection coefficient is proportional to the change in the second derivative of N^2 at the interface; V is much smaller than in (10.4.19) and vanishes more rapidly with increasing frequency. Taking into account the appropriate number of terms in the series (8.1.6), one can show that, when all the derivatives of N^2 up to $\partial^l N^2/\partial\zeta^l$ are continuous and when $\partial^{l+1} N^2/\partial\zeta^{l+1}$, $l = 0, 1, 2, \ldots$ is discontinuous, the reflection coefficient is proportional to the change in the $(l+1)$th derivative at the interface and is inversely proportional to $N_1^2(k_0 N_1)^{l+1}$. An analogous result for the case of motionless media with constant density was obtained in [10.48] and [Ref. 10.40, Chap. 3]. Such interfaces are called weak interfaces of the $(l+1)$th order. They arise when all derivatives of ϱ, c and v_0 with respect to z up to the lth are continuous and at least one of the quantities $\partial^{l+1}\varrho/\partial z^{l+1}$, $\partial^{l+1}c/\partial z^{l+1}$ or $\partial^{l+1}v_0/\partial z^{l+1}$ is discontinuous at some horizon.

When the medium parameters are infinitely differentiable, all the functions y_m in (8.1.7) are continuous and according to the WKB approximation there is no wave reflection. But, in general, infinitely differentiable profiles are not reflectionless (see, for example, (3.4.28) for the reflection coefficient from the Epstein layer considered in Chap. 3). Here we encounter a difference between an exact solution and an asymptotic one even when the asymptotic solution is derived with an infinite number of series terms. This difference could have been anticipated. For example, although the function $\exp(-w^{-2})$ is smooth, it is not equal to the sum of its Taylor series in the vicinity of the point $w = 0$. (This sum is identically equal to zero.) Quite analogously, the WKB approximation enables us to compute all the coefficients in the expansion of V in powers of k_0^{-1} but it treats a reflection coefficient of the type $\exp(-\alpha k_0 L)$, where $\alpha = \mathrm{const} > 0$ as if it were zero. It is thus clear that when the frequency tends to infinity the reflection coefficient from an infinitely differentiable profile without turning points vanishes more rapidly than any finite power of $1/k_0$. Under the somewhat more restrictive condition that the medium parameters be *analytic* functions of z, it can be shown [10.49] that $V = O[\exp(-\alpha k_0 L)]$ at $k_0 L \to \infty$.

We now continue tó consider reflection at weak interfaces. At a weak interface of first order the expressions for the reflection coefficient (10.4.19) and for the wave amplitudes (10.4.11) do not contain the functions $\varepsilon_j(\zeta)$. But for $\zeta \neq 0$ account of ε_j in (10.4.8) leads to corrections of the same order $(k_0 L)^{-1}$ as reflection at a weak interface. Using concepts introduced in Sect. 8.3, this fact may be interpretated as follows: the sum of first order waves arising in an inhomogeneous medium under incident wave propagation are comparable in amplitude with the wave reflected by a weak interface.

When one needs to account for reflection at a weak interface of the second order, the accuracy of (10.4.8) becomes insufficient to describe the sound field and one should use the more exact formulas given in (8.1.9). In general, when taking into account reflections at weak interfaces of the lth order, one should keep $l + 2$ terms in the series (8.1.6).

A different approach to the analysis of the field in the vicinity of a weak interface in a smoothly layered medium is to proceed from the Riccati equation (10.1.3) for the reflection coefficient and to apply the method of successive approximations presented in Sect. 10.3. Then, if the upper medium is homogeneous, expressions for the reflection coefficient analogous to those obtained above follow. See, e.g., [Ref. 10.42; Sect. 25.7], where reflection by a half-space with a linearly dependent square of the refraction index is considered by this method. In general, however, the two approaches lead to different results for a reflection coefficient determined inside an inhomogeneous medium. Consider an example.

It was shown in Sect. 10.1.2 that the reflection coefficient is a continuous function of ζ if the effective refraction index N is continuous. In particular, V is not discontinuous at weak interfaces. Thus, in the framework of the approach based on the Riccati equation there is no reflection at any boundary except for the usual interface at which N is discontinuous, and only inhomogeneous layers create reflected waves. This conclusion obviously contradicts (10.4.19, 20). The contradiction is still being debated, see [Ref. 10.42, Sect. 25.7] and [Ref. 10.5, Chap. 4], where contrary opinions are argued.

In fact, the contradiction between the approaches is due to a difference in terminology. It can be shown that both methods give identical expressions for the total sound field. But the *definitions* for the reflection coefficient in heterogeneous media differ. Indeed, in Sect. 10.1 incident and reflected waves were defined by using (10.1.1), which require that the vertical dependences of the acoustic pressure and the z-component of the particle displacement be proprotional. In the present section we use the sign of the z-component of the wave vector to distinguish between incident and reflected waves. Both definitions seem to be quite natural, but in inhomogeneous media they are consistent only *in the first approximation* of the WKB method. So, in the wave $\Phi_1^{(2)}(\zeta)$ given by (10.4.8), according to the exact expression (8.3.1), the vertical dependence of the z-component of the particle displacement is given by

$$f_1^{(2)} \equiv \partial \Phi_1^{(2)}/\partial \zeta = ik_0[N_1(1 + \varepsilon_1) - (2ik_0N_1)^{-1}\partial N_1/\partial \zeta]\Phi_1^{(2)} \quad . \tag{10.4.21}$$

This expression is consistent with the definition of a reflected wave, given in (10.1.1), only if terms in $O[(k_0L)^{-1}]$ are negligible compared to unity.

This illustrates once again the ambiguity incurred in separating a sound field into direct and inverse waves, as pointed out in Sect. 10.1.3. Finally, which of the two definitions of the reflection coefficient is used is insignificant since they differ only in inhomogeneous parts of the medium. In the present section, incident and reflected waves are defined according to the sign of their phases because this leads to a graphic description of changes occurring at a weak interface in high order terms of the high-frequency asymptotic expansion of the sound field.

To avoid confusion, it should be emphasized that the concept of waves which are reflected at usual and weak interfaces only is applicable provided $k_0L \gg 1$, i.e., when the medium varies slowly between the interfaces. If around an interface in a layer that is thin compared to the wavelength $N(\zeta)$ is smoothed so that it becomes infinitely differentiable, then according to the results of Sect. 10.2 the sound field is essentially unchanged. Though the interface disappears, equivalent reflection is supplied by the layer with smooth parameters since the typical scale of changes in N in this layer is not large compared to k_0^{-1}.

The effect of weak interfaces on a field generated by a *point source* calculated in the geometrical-acoustics approximation was studied in [10.44, 50–53]. Reference [10.54] shows that in the case of waveguide propagation the WKB approximation becomes valid up to much greater distances from the source when weak interfaces, which arise in the approximation of the sound speed profile, are taken into account.

10.4.3 Ray Interpretation

Now we shall consider the case when the WKB approximation is valid in each layer between the interfaces and only the leading term of the high-frequency asymptotic expansion of the sound field needs to be known. Then the general expressions (10.4.4, 5) can be significantly simplified. Letting $\varepsilon_j = 0$ in (10.4.8) and substituting the obtained $\Phi_j^{(1,2)}$ into (10.4.5), one finds the coefficients $a_{1,2,3,4}$ to within an accuracy to the factor $1 + O[(k_0 L)^{-1}]$. After simple transformations the impedance re-counting of (10.4.4) takes the form

$$Z_{in}^{(j)} = \frac{Z_u^{(j)}[Z_{in}^{(j-1)} - iZ_d^{(j)} \tan \varphi]}{Z_d^{(j)} - iZ_{in}^{(j-1)} \tan \varphi} \quad , \tag{10.4.22}$$

where

$$Z_u^{(j)} = \frac{\omega \varrho_1}{k_0 N_j(\zeta_j)} \quad , \quad Z_d^{(j)} = \frac{\omega \varrho_1}{k_0 N_j(\zeta_{j-1})} \quad , \quad \varphi = k_0 \int_{\zeta_{j-1}}^{\zeta_j} N_j(\zeta) d\zeta \quad . \tag{10.4.23}$$

The quantities $Z_u^{(j)}$ and $Z_d^{(j)}$ are equal to the plane-wave impedances in a homogeneous medium with the same parameters as the jth layer has close respectively to its upper or lower boundary. The physical meaning of φ is that it is a phase advance of the wave after a single pass through the layer.

Equation (10.4.22) is quite similar to the analogous result in (2.5.3) for a discretely layered medium. To the given level of approximation, a smooth inhomogeneous layer differs from a homogeneous layer with the same value of phase advance φ in the fact that the properties of the former, near to the upper and lower boundaries, are, in general, different and specified by two impedances $Z_d^{(j)}$ and $Z_u^{(j)}$ while the latter is specified by a unique impedance Z_j, which is used in (2.5.3).

According to (10.4.22) and (2.5.4) the reflection coefficient from a layer ($j = 2$) bounded by two homogeneous or smoothly layered half-spaces ($j = 1, 3$, Fig. 2.4) equals

$$V = \frac{Z_1 Z_u^{(2)} - Z_3 Z_d^{(2)} - i \tan \varphi(Z_u^{(2)} Z_d^{(2)} - Z_1 Z_3)}{Z_1 Z_u^{(2)} + Z_3 Z_d^{(2)} - i \tan \varphi(Z_u^{(2)} Z_d^{(2)} + Z_1 Z_3)} \quad , \tag{10.4.24}$$

where $Z_1 = \omega \varrho_1 / k_0 N_1(\zeta_1)$, $Z_2 = \omega \varrho_1 / k_0 N_3(\zeta_2)$. The reflection coefficient has poles at ξ values obeying the equation

$$\tan \varphi = \frac{-i(Z_1 Z_u^{(2)} + Z_3 Z_d^{(2)})}{Z_u^{(2)} Z_d^{(2)} + Z_1 Z_3} \quad . \tag{10.4.25}$$

At such ξ the sound pressure in this system is finite when there is no incident wave. The field may be interpreted to be a surface or leaky wave by an observer outside the layer (see Sect. 4.4) and is considered to be a normal mode [Ref. 10.40, Chap. 1, Sect. 4], [Ref. 10.42, Chap. 5], [Ref. 10.65, Chap. 4] when one is inside the layer. Equation (10.4.25) is a dispersion equation of these waves. When the layer has rigid ($Z_{1,3} \to \infty$) or pressure release ($Z_{1,3} \to 0$) boundaries the dispersion equation takes the form

$$k_0 \int_0^{\zeta_2} N_2(\zeta) d\zeta \equiv k_0 \int_{z_1}^{z_2} (n^2 \beta^2 - \xi^2 / k_0^2)^{1/2} dz = \pi l \quad , \quad l = 1, 2, \dots \quad . \tag{10.4.26}$$

225

Detailed analysis of normal modes in a smoothly layered medium is presented in [Ref. 10.42, Chap. 7] and [Ref. 10.55].

The result (10.4.22, 23) corresponds to a remarkably simple ray representation of wave propagation in smoothly layered media with interfaces. In this representation the rays are refracted without reflections in layers between interfaces; at the interface, where the effective refraction index changes discontinuously from N_1 to N_2, the incident ray is split into reflected and transmitted rays. The ratios of the amplitudes of the reflected and transmitted rays to the amplitude of the incident ray are equal to the plane wave reflection and transmission coefficients at a boundary of two homogeneous media with parameters N_1 and N_2. Let us apply these ray considerations to the calculation of the reflection coefficient of the layer, V. Summing up the complex amplitudes of rays with different numbers of reflections at the upper and lower boundaries and following the discussion given in Sect. 2.4 one again finds for V expression (2.4.17):

$$V = \frac{V_{32} + V_{21} \exp(2i\varphi)}{1 + V_{32} V_{21} \exp(2i\varphi)} \quad .$$

Here the reflection coefficients of a ray V_{32} at the interface of media 3 and 2 and V_{21} at the interface of media 2 and 1 are equal to

$$V_{32} = \frac{N_3(\zeta_2) - N_2(\zeta_2)}{N_3(\zeta_2) + N_2(\zeta_2)} \quad , \quad V_{21} = \frac{N_2(\zeta_1) - N_1(\zeta_1)}{N_2(\zeta_1) + N_1(\zeta_1)} \quad . \tag{10.4.27}$$

The phase advance φ is given by (10.4.23). It is not difficult to check that substitution of (10.4.27) into (2.4.17) gives the same result as (10.4.24) obtained above by the impedance re-counting procedure.

Postscript. To those readers who have stayed the course to the very end the authors would like to express their thanks. They hope that this book has been of value to all readers and look forward to meeting many again in their next book *Acoustics of Layered Media II: Point Sources and Bounded Beams*, published as volume 10 in Springer Series on Wave Phenomena in 1992. It presents many interesting transformations and applications of the plane-wave theory discussed in this volume. A new augmented and updated edition of the monograph is now in preparation.

References

Chapter 1

1.1 L.D. Landau, E.M. Lifshitz: *Course of Theoretical Physics*, Vol. 6, Fluid Mechanics (Pergamon, New York 1982) Sect. 1, 2

1.2 M.A. Isakovich: *Obschaya Akustika* (General acoustics) (Nauka, Moscow 1973)

1.3 O.A. Godin: Dokl. Akad. Nauk SSSR **293**, 63–67 (1987)
O.A. Godin: O Volnovom Uravnenii dlya Zvuka v Nestatsionarnoi Dvizhuscheisya Srede (On a wave equation for sound in a nonstationary moving medium) in *Akustika Okeanskoi Sredy* (Acoustics of oceanic medium), ed. by L.M. Brekhovskikh, I.B. Andreeva (Nauka, Moscow 1989) pp. 217–220

1.4 L.M.B.C. Campos: J. Sound Vibr. **110**, 41–57 (1986)

1.5 L. Brekhovskikh, V. Goncharov: *Mechanics of Continua and Wave Dynamics*, Springer Ser. Wave Phenom., Vol. 1 (Springer, Berlin, Heidelberg 1985)

1.6 E. Gossard, W. Hooke: *Waves in the Atmosphere* (Elsevier, New York 1975)

1.7 I. Tolstoy: *Wave Propagation* (McGraw-Hill, New York 1973)

1.8 W.S. Wladimirov: *Gleichungen der Mathematischen Physik* (Deutscher Verlag der Wissenschaften, Berlin 1972) Sect. 30
F.B. Jensen, W.A. Kuperman, M.B. Porter, H. Schmidt: *Computational Ocean Acoustics* (AIP Press, New York 1994) Sect. 2.3

1.9 V.G. Gavrilenko, L.A. Zelekson: Akusth. Zh. **23**, 867–872 (1977) [English transl.: Sov. Phys.-Acoust. **23**, 497–499 (1977)]

1.10 A.L. Fabrikant: Akust. Zh. **22**, 107–114 (1976) [English transl.: Sov. Phys.-Acoust. **22**, 56 (1976)]

1.11 L.M. Lyamshev: Akust. Zh. **28**, 367–374 (1982) [English transl.: Sov. Phys.-Acoust. **28**, 217–221 (1982)]

1.12 C. Yeh: J. Acoust. Soc. Am. **43**, 1454–1455 (1968)

1.13 O.A. Godin: Dokl. Akad. Nauk SSSR **276**, 579–582 (1984) [English transl.: Doklady - Earth Sci. **276**, 8–11 (1986)]
O.A. Godin: A new form of the wave equation for sound in a general layered fluid, in *Progress in Underwater Acoustics*, ed. by H.M. Merklinger (Plenum, New York 1987) pp. 337–349

1.14 L.D. Landau, E.M. Lifshitz: *Course of Theoretical Physics*, Vol. 7, Elasticity Theory, 2nd edn. (Pergamon, New York 1970)

1.15 J.F. Hook: J. Acoust. Soc. Am. **33**, 302–313 (1961)

1.16 V.Yu. Zavadskiy: *Vychisleniye Volnovykh Polei v Otkrytykh Oblastyakh i Volnovodakh* (Wave fields calculation in open regions and waveguides) (Nauka, Moscow 1972) Chap. 2

1.17 K. Aki, P.G. Richards: *Quantitative Seismology. Theory and Methods*, Vols. 1, 2 (Freeman, San Francisco 1980) Chap. 9

1.18 A.D. Pierce: *Acoustics – An Introduction to Its Physical Principles and Applications*, 2nd edn. (AIP Press, New York 1989)

1.19 L.M. Brekhovskikh, O.A. Godin: *Acoustics of Layered Media. II: Point Sources and Bounded Beams*, Springer Ser. Wave Phenom., Vol. 10 (Springer, Berlin,, Heidelberg 1992)

1.20 S.O. Coen: J. Acoust. Soc. Am. **70**, 172–175, 1473–1479 (1981)

Chapter 2

2.1 M.A. Isakovich: *Obschaya Akustika* (General acoustics) (Nauka, Moscow 1973) Chap. 4
2.2 L. Brekhovskikh, V. Goncharov: *Mechanics of Continua and Wave Dynamics*, 2nd edn. Springer Ser. Wave Phenom., Vol. 1 (Springer, Berlin, Heidelberg 1994) Sect. 12.1.4
2.3 L.M. Brekhovskikh: *Waves in Layered Media*, 2nd ed. (Academic, New York 1980)
2.4 T.E.W. Embleton, J.E. Pierey, G.A. Daigle: J. Acoust. Soc. Am. **74**, 1239–1244 (1983)
2.5 L.-E. Andersson, B. Lundberg: Wave Motion **6**, 398–406 (1984)
2.6 G.G. Steinmetz, J.J. Singh: J. Acoust. Soc. Am. **51**, 218–222 (1972)
2.7 L.M. Lyamshev: Dokl. Akad. Nauk SSSR **261**, 74–78 (1981) [English transl.: Sov. Phys.-Dokl. **26**, 1058 (1981)]
2.8 J.W. Miles: J. Acoust. Soc. Am. **29**, 226–228 (1957)
2.9 H.S. Ribner: J. Acoust. Soc. Am. **29**, 435–441 (1957)
2.10 A.A. Andronov, A.L. Fabrikant: "Zatukhaniye Landau, Vetrovye Volny i Svistok" ("Landau attenuation, wind waves, and whistle") in *Nelineinye Volny* (Nonlinear waves) (Nauka, Moscow 1979) pp. 68–104
2.11 N.G. Kikina, D.G.Sannikov: Akust. Zh. **15**, 543–546 (1969) [English transl.: Sov. Phys. Acoust. **15**, N 4 (1969]
2.12 L.M. Lyamshev: Akust. Zh. **28**, 367–374 (1982) [English transl.: Sov. Phys. Acoust. **28**, 217–221 (1982)]
2.13 C. Yeh: J. Acoust. Soc. Am. **43**, 1454–1455 (1968)
2.14 L.M. Lyamshev: Akusth. Zh. **6**, 505–507 (1960) [English transl.: Sov. Phys. Acoust. **6**, N 4 (1961)]
2.15 Yu.A. Stepanyantz, A.L. Fabrikant: Usp. Fiz. Nauk **159**, 83–123 (1989)
2.16 C.L. Morfey: J. Sound Vibr. **14**, 159–169 (1971)
     O.A. Godin: Physics-Doklady **41**, 580–584 (1996)
     O.A. Godin: Wave Motion **19**, 143–167 (1997)
2.17 L.M. Brekhovskikh, O.A. Godin: *Acoustics of Layered Media. II: Point Sources and Bounded Beams*, Springer Ser. Wave Phenom., Vol. 10 (Springer, Berlin, Heidelberg 1992)
2.18 H.A. Macleod: *Thin Film Optical Filters* (McGraw-Hill, New York 1986)
2.19 A. Thelen: *Design of Optical Interference Coatings* (McGraw-Hill, New York 1988)
2.20 J.A. Dobrovolski, A.V. Tikhonravov, M.K. Trubetskov, B.T. Sullivan, P.G. Verly: Appl. Opt. **35**, 644–658 (1996)
2.21 B.B. Kadomtsev, A.B. Mikhailovskii, A.V. Timofeev: Zh. Eksp. Teor. Fiz. **47**, 2266–2268 (1964) [English transl.: Sov. Phys. JETP **20**, 1517–1518 (1965)]
     L.A. Ostrovsky, S.A. Rybak, L.Sh. Tsimring: Usp. Fiz. Nauk **150**, 417–437 (1986) [English transl.: Sov. Phys.–Uspekhi **29**, 1040–1052 (1986)]
2.22 D.S. Jones, J.D. Morgan: Proc. Cambr. Philos. Soc. **72**, 465–488 (1972)

Chapter 3

3.1 L.M. Brekhovskikh: *Waves in Layered Media*, 2nd edn. (Academic, New York 1980)
3.2 V.L. Ginzburg: *Electromagnetic Wave Propagation in Plasmas*, 2nd edn. (Pergamon, New York 1970)
3.3 A.M. Goncharenko, V.A. Karpenko: *Osnovy Teorii Opticheskikh Volnovodov* (Fundamentals of the theory of optical waveguides) (Nauka i Tekhnika, Minsk 1983)
3.4 W.M. Ewing, W.S. Jardetzky, F. Press: *Elastic Waves in Layered Media* (McGraw-Hill, New York 1957)
3.5 J.R. Wait: *Electromagnetic Waves in Stratified Media* (Pergamon, New York 1970)
3.6 E. Kamke: *Handbook of Ordinary Differential Equations* (Chelsea, New York 1971)
3.7 O.A. Godin: Dokl. Akad. Nauk SSSR **255**, 1069–1072 (1980) [English transl.: Doklady - Earth Sci. **255**, 41–45 (1982)]
3.8 O.A. Godin: "Primery Rashcheta Otrazhenia Ploskoy Volny ot Sloistikh Sred" ("Examples of the calculation of the reflection of plane waves from layered media"), in *Voprosy Difraktsii Elektromagnitnykh Voln* (Problems in diffraction of electromagnetic waves) (MFTI, Moscow 1982) pp. 107–114

3.9   L.I. Slater: *Confluent Hypergeometric Functions* (Cambridge Univ. Press, Cambridge 1960)

3.10   M. Abramovitz, I.A. Stegun (eds.): *Handbook of Mathematical Functions with Formulas, Graphs and Tables*, Appl. Math. Ser., Vol. 55 (National Bureau of Standards, Washington 1964)

3.11   E.T. Whittaker, G.N. Watson: *A Course of Modern Analysis* (Cambridge Univ. Press, Cambridge 1952)

3.12   B.S. Westcott: Quart. J. Mech. Appl. Math. **23**, 431–440 (1979)

3.13   S.M. Rytov, F.S. Yudkevich: Zh. Eksp. Teor. Fiz. **10**, 887–902 (1940)

3.14   R. Iamada: J. Phys. Soc. Jpn. **10**, 71–79 (1955)

3.15   E.P. Masterov, V.N. Muromtseva: Akust. Zh. **6**, 335–339 [English transl.: Sov. Phys.-Acoust. **6**, N 3 (1960)]

3.16   L.D. Landau, E.M. Lifshitz: *Course of Theoretical Physics,* Vol. 3, Quantum Mechanics, Nonrelativistic Theory (Pergamon, New York 1972)

3.17   E.P. Masterov: Akust. Zh. **5**, 332–336 (1959) [English transl.: Sov. Phys.-Acoust. **5**, N 3 (1959)]

3.18   J. Heading: Proc. Cambr. Phil. Soc. **61**, 897–913 (1965)

3.19   J. Heading: Quart. J. Mech. Appl. Math. **22**, 75–86 (1969)

3.20   F.R. DiNapoli, R.I. Deavenport: Numerical models of underwater acoustic propagation, in *Ocean Acoustics,* ed. by J.A. DeSanto, Topics Curr. Phys., Vol. 8 (Springer, Berlin, Heidelberg 1979) pp. 79–157

3.21   P. Epstein: Proc. Nat. Acad. Sci. USA **16**, 627–637 (1930)

3.22   J.C.P. Miller: *The Airy Integral* (Cambridge Univ. Press, Cambridge 1946)

3.23   A.D. Smirnov: *Tablitsy Funktsyi Airy i Spetsial' nykh Virozhdennykh Gipergeometricheskikh Funktsyi* (Tables of Airy functions and special confluent hypergeometric functions) (AN SSSR, Moscow 1955)

3.24   P.B. Abraham, H.E. Moses: J. Acoust. Soc. Am. **71**, 1391–1399 (1982)

3.25   R.K. Bullough, P.J. Caudrey (eds.): *Solitons,* Topics Curr. Phys., Vol. 18 (Springer, Berlin, Heidelberg 1980)

3.26   G.L. Lamb, Jr.: *Elements of Soliton Theory* (Wiley, New York 1980)

3.27   P.B. Abraham, B. DeFaccio, H.E. Moses: Phys. Rev. Lett. **46**, 1657–1659 (1981)

3.28   I. Kay, H.E. Moses: Nuovo Cimento **3**, 276–304 (1956)
       I. Kay: Commun. Pure Appl. Math. **13**, 371–393 (1960)

3.29   M. Rasavy: J. Acoust. Soc. Am. **58**, 956–963 (1975)

3.30   K. Rawer: Ann. Phys. (Leipzig) **35**, 385–416 (1939)

3.31   J.W.S. Rayleigh: *The Theory of Sound,* Vol. 2 (Dover, New York 1945)

3.32   V.Yu. Zavadskiy: *Vychislenie Volnovykh Poley v Otkrytykh Oblastyakh i Volnovodakh* (Calculation of wave fields in open regions and waveguides) (Nauka, Moscow 1972)

3.33   M. Goldstein, E. Rice: J. Sound Vibr. **30**, 79–84 (1973)

3.34   S.P. Koutsoyannis, K. Karamcheti, D.G. Gialant: AIAA J. **18**, 1446–1454 (1980)

3.35   V.P. Goncharov: Izv. Akad. Nauk SSSR Fiz. Atmos. Okeana **20**, 312–314 (1984) [English transl.: Izv. Acad. Sci. USSR Atmos. Oceanic Phys. **20**, 325–327 (1984)]

3.36   I.P. Chunchuzov: Akust. Zh. **31**, 134–136 (1985) [English transl.: Sov. Phys.-Acoust. **31**, 78–79 (1985)]

3.37   L.M. Brekhovskikh, O.A. Godin: *Acoustics of Layered Media. II: Point Sources and Bounded Beams,* Springer Ser. Wave Phenom., Vol. 10 (Springer, Berlin, Heidelberg 1992)

3.38   R. Vasudevan, K. Venkatesan, G. Jagannathan: Nuovo Cimento (Suppl.) **5**, 621–641 (1967)

3.39   G.R. Gogate, M.L. Munjal: J. Acoust. Soc. Am. **92**, 2915–2923 (1992)

3.40   O.A. Godin: A new form of the wave equation for sound in a general layered fluid, in *Progress in Underwater Acoustics,* ed. by H.M. Merklinger (Plenum, New York 1987) pp. 337–349

3.41   I.P. Chunchuzov, G.A. Bush, S.N. Kulichkov: J. Acoust. Soc. Am. **88**, 455–461 (1990)

Chapter 4

4.1   D.L. Arenberg: J. Acoust. Soc. Am. **20**, 1 (1948)

4.2   L.D. Landau, E.M. Lifshitz: *Course of Theoretical Physics,* Vol. 7. Elasticity Theory, 2nd edn. (Pergamon, New York 1970)

4.3   L. Brekhovskikh, V. Goncharov: *Mechanics of Continua and Wave Dynamics*, 2nd edn. Springer Ser. Wave Phenom., Vol. 1 (Springer, Berlin, Heidelberg 1994)

4.4   J.G.J. Scholte: Monthly Notices Roy. Astron. Soc., Geophys. Suppl. **5**, 120–126 (1947)

4.5   G.B. Young, L.W. Braile: Bull. Seismol. Soc. Am. **66**, 1881–1885 (1976)

4.6   B.L.N. Kennet, N.J. Kerry, J.H. Woodhouse: Geophys. Roy. Astron. Soc. **52**, 215–229 (1978)

4.7   B.D. Tartakovskii: Zh. Tekh. Fiz. **21**, 1194–1201 (1951)

4.8   K. Ergin: Bull. Seismol. Soc. Am. **42**, 349–372 (1952)

4.9   W.T. Thomson: J. Appl. Phys. **21**, 89–93 (1950)

4.10   N.A. Haskell: Bull. Seismol. Soc. Am. **43**, 17–34 (1953)

4.11   F. Gilbert, G. Backus: Geophysics **31**, 326–332 (1966)

4.12   V.M. Babich, P.V. Krauklis, L.A. Molotkov: Akust. Zh. **30**, 693–695 (1984) [English transl.: Sov. Phys.-Acoust. **30**, N 5 (1984)]

4.13   B. Ursin: Geophysics **48**, 1063–1081 (1983)

4.14   K. Aki, P.G. Richards: *Quantitative Seismology. Theory and Methods*, Vols. 1, 2 (Freeman, San Francisco 1980)

4.15   L.A. Molotkov: *Matrichniy Metod v Teorii Rasprostraneniya Voln v Sloistykh Uprugikh i Zhydkikh Sredakh* (Matrix method in the theory of wave propagation in layered elastic and liquid media) (Nauka, Leningrad 1984)

4.16   V. Cerveny: Studia Geophysica et Geodaetica **18**, 59–68 (1974)

4.17   L.M. Brekhovskikh: *Waves in Layered Media*, 2nd ed. (Academic, New York 1980)

4.18   V.T. Grinchenko, V.V. Meleshko: *Garmonicheskiye Kolebaniya i Volny v Uprugikh Telakh* (Harmonic oscillations and waves in elastic media) (Naukova Dumka, Kiev 1981)

4.19   D.L. Folds, C.D. Loggins: J. Acoust. Soc. Am. **62**, 1102–1109 (1977)

4.20   G.R. Barnard, J.L. Bardin, J.M. Whiteley: J. Acoust. Soc. Am. **57**, 577–584 (1975)

4.21   I.A. Viktorov: *Rayleigh and Lamb Waves* (Plenum, New York 1967)

4.22   I.A. Victorov: *Zvukovye Poverkhnostnye Volny v Tverdykh Telakh* (Sound surface waves in Solids) (Nauka, Moscow 1981)

4.23   B.I. Vybornov: *Ul' trazvukovaya Defektoskopiya* (Ultrasonic defectoscopy) (Metallurgiya, Moscow 1974)

4.24   Surface acoustic waves devices and applications. Special Issue: Proc. IEEE **64**, No. 5 (1976)

4.25   E.A. Ash, E.G.S. Paige (eds.): *Rayleigh-Wave Theory and Application*, Springer Ser. Wave Phenom., Vol. 2 (Springer, Berlin, Heidelberg 1985)

4.26   J.W. Strutt (Rayleigh): Proc. London Math. Soc. **17**, 4–11 (1887)

4.27   L. Knopoff: Bull. Seismol. Soc. Am. **42**, 307–308 (1952)

4.28   L.R.F. Rose: Wave Motion **6**, 359–361 (1984)

4.29   V.P. Gogoladze: Tr. Seismol. Inst. AN SSSR **127**, 1–87 (1948)

4.30   F. Press, A.P. Crary, J. Oliver, S. Kutz: Trans. Am. Geophys. Union **32**, 166–172 (1951)

4.31   I. Tolstoy, C.S. Clay: *Ocean Acoustics. Theory and Experiment in Underwater Acoustics* (McGraw-Hill, New York 1966)

4.32   M.A. Biot: Bull. Seismol. Soc. Am. **42**, 81–93 (1952)

4.33   L.B. Felsen: Quasi-optic diffraction, in *Quasi-Optics*, ed. by J. Fox (Polytechnic Press, New York 1964) pp. 1–40

4.34   V.M. Kurtepov: Akust. Zh. **15**, 560–566 (1969) [English transl.: Sov. Phys.-Acoust. **15**, 484–489 (1970)]

4.35   R. Stoneley: Proc. Roy. Soc. London **A 106**, 416–428 (1924)

4.36   L. Cagniard: *Reflexion et Refraction des Ondes Seismiques Progressives* (Gouthier-Villars, Paris 1939)

4.37   K. Sezawa, K. Kanai: Bull. Earthquake Res. Inst. (Tokyo) **17**, 1 (1939)

4.38   S. Yamaguchi, Y. Sato: Bull. Earthquake Res. Inst. (Tokyo) **33**, 549 (1955)

4.39   A.S. Ginzburg, E. Strick: Bull. Seismol. Soc. Am. **48**, 51–63 (1958)

4.40   R.A. Phynney: Bull. Seismol. Soc. Am. **51**, 527–555 (1961)

4.41   V.I. Keilis-Borok: *Interferenzionnye Poverkhnostnye Volny* (Interference surface waves) (AN SSSR, Moscow 1960)

4.42   A.L. Levshin: *Poverkhnostnye i Kanalovye Seimicheskiye Volny* (Surface and guided seismic waves) (Nauka, Moscow 1973)

4.43   M.D. Cochran, A.F. Woeber, J.C. DeBreamecker: Rev. Geophys. Space Phys. **8**, 321–357 (1970)

4.44   G.S. Murty: J. Acoust. Soc. Am. **58**, 1094–1095 (1975)

4.45   A. Pilarski: Archiv. Acoust. **7**, 61–70 (1982)

4.46   A.H. Nayfeh, T.W. Taylor: J. Acoust. Soc. Am. **84**, 2187–2191 (1988)

4.47 L. Kazandjian: J. Acoust. Soc. Am. **97**, 2048–2051 (1995)
4.48 M. Ainslie, P.W. Burns: J. Acoust. Soc. Am. **98**, 2836–2840 (1995)
4.49 B.L.N. Kennett: *Seismic Wave Propagation in Stratified Media* (Cambridge Univ. Press, Cambridge 1983)
4.50 F.B. Jensen, W.A. Kuperman, M.B. Porter, H. Schmidt: *Computational Ocean Acoustics* (AIP Press, New York 1994) Sect. 4.3
4.51 M. Schoenberg: Wave Motion **6**, 303–320 (1984)
4.52 P. Chervenka, P. Challande: J. Acoust. Soc. Am. **89**, 1579–1589 (1991)
4.53 J. George: J. Acoust. Soc. Am. **80**, 1235 (1986); **90**, 3371 (1991)
4.54 D. Lévesque, L. Piché: J. Acoust. Soc. Am. **92**, 452–467 (1992)
4.55 G. Guidarelli, A. Marini: J. Acoust. Soc. Am. **94**, 1476–1481 (1993)
4.56 T. K. Owen: Progr. Appl. Mater. Res. **6**, 69–86 (1964)
4.57 D.M. Barnett, J. Lothe, S.D. Gavazza, M.J.P. Musgrave: Proc. Roy. Soc. (London) A **402**, 153–166 (1985)
4.58 S.V. Biryukov, Yu.V. Gulyaev, V.V. Krylov, V.P. Plesskii: *Surface Acoustic Wwaves in Inhomogeneous Media*, Springer Ser. Wave Phenom., Vol. 20 (Springer, Berlin, Heidelberg 1995)
4.59 S. Rokhlin, M. Hefets, M. Rosen: J. Appl. Phys. **51**, 3579–3582 (1981); **52**, 2847–2851 (1981)
4.60 L. Singher, Y. Segal, E. Segal, J. Shamir: J. Acoust. Soc. Am. **96**, 2497–2505 (1994)

Chapter 5

5.1 L.M. Brekhovskikh: Usp. Fiz. Nauk **50**, 539–576 (1953)
5.2 F.A. Fisher: Ann. Phys. (Leipzig) **2**, 211–224 (1948)
5.3 L.G. Chambers: Wave Motion **2**, 247–253 (1980)
5.4 W.S. Wladimirov: *Gleichungen der Mathematischen Physik* (Deutscher Verlag der Wissenschaften, Berlin 1972)
5.5 A.B. Arons, D.R. Yennie: J. Acoust. Soc. Am. **22**, 231–237 (1950)
5.6 L.-E. Andersson, B. Lundberg: Wave Motion **6**, 389–406 (1984)
5.7 P.R.Stepanishen, B. Strozeski: J. Acoust. Soc. Am. **71**, 9–21 (1982)
5.8 J.N. Tjøtta, S. Tjøtta: J. Acoust. Soc. Am. **73**, 826–834 (1983)
5.9 G.A. Korn, T.M. Korn: *Mathematical Handbook for Scientists and Engineers*, 2nd edn. (McGraw Hill, New York 1968)
5.10 F.G. Friedlander: Quart. J. Mech. Appl. Math. **1**, 376–384 (1948)
5.11 L.R.B. Duykers: J. Acoust. Soc. Am. **37**, 1052–1055 (1965)
5.12 B.F. Cron, A.H. Nuttal: J. Acoust. Soc. Am. **37**, 486–492 (1965)
5.13 G.L.Choy, P.G. Richards: Bull. Seismol. Soc. Am. **65**, 55–70 (1975)
5.14 G. Lingh: Acustica **5**, 257–262 (1955)
5.15 A.H. Nuttal, B.F. Cron: J. Acoust. Soc. Am. **40**, 1094–1107 (1966)
5.16 V.L. Ginzburg: *Electromagnetic Wave Propagation in Plasmas*, 2nd edn. (Pergamon, New York 1970)
5.17 L.A. Vainstein: Usp. Fiz. Nauk **118**, 339–367 (1976) [English transl.: Sov. Phys.-Usp. **19**, 189 (1976)]
5.18 L.M. Brekhovskikh: *Waves in Layered Media,* 2nd ed. (Academic, New York 1980)
5.19 V.V. Kurin, B.E. Nemtsov, B.Ya. Eidman: Akust. Zh. **31**, 62–68 (1985) [English transl.: Sov. Phys.-Acoust. **31**, 36–39 (1985)]
5.20 I. Tolstoy: J. Acoust. Soc. Am. **37**, 1153–1155 (1965)
5.21 M.G. Brown: J. Acoust. Soc. Am. **79**, 1367–1401 (1986)
5.22 M.G. Brown, F.D. Tappert: J. Acoust. Soc. Am. **80**, 251–255 (1986)
5.23 C.H. Chapman: Geophys. J. Roy. Astron. Soc. **54**, 481–518 (1978)
5.24 E. Heyman, L.B. Felsen: Wave Motion **7**, 335–358 (1985)
5.25 L. Tzang, J.A. Kong: J. Math. Phys. **20**. 1170–1182 (1979)
5.26 D.N. Towne: J. Acoust. Soc. Am. **44**, 65–76, 77–83 (1968)
5.27 K. Aki, P.G. Richards: *Quantitative Seismology. Theory and Methods*, Vols. 1, 2 (Freeman, San Francisco 1980)
5.28 I.D. Ivanov: Akust. Zh. **27**, 234–242 (1981) [Engl. transl.: Sov. Phys.-Acoust. **27**, 128–132 (1981)]
5.29 L.M. Brekhovskikh, O.A. Godin: *Acoustics of Layered Media. II: Point Sources and Bounded Beams*, Springer Ser. Wave Phenom., Vol. 10 (Springer, Berlin, Heidelberg 1992)
5.30 M.D. Verweij, A.T. de Hoop: Geophys. J. Int'l **108**, 731–754 (1990)
5.31 M.D. Verweij: J. Acoust. Soc. Am. **92**, 2223–2238 (1992)
5.32 J.H.M.T. van der Hijden: *Propagation of Transient Elastic Waves in Stratified Anisotropic Media* (Elsevier, Amsterdam 1987)

Chapter 6

6.1　O.A. Godin: Wave Motion **7**, 515–528 (1985)
　　O.A. Godin: Modifikatsiya Uravneniya Rasprostraneniya Zvuka v Sloistoi Srede (Modification of the equation governing sound propagation in layered media) in *Akusticheskiye Volny v Okeane* (Acoustic waves in the ocean), ed. by I.B. Andreeva, L.M. Brekhovskikh (Nauka, Moscow 1987) pp. 34–40

6.2　O.A. Godin: Physics -Doklady **41**, 580–584 (1996)
　　O.A. Godin: Wave Motion **25**, 143–167 (1997)

6.3　M.E. Goldstein: *Aeroacoustics* (McGraw-Hill, New York 1976)

6.4　K.P. Scharnhorst: J. Acoust. Soc. Am. **74**, 1883–1886 (1983)

6.5　L.D. Faddeev: Usp. Mat. Nauk **14**, 57–119 (1959)
　　L.D. Faddeev: Tr. Mat. Inst. Stekl. **73**, 314–336 (1964) [English transl.: Proc. Steklov Inst. Math. **73**, 139 (1964)]

6.6　V. Červeny, R. Ravindra: *Theory of Seismic Head Waves* (Univ. Toronto, Toronto 1971)

6.7　C.W. Frazier: Geophysics **35**, 197–219 (1970)

6.8　E.R. Lapwood, J.A. Hudson: Geophys. J. Roy. Astron. Soc. **40**, 255–268 (1975)

6.9　B.L.N. Kennett, N.J. Kerry, J.H. Woodhouse: Geophys. J. Roy. Astron. Soc. **52**, 215–229 (1978)

6.10　R.K. Bullough, P.J. Caudrey (eds.): *Solitons,* Topics Curr. Phys. (Springer, Berlin, Heidelberg 1980)

6.11　G.L. Lamb, Jr.: *Elements of Soliton Theory* (Wiley, New York 1980)

6.12　L.M. Brekhovskikh: Izv. Akad. Nauk SSSR Ser. Fiz. **13**, 505–545 (1949)

6.13　E.T. Whittaker, G.N. Watson: *A Course of Modern Analysis* (Cambridge Univ. Press, Cambridge 1952)

6.14　M. Lavrentiev, B. Chabat: *Méthodes de la Théorie des Fonctions d'une Variable Complexe*, 2nd edn. (Mir, Moscow 1977)

6.15　M. Abramovitz, I.A. Stegun (eds.): *Handbook of Mathematical Functions with Formulas, Graphs and Tables*, Appl. Math. Ser., Vol. 55 (National Bureau of Standards, Washington 1964)

6.16　L.M. Brekhovskikh: *Waves in Layered Media*, 2nd edn. (Academic, New York 1980)

6.17　W.M. Ewing, W.S. Jardetzky, F. Press: *Elastic Waves in Layered Media* (McGraw-Hill, New York 1957)

6.18　L.B. Felsen, N. Marcuvitz: *Radiation and Scattering of Waves* (Prentice-Hall, Englewood Cliffs, N.J. 1973)

6.19　E. Kamke: *Handbook of Ordinary Differential Equations* (Chelsea, New York 1971)

6.20　W. Kofink: Ann. Phys. (Leipzig) **1**, 119–132 (1947)

6.21　M. Razavy: J. Acoust. Soc. Am. **58**, 956–963 (1975)

6.22　J. Lekner: *Theory of Reflection of Electromagnetic and Particle Waves* (Nijhoff, Amsterdam 1987)

6.23　J. Lekner: J. Acoust. Soc. Am. **87**, 2325–2331 (1990)

6.24　J. Qu, J.D. Achenbach, R.A. Roberts: IEEE Trans. UFFC **36**, 280–286 (1989)

6.25　B.L.N. Kennett: *Seismic Wave Propagation in Stratified Media* (Cambridge Univ. Press, Cambridge 1983) Chap. 5

6.26　Yu.M. Aivazyan, V.A. Sozinov: Izv. VUZ. Radiofiz. **32**, 593–599 (1989) [English transl.: Radiophys. Quantum Electron. **32**, N 5 (1989)]

6.27　L.M. Brekhovskikh, O.A. Godin: *Acoustics of Layered Media. II: Point Sources and Bounded Beams*, Springer Ser. Wave Phenom., Vol. 10 (Springer, Berlin, Heidelberg 1992)

6.28　I. Kay, H.E. Moses: J. Appl. Phys. **27**, 1503–1508 (1956)

Chapter 7

7.1　L.D. Landau, E.M. Lifshitz: *Course of Theoretical Physics*, Vol. 6, Fluid Mechanics (Pergamon, New York 1982)

7.2　D.R. Bland: *The Theory of Linear Viscoelasticity* (Pergamon, New York 1960)

7.3　L.D. Landau, E.M. Lifshitz: *Course of Theoretical Physics*, Vol. 7, Elasticity Theory, 2nd edn. (Pergamon, New York 1970)

7.4　G.R. Barnard, J.L. Bardin, J.M. Whiteley: J. Acoust. Soc. Am. **57**, 577–584 (1975)

7.5　N.A. Trapeznikova: *Prognoz i Interpretatsiya Dinamiki Seismicheskikh Voln* (Forecast and interpretation of the dynamics of seismic waves) (Nauka, Moscow 1985)

7.6　A. Atalar: J. Acoust. Soc. Am. **70**, 1182–1183 (1981)

7.7   V.A. Vasil'ev: Akust. Zh. **23**, 223–227 (1977) [English transl.: Sov. Phys.-Acoust. **23**, 127–129 (1977)]
7.8   L. Brekhovskikh, V. Goncharov: *Mechanics of Continua and Wave Dynamics*, 2nd edn. Springer Ser. Wave Phenom., Vol. 1 (Springer, Berlin, Heidelberg 1994)
7.9   F.F. Legusha: Usp. Fiz. Nauk **144**, 509–522 (1984)
7.10 A.E.H. Love: *A Treatise on the Mathematical Theory of Elasticity*, 4th edn. (Cambridge Univ. Press, Cambridge 1927)
7.11 V.A. Krasil'nikov, V.V. Krylov: *Vvedeniye v Fizicheskuyu Akustiku* (Introduction to physical acoustics) (Nauka, Moscow 1984)
7.12 G.I. Petrashen: *Rasprostraneniye Voln v Anisotropnykh Uprugikh Sredakh* (Wave propagation in anisotropic elastic media) (Nauka, Leningrad 1980)
7.13 F.I. Fedorov: *Theory of Elastic Waves in Crystals* (Plenum, New York 1968)
7.14 E. Dieulesaint, D. Royer: The relationship between surface wave displacement and anisotropy on selected crystal structures, in *Rayleigh-Wave Theory and Application,* ed. by E.A. Ash, E.G.S. Paige, Springer Ser. Wave Phenom., Vol. 2 (Springer, Berlin, Heidelberg 1985) pp. 29–36
7.15 D. Royer, E. Dieulesaint: J. Acoust. Soc. Am. **76**, 1438–1444 (1984)
7.16 N.J. Vlaar: Bull Seismol. Soc. Am. **56**, 2053–2072 (1966)
7.17 V. Červeny: Geophys. J. Roy. Astron. Soc. **29**, 1–13 (1972)
7.18 G. Backus: J. Geophys. Res. **70**, 3429–3439 (1965)
7.19 M.L. Smith, F.A. Dahlen: J. Geophys. Res. **78**, 3321–3333 (1973)
7.20 D. J. Vezzetti: J. Acoust. Soc. Am. **78**, 1072–1080 (1985)
7.21 S. Crampin: Geophys. J. Roy. Astron. Soc. **21**, 387–402 (1970)
7.22 S. Crampin: Geophys. J. Roy. Astron. Soc. **49**, 9–27 (1977)
7.23 H. Takeuchi, M. Saito: Seismic surface waves, in *Seismology: Surface Waves and Earth Oscillations,* ed. by B.A. Bolt, Meth. Comp. Phys., Vol. II (Academic, New York 1972) pp. 217–295
7.24 T.C. Lim, G.W. Farnell: J. Acoust. Soc. Am. **45**, 845–851 (1968)
7.25 M.K. Balakirev, I.A. Gilinskii: *Volny v Piezokristallakh* (Waves in piezocrystals) (Nauka, Novosibirsk 1982)
7.26 V.E. Lyamov: *Polyarizatsyonnye Effekty i Anisotropiya Vzaimodeystviya Akusticheskikh Voln v Kristallakh* (Polarization effects and the anisotropy of the interaction of acoustic waves in crystals) (Moscow State University, Moscow 1983)
7.27 B.A. Auld: *Acoustic Fields and Waves in Solids,* Vols. 1, 2 (Wiley, New York 1973)
7.28 E.A. Ash, E.G.S. Paige (eds.): *Rayleigh-Wave Theory and Application,* Springer Ser. Wave Phenom., Vol. 2 (Springer, Berlin, Heidelberg 1985)
7.29 Yu.V. Riznichenko: Izv. Akad. Nauk SSSR. Ser. Geophys. Geograph. **13**, 115–128 (1949)
7.30 N.A. Shul'ga: *Osnovy Mekhaniki Sloistykh Sred Periodicheskoi Struktury* (Fundamentals of mechanics of layered media with periodic structure) (Naukova Dumka, Kiev 1981)
7.31 G.E. Backus: J. Geophys. Res. **67**, 4427–4440 (1962)
7.32 E. Behrens: J. Acoust. Soc. Am. **42**, 378–387 (1967)
7.33 S.M. Rytov: Akust. Zh. **2**, 71–83 (1956)
7.34 L.A. Molotkov: *Matrichniy Metod v Teorii Rasprostraneniya Voln v Sloistykh Uprugikh i Zhydkikh Sredakh* (Matrix method in the theory of wave propagation in layered elastic and liquid media) (Nauka, Leningrad 1984)
7.35 G.A. Korn, T.M. Korn: *Mathematical Handbook for Scientists and Engineers*, 2nd edn. (McGraw Hill, New York 1968)
7.36 L.M. Brekhovskikh: *Waves in Layered Media*, 2nd edn. (Academic, New York 1980)
7.37 L.A. Molotkov, A.E. Hilo: Zap. Nauchn. Semin. LOMI **128**, 130–138 (1983)
7.38 V.L. Berdichevskii: *Variatsyonnye Printsypy Mekhaniki Sploshnoi Sredy* (Variational principles in the mechanics of continua) (Nauka, Moscow 1983)
7.39 N.S. Bahvalov, G.P. Panasenko: *Osredneniye Protsessov v Periodicheskikh Sredakh* (Processes averaging in Periodic Media) (Nauka, Moscow 1984)
7.40 S.M. Kozlov: Dokl. Akad. Nauk SSSR, **236**, 1068–1071 (1977)
7.41 G. Guidarelli, A. Marini: J. Acoust. Soc. Am. **94**, 1476–1481 (1993)
7.42 M. Ainslie, P.W. Burns: J. Acoust. Soc. Am. **98**, 2836–2840 (1995)
7.43 M. Deschamps: J. d'Acoustique **3**, 251–261 (1990)
7.44 M.J. Anderson, P.G. Vaidya: J. Acoust. Soc. Am. **86**, 2385–2396 (1989)
7.45 A.D. Pierce: *Acoustics – An Introduction to Its Physical Principles and Applications*, 2nd edn. (AIP Press, New York 1989)
7.46 A.J. Rudgers: J. Acoust. Soc. Am. **88**, 1078–1094 (1990)

7.47 S. Sutcliffe: J. Acoust. Soc. Am. **74**, 357–358 (1983)
7.48 M.J.P. Musgrave: *Crystal Acoustics* (Holden Day, San Francisco 1970)
7.49 J.H.M.T. van der Hijden: *Propagation of Transient Elastic Waves in Stratified Anisotropic Media* (Elsevier Science, Amsterdam 1987)
7.50 D.M. Barnett, J. Lothe: Proc. Roy. Soc. London A **402**, 135–152 (1985)
7.51 S.V. Biryukov, Yu.V. Gulyaev, V.V. Krylov, V.P. Plesskii: *Surface Acoustic Waves in Inhomogeneous Media*, Springer Ser. Wave Phenom., Vol. 20 (Springer, Berlin, Heidelberg 1995)
7.52 A.H. Nayfeh: J. Acoust. Soc. Am. **86**, 2007–2012 (1989)
7.53 R.F. O'Doherty, N.A. Anstey: Geophys. Prosp. **19**, 430–458 (1971)
7.54 R. Burridge, M.V. de Hoop, K. Hsu, L. Le, A. Norris: J. Acoust. Soc. Am. **94**, 2884–2894 (1993)
7.55 Yu.A. Kravtsov, O.N. Naïda, A-A. Fuki: Usp. Fiz. Nauk **162**, 141–167 (1996) [English transl.: Phys.-Uspekhi **39**, 129–154 (1996)]
7.56 L. Thomsen: Geophysics **51**, 1954–1966 (1986)
7.57 C.M. Sayers: Geophys. J. Int'l **116**, 799–805 (1994)
7.58 T. Mensch, P. Rasolofosan: Geophys. J. Int'l **128**, 43–64 (1997)

Chapter 8

8.1   P. Debay: Ann. Phys. (Leipzig) **35**, 277 (1911)
8.2   S.L. Sobolev: Tr. Seismol. Inst. AN SSSR **6**, 1–57 (1930)
8.3   S.M. Rytov: Dokl. Akad. Nauk SSSR **18**, 263–266 (1938)
8.4   S.M. Rytov: Tr. Fiz. Inst. An SSSR **2**, 41–133 (1940)
8.5   V.M. Babič, V.S. Buldyrev: *Short-Wavelength Diffraction Theory: Asymptotic Methods*, Springer Ser. Wave Phenom., Vol. 4 (Springer, Berlin, Heidelberg 1991)
8.6   V.P. Maslov, M.V. Fedoriuk: *Semi-Classical Approximation in Quantum Mechanics* (Reidel, Dordrecht 1981)
8.7   A.H. Nayfeh: *Perturbation Methods* (Wiley, New York 1973)
8.8   M. Fedoryuk: *Methodes Asymptotiques pour les Equations Differentielles Ordinaires Linéaires* (Mir, Moscow 1987)
8.9   N. Frömann, P.O. Frömann: *JWKB Approximation. Contributions to the Theory* (North-Holland, Amsterdam 1965)
8.10 J. Heading: *An Introduction to Phase-Integral Methods* (Wiley, New York 1962)
8.11 Yu.A. Kravtsov, Yu.I. Orlov: *Geometrical Optics of Inhomogeneous Media*, Springer Ser. Wave Phen., Vol. 6 (Springer, Berlin, Heidelberg 1989)
8.12 H.J. Korsch, H. Laurent: J. Phys. B. **14**, 4213–4230 (1981)
8.13 E.A. Solov'ev: Pis. Zh. Eksp. Teor. Fiz. **39**, 84–86 (1984)
8.14 F.W.J. Olver: Proc. Cambr. Phil. Soc. **57**, 790–810 (1961)
8.15 F.W.J. Olver, F. Stengler: J. SIAM Numer. Anal. B **2**, 244–249 (1965)
8.16 L.M. Brekhovskikh: *Waves in Layered Media*, 2nd edn. (Academic, New York 1980)
8.17 L.M. Brekhovskikh: Izv. Akad. Nauk SSSR Fiz. Atmos. Okeana **4**, 1291–1304 (1968) [English transl.: Izv. Acad. Sci. USSR. Atmos. Oceanic Phys. **4**, N 12 (1968)]
8.18 A.L. Virovlyanskii: Izv. VUZ. Radiofiz. **27**, 1592–1594 (1984)
8.19 J.W.S. Rayleigh: *The Theory of Sound*, Vol. 2 (Dover, New York 1945)
8.20 D.I. Blokhintsev: *Acoustics of an Inhomogeneous Moving Medium* (Brown Univ. Tren Providence, R.I. 1956)
8.21 O.S. Golod, N.S. Grigor'eva: Akust. Zh. **28**, 758–762 (1982) [English transl.: Sov. Phys.-Acoust. **28**, 449–451 (1982)]
8.22 V.E. Ostashev: Akust. Zh. **31**, 225–229 (1985) [English transl.: Sov. Phys.-Acoust. **31**, 130–132 (1985)]
8.23 V.A. Polyanskaya: Akust. Zh. **31**, 628–632 (1985) [English transl.: Sov. Phys.-Acoust. **31**, 376–379 (1985)]
8.24 L.A. Chernov: Akust. Zh. **4**, 299–306 (1958) [English transl.: Sov. Phys.-Acoust. **4**, 311–318 (1958)]
8.25 C.I. Chessel: J. Acoust. Soc. Am. **53**, 83–87 (1973)
8.26 A.D. Gorman, R. Wells: J. Acoust. Soc. Am. **73**, 363–365 (1983)
8.27 S.V. Chibisov: Izv. AN SSSR. Ser. Geograf. Geofiz. N 1, 33–118; N 2, 207–222; N 4, 475–520 (1940)
8.28 E.T. Kornhauser: J. Acoust. Soc. Am. **25**, 945–949 (1953)
8.29 B.K. Newhall, M.J. Jacobson, W.L. Siegmann: J. Acoust. Soc. Am. **67**, 1997–2010 (1980)
8.30 T.B. Sanford: J. Acoust. Soc. Am. **56**, 1118–1121 (1974)

8.31 R.J. Thompson: J. Acoust. Soc. Am. **51**, 1675–1682 (1972)
8.32 R.J. Thompson: J. Acoust. Soc. Am. **55**, 729–737 (1974)
8.33 P. Ugincius: J. Acoust. Soc. Am. **51**, 1759–1763 (1972)
8.34 C.H.E. Warren: J. Sound Vibr. **1**, 175–178 (1964)
8.35 V.E. Ostashev: Izv. AN SSSR. Fiz. Atmos. Okeana **21**, 358–373 (1985) [English transl.: Izv. Acad. Sci. USSR. Atmos. Oceanic Phys. **21**, 274–285 (1985)]
8.36 E.H. Brown, F.F. Hall: Rev. Geophys. Space Phys. **16**, 47–110 (1978)
8.37 H. Bremmer: Physica **15**, 593 (1949)
8.38 H. Bremmer: Commun. Pure Appl. Math. **4**, 105–115 (1951)
8.39 R. Bellman, R. Kalaba: Proc. Nat. Acad. Sci. USA **44**, 317 (1958)
8.40 F.V. Atkinson: J. Math. Anal. Appl. **1**, 255–276 (1960)
8.41 S.H. Gray: Wave Motion **5**, 249–255 (1983)
8.42 V.A. Bailey: Phys. Rev. **96**, 865–868 (1954)
8.43 N. Fröman, P.O. Fröman: Ann. Phys. (N.Y.) **163**, 215–226 (1985)
8.44 E. Bahar: J. Math. Phys. **8**, 1735–1746 (1967)
8.45 E. Bahar: Radio Sci. **15**, 573–579 (1980)
8.46 B.Z. Katsenelenbaum: Radiotekh. Electron. **22**, 2414–2417 (1977)
8.47 B.Z. Katsenelenbaum: Radiotekh. Electron. **27**, 1451 (1982)
8.48 S.L. Ziglin: Radiotekh. Electron. **24**, 173–175 (1979)
8.49 S.L. Ziglin: Radiotekh. Electron. **24**, 2131–2133 (1979)
8.50 S.L. Ziglin: Radiotekh. Electron. **29**, 836–842 (1984)
8.51 C.E. Hecht, J.E. Mayer: Phys. Rev. **106**, 1156–1161 (1957)
8.52 N. Fröman: Ann. Phys. (N. Y.) **61**, 451–464 (1970)
8.53 E.R. Floyd: J. Math. Phys. **17**, 880–884 (1976)
8.54 E.R. Floyd: J. Acoust. Soc. Am. **60**, 801–809 (1976); **75**, 803–808 (1984); **79**, 1741–1747 (1986); **80**, 877–887 (1986)
8.55 R. Zhang: Chi. J. Acoust. **1**, 23–24 (1982)
     R. Zhang, Y. He, H. Liu: Chi. J. Acoust. **13**, 1–12 (1994)
8.56 N.E. Mal'tsev: Dokl. Akad. Nauk SSSR **271**, 1108–1111 (1983)
     A.S. Aralkin, N.E. Mal'tsev: Akust. Zh. **35**, 577–583 (1989) [English transl.: Sov. Phys.-Acoust. **35**, N 4 (1989)]
     N.E. Mal'tsev: J. Math. Phys. **35**, 1387–1398 (1994)
8.57 A. Beilis: J. Acoust. Soc. Am. **74**, 171–180 (1983)
8.58 J.G. Taylor: J. Math. Anal. Appl. **85**, 79–89 (1982)
8.59 R.M. Jones, T.M. Georges, J.P. Riley: IEEE Trans. GE-**22**, 633–640 (1984)
8.60 J.A. Mercer: J. Acoust. Soc. Am. **84**, 999–1006 (1988)
8.61 F. Walkden, M. West: J. Acoust. Soc. Am. **84**, 321–326 (1988)
8.62 M.M. Boone, E.A. Vermaas: J. Acoust. Soc. Am. **90**, 2109–2117 (1991)
8.63 O.A. Godin, D.Yu. Mikhin, S.Ya. Molchanov: Izv. AN SSSR. Fiz. Atmos. Okeana **27**, 139–150 (1991) [English transl.: Izv. Acad. Sci. USSR. Atmos. Ocean Phys. **27**, N 2 (1991)]
     O.A. Godin, D.Yu. Mikhin, S.Ya. Molchanov: Izv. Akad. Nauk. Fiz. Atmos. Okeana **28**, 1146–1158 (1992); **29**, 194–201 (1993) [English transl.: Izv. Atmos. Ocean Phys. **28**, N 12 (1992); **29**, N 2 (1993)]
8.64 D.Yu. Mikhin, O.A. Godin, O. Boebel, W. Zenk: J. Atmos. Ocean Techn. **14**, 938–949 (1997)
8.65 L.M. Brekhovskikh, O.A. Godin: *Acoustics of Layered Media. II: Point Sources and Bounded Beams*, Springer Ser. Wave Phenom., Vol. 10 (Springer, Berlin, Heidelberg 1992)
8.66 J. Corones: J. Math. Anal. Appl. **50**, 361–372 (1975)
8.67 M.V. de Hoop: J. Math. Phys. **37**, 3246–3282 (1996)

Chapter 9

9.1 R.E. Langer: Trans. Am. Math. Soc. **48**, 461–490 (1949)
9.2 R.E. Langer: Commun. Pure Appl. Math. **3**, 427–438 (1951)
9.3 W.S. Wladimirov: *Gleichungen der Mathematischen Physik* (Deutscher Verlag der Wissenschaften, Berlin 1972)
9.4 R. Lynn, J.B. Keller: Commun. Pure Appl. Math. **23**, 379–408 (1970)
9.5 J.M. Arnold: J. Acoust. Soc. Am. **69**, 17–24 (1981)

9.6   W.A. Wasov: *Asymptotic Expansions for Ordinary Differential Equations* (Wiley, New York 1965)

9.7   A.A. Dorodnitsin: Usp. Mat. Nauk **7**, 3–96 (1952)

9.8   A.H. Nayfeh: *Perturbation Methods* (Wiley, New York 1973)

9.9   F.W.J. Olver: *Asymptotics and Special Functions* (Academic, New York 1974)

9.10 M. Fedoryuk: *Methodes Asymptotiques pour les Equations Differentielles Ordinairs Linéaires* (Mir, Moscow 1987)

9.11 F.W.J. Olver: Phil. Trans. Roy. Soc. London A 250, 479–517 (1958)

9.12 F.W.J. Olver: Proc. Cambr. Phil. Soc. **57**, 790–810 (1961)

9.13 F.W.J. Olver, F. Stengler: J. SIAM Numer. Anal. B **2**, 244–249 (1965)

9.14 V.S. Buldyrev, S.Yu. Slavyanov: Vestnik LGU **22**, 70–84 (1968)

9.15 Yu.A. Kravtsov: Izv. VUZ. Radiofiz. **8**, 659–667 (1965) [English transl.: Radiophys. Quantum Electron. **8**, N4 (1965)]

9.16 Yu.I. Orlov: Izv. VUZ. Radiofiz. **9**, 1036–1038 (1966) [English transl.: Radiophys. Quantum Electron. **9**, N5 (1966)]

9.17 V.A. Fock: Radiotekh. Elektron **1**, 560–574 (1956)

9.18 N.D. Kazarinoff: Arch. Rat. Mech. Anal. **2**, 129–150 (1958–1959)

9.19 R.E. Langer: Trans. Am. Math. Soc. **90**, 113–142 (1959)

9.20 L.D. Landau, E.M. Lifshitz: *Course of Theoretical Physics*, Vol. 3, Quantum Mechanics. Nonrelativistic Theory (Pergamon, New York 1972)

9.21 L.B.Felsen, N. Marcuvitz: *Radiation and Scattering of Waves* (Prentice-Hall, Englewood Cliffs, N.J. 1973)

9.22 M. Abramovitz, I.A. Stegun (eds.): *Handbook of Mathematical Functions with Formulas, Graphs and Tables*, Appl. Math. Ser., Vol. 55 (National Bureau of Standards, Washington 1964)

9.23 J.C.P. Miller: *Tables of Weber Parabolic Cylinder Functions* (Her Majesty's Stationary Office, London 1955)

9.24 E.T. Whittaker, G.N. Watson: *A Course of Modern Analysis* (Cambridge Univ. Press, Cambridge 1952)

9.25 M. Lavrentiev, B. Chabat: *Méthodes de la Théorie des Fonctions d'une Variable Complexe*, 2nd edn. (Mir, Moscow 1977)

9.26 F.W.J. Olver: J. Res. NBS 63 B, 131–169 (1959)

9.27 E.L. Murphy: J. Acoust. Soc. Am. **43**, 610–618 (1968)

9.28 V.A. Kaloshin, Yu.I. Orlov: Radiotekh. Elektron. **18**, 2028–2033 (1973)

9.29 Yu.A. Kravtsov: Izv. VUZ. Radiofiz. **10**, 1283–1305 (1967) [English transl.: Radiophys. Quantum Electron **10**, N9–10 (1967)]

9.30 J. Heading: J. Lond. Math. Soc. **37**, 195–208 (1962)

9.31 E. Kamke: *Handbook of Ordinary Differential Equations* (Chelsea, New York 1971)

9.32 A.A. Andronov, A.L. Fabricant: Zatukhaniye Landau, Vetrovye Volny i Svistok (Landau attenuation, wind waves, and whistle), in *Nelineynye Volny* (Nonlinear waves) (Nauka, Moscow 1979) pp. 68–104

9.33 L.D. Landau, E.M. Lifshitz: *Course of Theoretical Physics*, Vol. 6, Fluid Mechanics (Pergamon, New York 1982)

9.34 V.G. Gavrilenko, L.A. Zelekson: Akust. Zh. **23**, 867–872 (1977) [English transl.: Sov. Phys.-Acoust. **23**, 497–499 (1977)]

9.35 A.L. Fabrikant: Akust. Zh. **22**, 107–114 (1976) [English transl.: Sov. Phys.-Acoust. **22**, 56–59 (1976)]

9.36 P.I. Kolyhalov: Dokl. Akad. Nauk SSSR **280**, 95–98 (1985)

9.37 B.B. Kadomtsev, A.B. Mikhailovskii, A.V. Timofeev: Zh. Eksp. Teor. Fiz. **47**, 2266–2268 (1964) [English transl.: Sov. Phys. JETP **20**, 1517–1518 (1965)] L.A. Ostrovsky, S.A. Rybak, L.Sh. Tsimring: Usp. Fiz. Nauk **150**, 417–437 (1986) [English transl.: Sov. Phys. Usp. **29**, 1040–1052 (1986)]

9.38 O.A. Godin, A.V. Mokhov: Akust. Zh. **37**, 58–64 (1991) [English transl.: Sov. Phys.-Acoust. **37**, N 1 (1991)]

9.39 L.A. Zelekson, V.D. Pikulin: Izv. VUZ. Radiofiz. **32**, 696–700 (1989) [English transl.: Radiophys. Quantum Electron. **32**, N 6 (1989)]

9.40 O.A. Godin: Dokl. Akad. Nauk SSSR **304**, 79–83 (1989) [English transl.: Sov. Phys.-Doklady **34**, 43 (1989)]

Chapter 10

10.1 J.R. Pierce: Bell System. Techn. J. **32**, 263 (1943)
10.2 F. Press, M. Ewing: Trans. Am. Geophys. Union **32**, 673 (1951)
10.3 G. Eckart: Acustica **2**, 256 (1952)
10.4 C.O. Hines: Quart. Appl. Math. **11**, 9–31 (1953)
10.5 V.L. Ginzburg: *Electromagnetic Wave Propagation in Plasmas*, 2nd edn. (Pergamon, New York 1970)
10.6 J.W.S. Rayleigh: Proc. Roy. Soc. A **86**, 207–266 (1912)
10.7 I. Tolstoy: J. Acoust. Soc. Am. **27**, 274–277 (1955)
10.8 V.Yu. Zavadskiy: *Vychisleniye Volnovykh Polei v Otkrytykh Oblastyakh i Volnovodakh* (Wave fields calculation in open regions and waveguides) (Nauka, Moscow 1972)
10.9 L.D. Presnyakov, I.I. Sobelman: Izv. VUZ. Radiofiz. **8**, 57–63 (1965) [English transl.: Radiophys. Quantum Electron **8**, N 1 (1965)]
10.10 V.A. Bailey: Phys. Rev. **96**, 865–868 (1954)
10.11 M.B. Lesser: J. Acoust. Soc. Am. **47**, 1297–1302 (1970)
10.12 J.R. Wait: *Electromagnetic Waves in Stratified Media* (Pergamon, New York 1970)
10.13 J.A. Ware, K. Aki: J. Acoust. Soc. Am. **45**, 911–921 (1969)
10.14 R.N. Gupta: J. Acoust. Soc. Am. **39**, 255–260 (1966)
10.15 W.R. Hoover, A. Nagl, H. Uberall: Resonances in acoustic bottom reflection and their relation to the ocean bottom properties, in *Bottom-Interacting Ocean Acoustics*, ed. by W.A. Kuperman, F.B. Jensen (Plenum, New York 1980) pp. 209–224
10.16 N.M. Eryshov: Radiotekh. Electron. **26**, 457–462 (1981)
10.17 N. Fröman, P.O. Fröman: Ann. Phys. (N.Y.) **163**, 215–226 (1985)
10.18 J. Lekner: *Theory of Reflection of Electromagnetic and Particle Waves* (Nijhoff, Amsterdam 1987)
10.19 L.M. Brekhovskikh: Izv. Akad. Nauk SSSR. Ser. Fiz. **13**, 505–545 (1949)
10.20 L.M. Brekhovskikh: Zh. Tekh. Fiz. **19**, 1126–1135 (1949)
10.21 O.A. Godin: A new form of the wave equation for sound in a general layered fluid, in *Progress in Underwater Acoustics*, ed. by H.M. Merklinger (Plenum, New York 1987) pp. 337–349
10.22 O.A. Godin: Wave Motion **7**, 515–528 (1985)
10.23 G. Franceschetti: IEEE Trans AP-**12**, 754 (1964)
10.24 J.M. McKisik, D.P. Hamm: J. Acoust. Soc. Am. **59**, 294–304 (1976)
10.25 P.E. Krasnushkin: Dokl. Akad. Nauk SSSR **252**, 332–335 (1980)
10.26 V.V.Tyutekin: Akust. Zh. **30**, 373–379 (1984) [English transl.: Sov. Phys.-Acoust. Zh. **30**, 220 (1984)]
10.27 J. Casti, R. Kalaba: *Imbedding Methods in Applied Mathematics* (Addison-Wesley, London 1973)
10.28 V.I. Klyatskin: *Metod Pogruzheniya v Teorii Rasprostraneniya Voln* (Imbedding method in the theory of wave propagation) (Nauka, Moscow 1986)
10.29 S.A. Schelkunoff: Commun. Pure Appl. Math. **4**, 117–128 (1951)
10.30 F.W. Sluijter: J. Opt. Soc. Am. **60**, 8–10 (1970)
10.31 V.P. Maslov: Akust. Zh. **27**, 428–433 (1981) [English transl.: Sov. Phys.-Acoust. **27**, 235–238 (1981)]
10.32 W.S. Wladimirow: *Gleichungen der Mathematischen Physik* (Deutscher Verlag der Wissenschaften, Berlin 1972)
10.33 R.H. Lang, J. Shmoys: J. Acoust. Soc. Am. **48**, 242–252 (1970)
10.34 A.A. Andronov, A.L. Fabricant: Zatukhanie Landau, Vetrovye Volny i Svistok (Landau attenuation, wind waves, and whistle), in *Nelineinye Volny* (Nonlinear waves) (Nauka, Moscow 1979) pp. 68–104
10.35 O.A. Godin: Izv. Akad. Nauk SSSR. Fiz. Atmos. Okeana **21**, 1252–1259 (1985) [English transl.: Izv. Acad. Sci. USSR. Atmos. Oceanic Phys. **21**, N 12 (1985)]
10.36 Yu.V. Sidorov, M.V. Fedoryuk, M.I. Shabunin: *Lectures on the theory of functions of a complex variable* (Mir, Moscow 1985)
10.37 E. Kamke: *Handbook of Ordinary Differential Equations* (Chelsea, New York 1971)
10.38 N. Fröman, P.O. Fröman: *JWKB Approximation. Contributions to the Theory* (North-Holland, Amsterdam 1965)
10.39 V.P. Maslov: Akust. Zh. **27**, 914–918 (1981) [English transl.: Sov. Phys.-Acoust. **27**, N 6 (1981)]
10.40 L.B. Felsen, N. Marcuvitz: *Radiation and Scattering of Waves* (Prentice-Hall, Englewood Cliffs, N.J. 1973)
10.41 S.M. Rytov, Yu.A. Kravtsov, V.I. Tatarskiy: *Vvedenie v Statisticheskuyu Radiofiziku. Sluchainye Polya* (Nauka, Moscow 1978) [Engl. transl.: *Introduction to Statistical Radiophysics*, Vol. 3 (Springer, Berlin, Heidelberg 1989)]

10.42 L.M. Brekhovskikh: *Waves in Layered Media*, 2nd edn. (Academic, New York 1980)

10.43 S.N. Stolyarov, Yu.A. Filatov: Radiotekh. Elektron. **28**, 2330–2335 (1983)

10.44 L.M. Brekhovskikh: Elementy Teorii Zvukovogo Polya v Okeane (Elements of sound field theory in the ocean) in *Akustika Okeana* (Ocean acoustics), ed. by L.M. Brekhovskikh (Nauka, Moscow 1974) pp. 79–162

10.45 F.R. DiNapoli, R.I. Deavenport: Numerical models of underwater acoustic propagation, in *Ocean Acoustics*, ed. by J.A. DeSanto, Topics Curr. Phys., Vol. 8 (Springer, Berlin, Heidelberg 1979) pp. 79–157

10.46 N.E. Mal'tsev: Matematicheskoye Modelirovaniye Zvukovykh Polei v Okeane (Mathematical modeling of sound fields in the ocean), in *Akustika Okeana. Sovremennoe Sostoyaniye* (Ocean acoustics. State of the art) (Nauka, Moscow 1982) pp. 5–24

10.47 K. Aki, P.G. Richards: *Quantitative Seismology. Theory and Methods*, Vols. 1, 2 (Freeman, San Francisco 1980)

10.48 R.M. Lewis, J.B. Keller: Asymptotic methods for partial differential equations: The reduced wave equation and Maxwell's equations, New York Univ. Res. Rep. **EM-194** (1964)

10.49 R.E. Meyer: J. Math. Phys. **17**, 1039–1041 (1976)

10.50 N.S. Ageeva: Zvukovoye Pole Sosredotochennogo Istochnika v Okeane (The sound field of a concentrated source in the ocean), in *Akustika Okeana* (Ocean acoustics) ed. by L.M. Brekhovskikh (Nauka, Moscow 1974) pp. 163–229

10.51 M.J. Mezzino: J. Acoust. Soc. Am. **53**, 581–589 (1973)

10.52 M.A. Pedersen: J. Acoust. Soc. Am. **33**, 465–474 (1961)

10.53 M.A. Pedersen, D.F. Gordon: J. Acoust. Soc. Am. **41**, 419–438 (1967)

10.54 O.A. Godin, I.V. Prokopyuk: Akust. Zh. **34**, 49–54 (1988) [English transl.: Sov. Phys.-Acoust. **34**, N 1 (1988)]

10.55 D.S. Ahluwalia, J.B. Keller: Exact and asymptotic representation of the sound field in a stratified ocean, in *Wave Propagation and Underwater Acoustics*, ed. by J.B. Keller, J.S. Papadakis, Lect. Note Phys., Vol. 70 (Springer, Berlin, Heidelberg 1977) pp. 14–84

10.56 H. Gingold, J. She, W.E. Zorumski: J. Acoust. Soc. Am. **91**, 1262–1269 (1992); **93**, 599–604 (1993)

10.57 I.V. Zuev, A.V. Tikhonravov: Zh. Vych. Mat. Mat. Fiz. **33**, 428–438 (1993)

10.58 J. Lekner: J. Acoust. Soc. Am. **86**, 2359–2362 (1989); **87**, 2325–2331 (1990)

10.59 L. Fishman, A.K. Gautesen, Z. Sun: An exact, well-posed, one-way reformulation of the Helmholtz equation with application to direct and inverse wave propagation modeling, in *New Perspectives on Problems in Classical and Quantum Physics*, ed. by A.W. Saenz, P.P. Delsanto (Gordon and Breach, Newark 1997)

10.60 Y.Y. Lu, J.R. McLaughin: J. Acoust. Soc. Am. **100**, 1432–1446 (1989)

10.61 R. Bellman, R. Vasudevan: *Wave Propagation, an Invariant Imbedding Approach* (Reidel, Dordrecht 1986)

10.62 J.P. Corones, G. Kriestensson, P. Nelson, D. Seth: *Invariant Imbedding and Inverse Problems* (SIAM, Philadelphia 1992)

10.63 O.A. Godin: Modifikatsiya Uravneniya Rasprostraneniya Zvuka v Sloistoi Srede (Modification of the equation governing sound propagation in layered media) in *Akusticheskiye Volny v Okeane* (Acoustics waves in the ocean) ed. by I.B. Andreeva, L.M. Brekhovskikh (Nauka, Moscow 1987) pp. 34–40

10.64 F.B. Jensen, W.A. Kuperman, M.B. Porter, H. Schmidt: *Computational Ocean Acoustics* (AIP Press, New York 1994) Sect. 4.3

10.65 L.M. Brekhovskikh, O.A. Godin: *Acoustics of Layered Media. II: Point Sources and Bounded Beams*, Springer Ser. Wave Phenom., Vol. 10 (Springer, Berlin, Heidelberg 1992)

Subject Index

Absolutely rigid boundary 4–5, 12, 15, 25, 106, 137, 206, 225

Absorption
— of elastic waves in solid 147–148, 163–164
— of sound 6, 45, 70, 129, 132, 144–146, 149, 163–164, 203

Airy equation 65, 184

Airy functions 49, 65–69, 70, 85, 184, 196, 197, 219

Airy integral 66

Analogy between SH waves and sound 15, 162

Angle of boundary's complete transparency 22, 30, 37, 83, 139–143

Angle of incidence 20

Angle of polarization exchange 88–91, 107

Anisotropic medium 144, 150–158, 161–164

Antiwaveguide 47, 50, 53

Asymptotics
—, local 185, 196, 198
— of sound field in the vicinity of turning horizon 183–186
—, uniform 184, 188, 198

Born approximation 215–216

Boundary, ideal 5

Boundary conditions 4–5, 10, 112, 147
—, electrical 157
— of the first kind 5
— in fluid at rest 5, 10, 21, 149
— in moving fluid 7–8, 10, 35
— of the second kind 5
— in solids 12, 15, 16, 99, 147, 153
— of the third kind 25

Branch point 136–139

Causality principle 123, 125, 194

Characteristic equation 107,
  see also Dispersion relation

Characteristic (or wave) impedance 21, 22

Christoffel equation 150, 156

Complete transparency 22, 30, 39, 83, 139–143

Complex geometrical acoustics 191

Confluent hypergeometric functions 44, 53,
  see also Whittaker functions

Continuity equation 1

Convective time derivative
  see Material time derivative

Critical angle of total reflection 22–23, 48, 90–91

Cylindrical functions 44, 49–51, 67–68, 85, 138

Decay of plane waves 144–147

Decay of surface and "leaky" waves 110–111

Density of sound waves' energy 18

Dielectric permittivity tensor 155

Diffracted rays 191

Discretely layered medium 17, 34, 102, 153

Dispersion relation 8, 107, 146, 147, 225
— of normal modes 225
— of sound moving medium 8, 173
— of surface waves 106, 107, 109–110, 225

Dispersive media 6, 123, 146, 147

Displacement of particles 11, 35

Displacement-stress vector 99, 153, 154, 162

Elastic modulus tensor 150, 151, 156

Elastic waves
—, horizontal polarization (SH) 15, 129, 138, 152, 154
— in solid, equations for 1, 11–16, 147, 150
—, vertical polarization (P–SV) 15, 87–118, 129–132, 138, 148–149, 152–154

Electric potential 155

Energy conservation law 39, 64, 93, 97, 116–117, 129, 133, 171

Energy flux 18–19, 24, 38, 49, 64, 95–97, 109, 111, 117, 127–129, 148–149, 171, 172, 191, 194, 201, 203, 222

Energy reflection coefficient 97, 191, 197, 200, 202–204

Energy transmission coefficient 24, 97, 127–129, 191, 197, 200, 203

Euler equation 1–3, 144

Excitation coefficient for plane waves in solid 92–95, 97, 131,
  see also Transmission coefficient, Scattering matrix

Fast Fourier Transformation (FFT) 122

239

Springer
and the
environment

At Springer we firmly believe that an international science publisher has a special obligation to the environment, and our corporate policies consistently reflect this conviction.

We also expect our business partners – paper mills, printers, packaging manufacturers, etc. – to commit themselves to using materials and production processes that do not harm the environment. The paper in this book is made from low- or no-chlorine pulp and is acid free, in conformance with international standards for paper permanency.